Theory of
Meson Interactions
with Nuclei

Theory of Meson Interactions with Nuclei

JUDAH M. EISENBERG

Tel Aviv University
Tel Aviv, Israel

DANIEL S. KOLTUN

University of Rochester
Rochester, New York, U.S.A.

A WILEY-INTERSCIENCE PUBLICATION
JOHN WILEY & SONS
New York · Chichester · Brisbane · Toronto

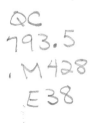

Library of Congress Cataloging in Publication Data:
Eisenberg, Judah M
   Theory of meson interactions with nuclei.

   "A Wiley-Interscience publication."
   Bibliography: p.
   Includes index.
   1. Mesons—Scattering.   2. Scattering (Nuclear
physics) I. Koltun, Daniel S., joint author.
II. Title.

QC793.5.M428E38   539.7′2162   79-24653
ISBN 0-471-03915-2

Printed in the United States of America

10 9 8 7 6 5 4 3 2 1

To our wives, Judith and Miriam

Preface

This book is an introduction to the theory of the interactions of π and K mesons with atomic nuclei. The main emphasis is on scattering of these mesons by nuclei, and the reactions that are thereby induced, including the absorption or production of mesons. This subject has always formed part of the border between nuclear and elementary particle physics; at one point these two branches were considered to be one field of inquiry. Nuclei provide a natural "laboratory" for the study of mesons as elementary particles, and have been used as such since the discovery of these particles. The use of meson reactions to study nuclei has developed more recently, having received new impetus through the construction of accelerators (meson factories) capable of producing meson beams of high intensity and good energy resolution, suitable for nuclear structure investigation.

Pions and kaons are the only mesons suitable for scattering reactions, since they are sufficiently stable to be collected in beams of laboratory dimensions. The pion is the best known meson and is most readily available for nuclear scattering; it is therefore our primary example in this book. Kaons are not as well studied at present, but they do present both interesting problems and possibilities for the investigation of nuclei and of hypernuclei. Other mesons (e.g., the vector mesons $\rho, \omega, \ldots,$ etc.) also are involved in nuclear reactions, as secondary, rather than as beam, particles; these are discussed in the appropriate contexts. All mesons, stable or otherwise, also play a role in nuclear dynamics, as constituents of stable nuclei that contribute to nuclear binding and to rates for electromagnetic and weak transitions. Although this is not the main topic of our book, we include enough discussion to introduce the reader to the subject and to connect it to the central themes.

The aim of this book is to present a theoretical framework for the analysis of the nuclear scattering and reactions of pions and kaons. The central method presented is the theory of multiple scattering, developed by Watson and others, for the scattering of fast particles from complex targets. This approach seems to give the most complete and flexible dynamical formulation, which is consistent with the usual description of nuclei in terms of Schroedinger mechanics. With this in mind, we give considerable space to the development and applications of the multiple scattering method, as well as to its limitations. It is in terms of this approach that we explain practical methods for interpretation of meson scattering experiments and show to what extent the known features of such experiments may be understood. However, in order to treat reactions like absorption or production of mesons, in which meson degrees of freedom are explicitly involved, we must develop methods that go beyond the theory of multiple scattering. Alternate approaches to scattering theory are also discussed.

The book is organized as follows. The first three chapters introduce the general subject, and then discuss in detail the two- and three-body problems: meson plus nucleon, nucleon plus nucleon, and meson plus deuteron. Here we introduce theoretical methods, some of which are specific to these elementary systems, and some of which are of direct use later. The next two chapters cover multiple scattering theory: Chapter 4 develops the theoretical tools for elastic and inelastic scattering, and Chapter 5 explains how these methods are applied to a variety of experimental cases. Chapter 6 deals with the inclusion of meson degrees of freedom in scattering reactions, and extends multiple scattering theory to include these. Applications to the problems of meson absorption and production in nuclear targets follow in Chapter 7. The last chapter introduces the theory of mesons in stable nuclei, to show the connection of the main material of the book to the subject of nuclear dynamics. Most chapters begin at a relatively elementary level, and introduce, in fairly general terms, the physics of the subject at hand. More advanced methods and detailed applications appear later in the development of each topic. Thus each chapter is relatively self-contained.

We would expect this material to be useful to advanced students and research physicists in nuclear or elementary particle physics. We consider the book to have a core of practical matters including theoretical methods and a discussion of phenomenology. This consists of Chapters 1, 2, 4, 5, and 7, which may be easiest to approach first. Among other uses, this core may serve as a guide for experimenters in meson-nucleus physics to the uses and limitations of current theoretical tools. The other chapters (3, 6, and 8) contain more specialized treatments of subjects of more theoretical interest.

The reader is expected to know quantum mechanics, including some results of the quantum theory of scattering. He should have some familiarity with nuclear physics, but may perhaps need less knowledge of elementary particles. This background should be sufficient for a reading of the core material, which is almost self-contained. Some of the more advanced material requires a knowledge of the elements of the quantum theory of fields.

We are happy to acknowledge continuing and useful interaction with a number of physics colleagues, especially H. Feshbach, A. Gal, F. Lenz, E. Levin, T. Mizutani, E. J. Moniz, F. Myhrer, J. V. Noble, D. Schneider, H. J. Weber, and A. I. Yavin. We thank D. Agassi, I. J. R. Aitchison, and M. Reiner for reading parts of the text in draft. Special thanks go to Mrs. E. Hughes for her excellent typing of the manuscript, to Mrs. B. Merkel and Ms. E. Hogan for their assistance with typing, to Mr. A. Covert for the drawings, and to Mr. A. Knapp for photographs.

One of us (D. S. K.) expresses his gratitude for the hospitality of the Department of Physics and Astronomy of Tel Aviv University, where this book was planned and begun during his residence as Visiting Professor in 1976 and 1977. Financial assistance during this year by the John Simon Guggenheim Memorial Foundation is also gratefully acknowledged. This work was also supported in part by the U.S. Department of Energy under a research contract with the University of Rochester.

JUDAH M. EISENBERG
DANIEL S. KOLTUN

Tel Aviv, Israel
Rochester, New York, U.S.A.
November 1979

Contents

1 Introduction: Mesons and Nuclei, 1

2 Meson Interactions with Nucleons, 7

2.1 Transitions and Scattering Amplitudes, 7
2.2 Information from πN and KN Scattering, 17
2.3 Potential Theory, 27
2.4 Pion-Nucleon Dynamics, 34
2.5 Nucleon-Nucleon Interaction, 57

3 Mesons and Deuterons: Three-body Systems, 62

3.1 Introduction to Multiple Scattering, 62
3.2 Faddeev Equations for Three-Body Systems, 72
3.3 Applications of the Faddeev Equations, 86
3.4 More Ambitious Approaches to Meson-Deuteron Scattering, 98
3.5 Meson Absorption, 106

4 Mesons and Nuclei: Multiple Scattering Theory, 112

4.1 Optics Approach to Meson-Nucleus Scattering, 113
4.2 The Watson Series: Theoretical Development, 127
4.3 Applications of the Watson Theory: Explicit Forms and Estimates of Convergence, 154
4.4 Glauber Theory, 177

5 Meson Scattering by Nuclei: Applications of Multiple Scattering Theory, 190

5.1 Elastic Scattering of Pions, 191
5.2 Total Cross Sections and Forward Scattering Amplitudes, 227
5.3 Mesic Atoms, 233
5.4 Scattering with Exchange of Quantum Numbers, 240

xi

5.5 Inelastic Scattering to Bound and Quasi-Bound States, 251

5.6 Knockout Reactions, 255

6 Field Theory and Coupled Channel Methods, 261

6.1 Field Theory Aspects of Scattering, 262

6.2 Scattering Theory for Many Degrees of Freedom, 273

6.3 Multiple Scattering, 284

6.4 Propagation of Resonances, 295

7 Absorption and Production of Mesons, 305

7.1 General Description, 305

7.2 Transition Amplitudes: Microscopic Theory, 313

7.3 Applications to Reactions, 326

8 Meson Interactions and Nuclear Structure, 347

8.1 Outline of Theory, 348

8.2 Examples of Mesic Features in Nuclear Structure, 362

References, 379

Index, 395

1 Introduction
— Mesons and Nuclei

The study of the strong interactions in nuclear and particle physics, and of the role of the meson in these interactions, began with the work of Yukawa in 1935. The early ideas about nuclear forces reflected the experience derived from studies of electrodynamics. Yukawa suggested that the nuclear forces would be mediated by the virtual exchange of a meson, whose mass he estimated from the roughly known range of the nucleon-nucleon interaction. The physical, on-shell manifestation of this particle was observed (about 1948) in nuclear bombardments. We now know it as the π-meson or pion, with mass $m_\pi \approx 140$ MeV. It is the lightest known strongly interacting particle and thus carries the part of the nuclear force with longest range. It has spin zero and, as a boson, can be absorbed or created individually in nuclear interactions; this feature is, indeed, intrinsic to its role as a mediator of the nuclear force. The pion has isospin one, appearing in the three charge states π^+, π^0, and π^-. As a consequence it is responsible for part of the charge-exchange in the nucleon-nucleon interaction. The pion in interaction with the nucleon also generates many resonances, or excited baryon states, whose properties are important to our understanding of strong-interaction physics.

In subsequent years, a variety of more massive mesons carrying other intrinsic quantum numbers have been discovered. Several of the lighter of these mesons are presumed to play a determinative role in the nuclear force (see Table 1.1). The heavier ones may contribute within the range of the core of very strong nuclear repulsion ($r < 0.5$ fm), which remains, in large measure, *terra incognita*. By the early 1950s K-mesons had also been discovered which were produced in association with a whole new family of strange baryons, the hyperons; of these, the Λ-particle in particular can be relatively easily bound in nuclei, yielding a species of hypernuclei with its own spectroscopy and dynamics.

By the very nature of a strongly interacting system, one must expect that all these hadrons may couple to nuclei, though, of course, one expects

1

Table 1.1 Meson Properties (Par–78)

Meson	Mass (MeV)	J^π	T	Γ (MeV)	Presumed Role in Nucleon-Nucleon Force
π	139.6	0^-	1	very narrow	mediates the longest range part of the nucleon-nucleon force; strong attractive tensor part
η	548.8	0^-	0	0.85×10^{-3}	does not seem to couple strongly to nucleons
ρ	773	1^-	1	152	assumed to counteract partially the attractive pion tensor force
ω	782.7	1^-	0	10	assumed to provide strong, short-range repulsion
$\sigma_{0,1}$	seek $\sim$550 MeV but lightest candidate is now apparently ε at $\sim$1200 MeV	0^+	0, 1	to the extent that this description makes sense, the width must apparently be very large ($\sim$300 MeV?)	simulates two-pion exchange in $T=0$ or 1 state, but no narrow resonance of sufficiently low mass has yet been seen

the states of large mass to mix less. In the early development of nuclear theory, the primary aspect of this coupling, namely, the mesic exchange that produces the nuclear force, was treated by eliminating the mesic degree of freedom in favor of a two-nucleon potential with exchange forces. More recently, however, it has become increasingly necessary to address the explicit role of the mesons and, for that matter, of isobar resonance admixtures, in producing many nuclear properties, especially at medium energies (very roughly from about 100 to 1000 MeV). Moreover, the stable mesons, π and K, have been formed into beams for the hadronic probing of the nucleus, for which purpose they are an allied tool to the use of nucleonic probes. One distinctive feature in this context is their boson nature which allows them to be absorbed in nuclear reactions, thus imparting their rest mass energy to the nuclear system producing quite high-lying excitations on the nuclear scale.

The central purpose of this book is to study theoretical methods for understanding the scattering, absorption, and production reactions of real mesons with nuclei. It is a significant part of our purpose to try to understand the extent to which this may be relevant to conventional nuclear structure studies of static proton and neutron density distributions, multiparticle correlations, and dynamic transition densities. Naturally, what we learn about mesic-nucleus interactions is highly pertinent to the investigation of the intrinsic role of mesons in nuclei, and we try to sketch this aspect of the field briefly in the closing chapter in order to acquaint the reader with the large literature on this subject. We also see that for a variety of purposes the nucleus can serve as a convenient study tool for properties of the elementary particles and, in particular, the mesons. This is especially the case when features of absorption or production are involved, since such processes are usually carried out in systems involving at least two nucleons and thus fall into the domain of nuclear physics. We constantly find ourselves looking at a range of phenomena and problems that are a complete mixture of the many-nucleon aspects of our subject and its mesic elements. This is in many ways an intrinsic ingredient of strongly interacting, many-body systems and forms one of its most intriguing features.

At the heart of our subject lies a source of some considerable difficulty and tension: On the one hand, it is inevitable in the medium-energy range and with the absorption and production of mesons that relativistic considerations enter importantly in the mesic features of our subject. On the other hand, the description of the mesic reaction mechanisms must be wedded to our knowledge of nuclear structure. For the latter point there hardly exists at present a systematic relativistic theory. We have therefore chosen to use as our primary tool the nonrelativistic multiple-scattering formalism, which has been rather thoroughly developed by Watson and others and which lends itself readily to nuclear applications. This formalism allows for the systematic development of the usual lowest-order result for the optical potential (as the product of the basic projectile-nucleon amplitude t and the nuclear density $\mathcal{V} = t\rho$) and—most importantly—of the higher-order corrections. Within the domain of potential scattering it is a complete, consistent, and exact theory. We occasionally discuss relativistic theories (as, for example, in Chapters 2, 6, and 8), usually in a somewhat phenomenological and hybrid fashion. By the same token, our interest in relating medium-energy reactions to nuclear structure theory leads us to take relatively little notice of detailed theories concerning the internal structure of the mesons and baryons, say, those which assume the particles are made up of two and three quarks, respectively. Occasionally, however, the quark model is useful for our purposes, in determining elementary couplings that enter in nuclear applications.

We roughly limit ourselves to energies below 1000 MeV in discussing the medium-energy range. This is partly a restriction of the theory, which follows on the recognition of the lack of a relativistic theory of nuclear structure and our inability to cope with the many hadronic channels that couple in at higher energies. Thus the nucleon mass sets a crude scale of the relevant upper limit of energy. This limitation also follows from a central experimental consideration: the size and binding potentials of nuclei dictate a scale of excitations with level separations on the order of 1 MeV or less. To form a flexible tool for nuclear studies, particle beams must therefore be able to achieve a comparable energy resolution, and this effectively limits the upper range of energies that can be realistically exploited. Moreover, the relatively large nuclear size means that scattering cross sections tend to fall rapidly at medium energies. This, on top of the intrinsically small cross sections for some of the processes we consider [e.g., mesic charge exchange, strangeness exchange, (p, π) reactions] and the need for high resolution, forces the use of very high intensity beams, which again implies a practical limitation in the accessible range of energies. At present, there exist three major facilities that were built to meet the criteria of medium energies and high beam intensities for use, at least in large measure, in nuclear structure studies. (They are also naturally useful for the study of intrinsic meson properties, such as rare decay modes.) These so-called "meson factories" include the Los Alamos Meson Physics Facility (LAMPF) in the United States, the Swiss Institute for Nuclear Research (SIN) in Switzerland, and the Tri-University Meson Facility* (TRIUMF) in Canada. The first of these accelerates protons to 800 MeV and the last two reach proton energies of between 500 and 600 MeV. All three accelerators are constructed to produce relatively intense secondary beams of pions. Beams of kaons have been used for nuclear study at the Brookhaven National Laboratory (BNL) in the United States and at the European Center for Nuclear Research (CERN) in Switzerland; these kaon beams are substantially less intense than those for pions, making nuclear experiments with them unusually challenging. As a result, much more information has thus far been accumulated on pion-nucleus interactions and hence we focus mainly on these.

These accelerators had been in operation one or two years as this book was being written, and in that short time produced a great deal more data than had been eked out during the preceding years at more conventional facilities. Although we refer occasionally both to the modern and to the earlier data, our primary purpose is to provide the theoretical framework for investigating meson-nucleus interactions. Toward that end we focus primarily on theoretical tools that we believe are of lasting value in such studies and develop these with sufficient generality that they may readily

---

*Later expanded to include a fourth university.

serve as a base on which to set future developments and improvements. We also intend the role of this book to be quite distinct from that of a review article on meson-nucleus physics (see, for example, Kol–69 and Huf–75a) or a text on multiple-scattering theory (e.g., Gol–64). It does not explore all aspects of meson-nucleus theory or of multiple-scattering theory, but rather lays out the broad lines of the multiple-scattering formalism as required for applications to meson-nucleus scattering, absorption, and production reactions.

The general features of the multiple-scattering theory as we develop it (Chapters 3 and, especially, 4) are directly applicable also for nucleon-nucleus studies in the intermediate-energy range.* Such studies share with the meson-nucleus studies the objective of learning about nuclear behavior, especially hadronic distributions at large momentum transfers or at small distance intervals; they differ, of course, in that they do not permit actual absorption of the probe, since baryon number must be conserved. Relatedly, the nucleon-nucleus processes probably achieve a more convincing description in terms of the nonrelativistic theory, since it is easier to believe that something approximating a conventional potential description exists for the NN interaction. The mesons, on the other hand, demand a treatment that takes into account their absorption and reemission on nucleons, and the πN (or KN) crossing properties, all of which may not be well simulated by a potential. Their very differences from the nucleon (smaller mass, spinlessness, unit isospin for the pion, and nonzero strangeness for the kaon) open possibilities and alternatives for the mesons as probes that are not available to the nucleon.

All of these hadronic probes at medium energies have in common a capability of examining hadronic behavior in the nucleus at high momentum transfers ($\gtrsim 400$ MeV/c, say, to pick a number pertinent to 90° πA scattering near the first πN resonance). In this respect they complement electron scattering measurements that are extremely suitable for probing the nuclear electromagnetic distribution—though only this distribution——at these momentum transfers. They also focus on nuclear behavior at momentum components and distance intervals [$\hbar$ (400 MeV/c)$^{-1} \simeq 0.5$ fm] where meson and baryon resonance effects are likely to manifest themselves explicitly. In this sense they address very directly the central feature of nuclear physics, namely, the consequence of strong-interaction coupling in a many-particle system.

A few remarks on notations and conventions. The nonrelativistic, Schroedinger energy of a system is denoted in this book by E, the physical

---

*We nevertheless often use the words "meson" or "pion" as shorthand notation for the projectile, even in more general contexts.

system and coordinate system to which it pertains being specified in the relevant context. In the relativistic situation, we also denote the total energy by E which is the sum of single-particle relativistic energies $\varepsilon(k) = \sqrt{k^2 + m^2}$ for three-momentum $\mathbf{k}$ and particle mass m; for mesons this is sometimes labeled ω_k, as in covariant formalisms where $k_\mu = (\omega_k, \mathbf{k})$. Three-vectors are shown in boldface type (thus $\mathbf{k}$), and vectors in isospin space carry an arrow over them (as $\vec{I}$; in our usage, the proton has $t_3 = +\frac{1}{2}$). Meson masses are referred to as m_α, where $\alpha = \pi, K, \ldots$ indicates the meson. For the pion, this is sometimes simply known as m without the subscript. The nucleon mass is denoted by M. Last, our plane waves are normalized such that $\langle \mathbf{k'} | \mathbf{k} \rangle = (2\pi)^3 \delta(\mathbf{k'} - \mathbf{k})$.

2 Meson Interactions with Nucleons

We begin with this subject because most of our ideas for dealing with the interactions of mesons with nuclei are based on what we know of the interactions with single nucleons. From the point of view of this book, nucleons are the primary constituents of nuclei; in the Schroedinger formulation, which we use in much of our treatment, nucleons are the *only* constituents. Therefore, we treat at some length the subject of meson interactions with individual nucleons. Since most of the information we have on this subject is derived from the study of meson-nucleon scattering reactions, we begin with a brief review of scattering theory. This provides us with definitions and notation used in the rest of the book. Then we summarize the information obtained from πN and KN scattering. Following this we discuss the dynamics of meson-nucleon scattering. Last we treat briefly the role of mesons in interactions among nucleons.

2.1 Transitions and Scattering Amplitudes

2.1a On-Energy-Shell Amplitudes; Partial Waves

Scattering reactions are described in terms of probability amplitudes relating the initial and final asymptotic states of the combined system of projectile and target. If we denote these states by a and b, respectively, we may define two matrix amplitudes: the S-matrix S_{ba} and the transition or t-matrix $\langle b|T|a \rangle$, which are related as follows:

$$S_{ba} = \delta_{ba} - 2\pi i\, \delta(E_b - E_a)\langle b|T|a \rangle. \qquad [1]$$

The labels a and b stand for the quantum numbers of the states asymptotically, that is, at a large enough distance that the projectile and target do not interact. Then, for example, we may specify the momentum of each

7

system, and the spin and isospin projections. For a meson plus nucleon, we may write

$$|a\rangle = (2\pi)^{-3}|\mathbf{k}, \mathbf{p}, \alpha\rangle \qquad [2]$$

where $\mathbf{k}$ and $\mathbf{p}$ are the linear momentum of the meson and the nucleon, respectively, and α stands for the spin, isospin (and strangeness) quantum numbers.* (For π and K mesons, which are spinless, α specifies the spin projection m_s of the nucleon and the isospin states t_3 (meson) and τ_3 (nucleon).)

The S-matrix in Eq. 1 is defined only for the final state energy E_b equal to the initial state energy E_a, where these energies are the sum of the internal and kinetic energies of the projectile and target. This is suitable for the description of the scattering of an isolated system (e.g., meson plus nucleon) for which energy is conserved; the S-matrix is referred to as being "on-energy-shell," with $E = E_a = E_b$. The S-matrix is unitary,

$$\sum_c S_{cb}^* S_{ca} = \delta_{ba}, \qquad [3]$$

where the sum is over all possible asymptotic states (channels) with energy $E_c = E_b = E_a = E$. The relation given by Eq. 3 expresses the conservation of probability (flux) in scattering processes.

The t-matrix defined in Eq. 1 is also "on-energy-shell," although we shortly extend the definition "off-energy-shell" as well. Again, for an isolated system, the total momentum is conserved, so we may write

$$\langle b|T|a\rangle = T_{ba}\,\delta(\mathbf{P}_b - \mathbf{P}_a), \qquad [4]$$

where $\mathbf{P}_a = \mathbf{k} + \mathbf{p}$. The cross section to some particular set of final states b, for a given initial state a, is given by

$$\Delta\sigma = \frac{(2\pi)^4}{v_a}\sum_b |T_{ba}|^2 \qquad [5]$$

where v_a is the initial relative velocity of the projectile and target, and the sum (or integral) b is defined by the choice of final states measured (e.g., angles, energies, spin projections of final state particles).

For elastic scattering, it is convenient to work in the center of mass (c.m.) frame ($\mathbf{P}_a = \mathbf{P}_b = 0$), with initial and final states given in Eq. 2

---

*We normalize plane wave states $\langle \mathbf{k}'|\mathbf{k}\rangle = (2\pi)^3\,\delta(\mathbf{k}' - \mathbf{k})$.

specified by the meson momentum κ, κ', respectively:

$$|a\rangle = (2\pi)^{-3/2}|\kappa, -\kappa, \alpha\rangle$$
$$|b\rangle = (2\pi)^{-3/2}|\kappa', -\kappa', \beta\rangle. \qquad [6]$$

The total energies in this frame are

$$E_a = \varepsilon_m(\kappa) + \varepsilon_M(\kappa)$$
$$E_b = \varepsilon_m(\kappa') + \varepsilon_M(\kappa') \qquad [7]$$

where $\varepsilon_m(\kappa) = (\kappa^2 + m^2)^{1/2}$ is the relativistic energy. Equation 7 defines the "energy shells" or surfaces in momentum space; the "on-energy-shell" relation $E_a = E_b$ means the surfaces coincide, and the magnitudes $\kappa = \kappa'$ are equal. We define the *elastic scattering amplitude* as

$$f_{\beta\alpha} = -(2\pi)^2 m^* T_{ba} \equiv -\frac{m^*}{2\pi}\langle \kappa', \beta | T | \kappa, \alpha \rangle \qquad [8]$$

where m^* is the reduced mass for the meson and nucleon: $m^* = [\varepsilon_m(\kappa)^{-1} + \varepsilon_M(\kappa)^{-1}]^{-1}$. The *differential cross section* for elastic scattering (unpolarized) is given by

$$\frac{d\sigma}{d\Omega} = \frac{1}{2}\sum_{\text{spins}} |f_{\beta\alpha}|^2 \qquad [9]$$

where the sum is over the spin projections of the nucleon in the initial and final states. The amplitude given in Eq. 8 and the cross section given in Eq. 9 depend on the energy $E_a = E_b = E$, the directions $\hat{\kappa}, \hat{\kappa}'$, and (before summing) the nucleon spin projections. If we introduce the Pauli spin matrices σ for the nucleon, we may represent the amplitude in Eq. 8 as a spin matrix. Invariance under rotations in space and parity inversions limits the form of $f_{\beta\alpha}$ to

$$f_{\beta\alpha}(E, \hat{\kappa}, \hat{\kappa}') = f(E, \theta) + g(E, \theta)i\sigma \cdot \mathbf{n}$$

$$\mathbf{n} = \frac{(\kappa \times \kappa')}{|\kappa \times \kappa'|} \qquad [10]$$

where θ is the angle between $\hat{\kappa}$ and $\hat{\kappa}'$. (The f and g in Eq. 10 are functions of E and θ, and still depend on isospin labels, which we have suppressed for the moment.) The differential cross section given in Eq. 9 can now be written using Eq. 10 as

$$\frac{d\sigma}{d\Omega}(E,\theta) = |f(E,\theta)|^2 + |g(E,\theta)|^2. \qquad \textbf{[11]}$$

The amplitude in Eq. 10 can be expanded in partial waves by expanding the f and g separately:

$$f(E,\theta) = \sum_{l=0}^{\infty} \{(l+1)f_{l+}(E) + lf_{l-}(E)\} P_l(\cos\theta)$$

$$\qquad\qquad\qquad\qquad\qquad\qquad\qquad \textbf{[12]}$$

$$g(E,\theta) = \sum_{l=0}^{\infty} \{f_{l+}(E) - f_{l-}(E)\} P_l'(\cos\theta)\sin\theta$$

in terms of the Legendre functions $P_l(x)$, $x = \cos\theta$, and their x-derivatives $P_l'(x) = (d/dx)P_l(x)$. The partial-wave amplitudes $f_{l\pm}(E)$ are defined for orbital angular momentum l and total angular momentum $J = l \pm \frac{1}{2}$ of the meson-nucleon system (in c.m.); due to invariance under rotations and parity, these are both conserved quantum numbers. This means that the S-matrix in Eq. 1 is also diagonal in l and J. Using the unitary condition given by Eq. 3 and the definitions in Eqs. 1, 4, 8, 10, and 12, one finds that the $f_{l\pm}(E)$ can be expressed in terms of two *real* quantities: the phase shift $\delta_{l\pm}(E)$ and the inelasticity parameter $\eta_{l\pm}(E)$, in the form

$$f_{l\pm}(E) = \frac{(\eta_{l\pm}(E)\exp 2i\delta_{l\pm}(E) - 1)}{2i\kappa}, \qquad \textbf{[13a]}$$

with κ on-energy-shell given by Eq. 7. The parameter η measures the fraction of flux lost from the elastic channel for each l and J; $\eta = 1$ holds for no inelasticity (all flux conserved in the elastic channel) and $0 \leqslant \eta < 1$ otherwise. For purely elastic scattering ($\eta = 1$) Eq. 13a takes the usual form

$$f_{l\pm}(E) = \frac{\exp[i\delta_{l\pm}(E)]\sin\delta_{l\pm}(E)}{\kappa}, \qquad \text{(elastic).} \quad \textbf{[13b]}$$

In the limit of very low energy, the partial-wave amplitudes $f_{l\pm}(E) \propto \kappa^{2l}$ based on centrifugal effects alone. For $l = 0$ and 1 it is conventional to introduce the *scattering lengths* and *volumes*:

$$a = \lim_{\kappa \to 0} f_{0+}(E) \qquad s\text{-wave} \qquad\qquad \textbf{[14a]}$$

$$a_{j=1\pm 1/2} = \lim_{\kappa \to 0} \frac{f_{1\pm}(E)}{\kappa^2} \qquad p\text{-wave.} \qquad \textbf{[14b]}$$

These quantities may be complex if there is inelasticity at zero energy (e.g., $K^-p \to \pi^0 \Lambda$); otherwise they are real.

We have omitted the isospin labels on scattering amplitudes, starting with Eq. 10. Rather than keeping the uncoupled isospin projections of the meson and nucleon separately, it is more convenient to deal with the total isospin, which is a conserved quantity (to a good approximation) for meson-nucleon reactions. If we introduce the isospin matrix $\vec{I}_m$ for the meson,* and the (Pauli) isospin matrix $\vec{\tau}$ for the nucleon, then the total isospin is given by a matrix (operator) $\vec{T} = \vec{I}_m + \frac{1}{2}\vec{\tau}$, with isospin quantum number T. We may write the scattering amplitude given in Eq. 10 or any partial wave component $f_{l\pm}$ (Eq. 12) as a sum:

$$f = \sum_T f_T Q_T, \qquad [15]$$

where Q_T project states of definite isospin T.

Pions have unit isospin, and nucleons one-half unit, so that $T = \frac{1}{2}$ or $\frac{3}{2}$, and

$$
\begin{aligned}
Q_{1/2} &= \tfrac{1}{3}\left(1 - \vec{I}_\pi \cdot \vec{\tau}\right), \\
Q_{3/2} &= \tfrac{1}{3}\left(2 + \vec{I}_\pi \cdot \vec{\tau}\right).
\end{aligned}
\qquad [16a]
$$

We may write the πN amplitudes using Eqs. 15 and 16a as

$$
\begin{aligned}
f &= \left(\tfrac{1}{3}f_{1/2} + \tfrac{2}{3}f_{3/2}\right) + \left(\tfrac{1}{3}f_{3/2} - \tfrac{1}{3}f_{1/2}\right)\vec{I}_\pi \cdot \vec{\tau} \\
&\equiv f^{(+)} - f^{(-)}\vec{I}_\pi \cdot \vec{\tau}
\end{aligned}
\qquad (\pi N) \qquad [16b]
$$

where the $(\pm)$ in the last line is convention: in terms of tensors in isospin, $f^{(+)}$ is the coefficient of the symmetric (isoscalar) part, and $f^{(-)}$ is the coefficient of the antisymmetric (isovector) part. Then we may write, for example, the amplitudes for the following charge modes in terms of isospin amplitudes:

$$
\begin{aligned}
f(\pi^+ p \to \pi^+ p) &= f_{3/2} = f^{(+)} - f^{(-)} \\
f(\pi^- p \to \pi^- p) &= \tfrac{1}{3}f_{3/2} + \tfrac{2}{3}f_{1/2} = f^{(+)} + f^{(-)} \\
f(\pi^- p \to \pi^0 n) &= \frac{\sqrt{2}}{3}\left(f_{3/2} - f_{1/2}\right) = -\sqrt{2}\, f^{(-)}.
\end{aligned}
\qquad [16c]
$$

---

*We use the arrow notation $\vec{a}$ only for vectors in isospin, with the isoscalar product $\vec{a} \cdot \vec{b}$.

Kaons have one-half unit isospin, so $T=0$ or 1, and

$$Q_0 = \frac{1-2\vec{I}_K \cdot \vec{\tau}}{4}$$

$$Q_1 = \frac{3+2\vec{I}_K \cdot \vec{\tau}}{4}$$

[17a]

and

$$f = \left(\tfrac{1}{4}f_0 + \tfrac{3}{4}f_1\right) + \tfrac{1}{2}(f_1 - f_0)\vec{I}_K \cdot \vec{\tau} \qquad \text{(KN)}.$$

[17b]

Note that the (K^+, K^0) form an isospin doublet with *strangeness* $S=1$, while $(\overline{K}^0, K^-)$ form an isospin doublet with strangeness $S=-1$. Since strangeness is conserved in strong interactions, Eq. 17b applies to these doublets separately; for example,

$$
\begin{aligned}
f(K^+ p \to K^+ p) &= f_1 & (S=1) \\
f(K^+ n \to K^+ n) &= \tfrac{1}{2}(f_0 + f_1) & (S=1) \\
f(K^- p \to K^- p) &= \tfrac{1}{2}(f_0 + f_1) & (S=-1) \\
f(K^- p \to \overline{K}^0 n) &= \tfrac{1}{2}(f_1 - f_0) & (S=-1).
\end{aligned}
$$

[18]

Differential cross sections (Eq. 11) for elastic and charge exchange meson-nucleon scattering, for example, in the modes in Eqs. 16 and 18, provide most of the information we have about the on-energy-shell amplitudes $f_{\beta\alpha}$ or T_{ba} in Eq. 8. Spin polarization experiments are also necessary to separate the f and g parts of Eq. 10. The information extracted from experiment is discussed in Section 2.2.

2.1b Off-Energy-Shell, Kinematics, Relativity

As we have seen, the on-energy-shell amplitudes S_{ba}, $\langle b|T|a \rangle$ (Eq. 1), and T_{ba} (Eqs. 4 and 8) determine the scattering properties of an isolated system of meson and nucleon, such as the differential cross section given in Eqs. 9 and 11. These properties only specify the asymptotic aspects of the system, that is, where projectile and target are at a large relative distance. If we wish to discuss the dynamics of the system, we also need to specify behavior at shorter relative distance where interactions take place. We may want to do this not only for an isolated meson-nucleus system but where that system constitutes part of a larger system, as when a meson scatters from a nucleus. In principle the description of the interacting region requires full information on the interaction, the dynamical equation of motion, and the form of the solutions of that equation. This is often more than we know directly; these are usually inferred by studying the meson-nucleon scattering, from which we know (in principle) the on-energy-shell t-matrix. An economical way of expressing the dynamical aspects of the meson-nucleon interaction is through the extension of the t-matrix "off-energy-shell." This approach proves particularly useful for the study of meson scattering from nuclei: in the multiple scattering theory, which we develop in the next few chapters, nuclear scattering is expressed in terms of the meson-nucleon t-matrices.

Let us discuss the off-energy-shell extensions. First we introduce the "completely off-energy-shell" t-matrix $\langle b|T(E)|a\rangle$. For this amplitude, E_b need not equal E_a, and we introduce a third energy variable that need not equal either E_a or E_b. Then we also define the two "half off-energy-shell" t-matrices by setting $E = E_a$ or $E = E_b$ in the "completely off-energy-shell" t-matrix; we obtain $\langle b|T(E_a)|a\rangle$ and $\langle b|T(E_b)|a\rangle$. These three amplitudes become equal on-energy-shell, $E = E_a = E_b$, and all are equal to $\langle b|T|a\rangle$, which was defined in Eq. 1. In summary, we have the following:

$$\langle b|T(E)|a\rangle \quad \text{(completely off-energy-shell), } E \neq E_a \neq E_b,$$

$$\begin{matrix} \langle b|T(E_a)|a\rangle \\ \langle b|T(E_b)|a\rangle \end{matrix} \quad \text{(half off-energy-shell), } E_b \neq E_a. \qquad [19]$$

The various off-energy-shell t-matrices in Eq. 19 are each generated by an equation of motion, which depends on the interaction. We postpone the specific forms until Sections 2.3 and 2.4 where we discuss dynamics. At this point we simply assume that there is dynamical information in the terms given in Eq. 19. The next task is to consider the kinematic variables that specify the off-energy-shell amplitudes in the light of the symmetries of the meson-nucleon system.

It is sufficient to discuss the completely off-energy-shell case, since the other cases are obtained by equating energy parameters. For an arbitrary reference frame, we may represent the asymptotic states $|a\rangle, |b\rangle$, in the form given by Eq. 2. The t-matrix may be written

$$\langle \mathbf{k}', \mathbf{p}', \beta | T(E) | \mathbf{k}, \mathbf{p}, \alpha \rangle \qquad [20]$$

in terms of the meson momenta $\mathbf{k}$, $\mathbf{k}'$, nucleon momenta $\mathbf{p}$, $\mathbf{p}'$, spin and isospin projections α, β, and the off-shell energy parameter E (see Fig. 2.1). The energies E_a and E_b are determined by k, p and k', p', respectively. If we assume a nonrelativistic limit, $E_a = m + M + (k^2/2m) + (p^2/2M)$, and the case is similar for E_b.

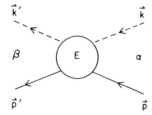

Fig. 2.1 Representation of t-matrix, Eq. 20, indicating kinematic variables. Dashed line refers to meson, solid line to nucleon.

Translational invariance means conservation of total momentum $\mathbf{P} = \mathbf{k} + \mathbf{p} = \mathbf{k}' + \mathbf{p}'$, as in Eq. 4. Choosing the relative momenta

$$\boldsymbol{\kappa} = \mathbf{k} - \left(\frac{m}{m+M} \right) \mathbf{P}, \quad \boldsymbol{\kappa}' = \mathbf{k}' - \left(\frac{m}{m+M} \right) \mathbf{P}, \qquad [21]$$

we may express Eq. 20 in terms of $\boldsymbol{\kappa}$, $\boldsymbol{\kappa}'$, and $\mathbf{P}$. However, invariance under change of frame (Galilei invariance, for the nonrelativistic limit) eliminates the $\mathbf{P}$-dependence; thus we may write

$$\langle \mathbf{k}', \mathbf{p}', \beta | T(E) | \mathbf{k}, \mathbf{p}, \alpha \rangle = (2\pi)^3 \delta(\mathbf{k}' + \mathbf{p}' - \mathbf{k} - \mathbf{p}) \langle \boldsymbol{\kappa}', \beta | T(E_{\text{c.m.}}) | \boldsymbol{\kappa}, \alpha \rangle$$
$$[22]$$

where the t-matrix is now expressed in terms of variables measured in the c.m.: $\boldsymbol{\kappa}$, $\boldsymbol{\kappa}'$, and $E_{\text{c.m.}} = E - [P^2/2(m+M)]$ (and α, β). The energy $E_a = (m+M) + (\kappa^2/2m^*)$ with the reduced mass $m^* = mM/(m+M)$ (compare Eq. 7) and the case is similar for E_b.

The argument from invariance under rotations and parity inversions is similar to that for the on-energy-shell case, Eqs. 8–10, and allows us to write the completely off-energy-shell t-matrix (c.m.) in the form

$$\langle \boldsymbol{\kappa}' | T(E_{\text{c.m.}}) | \boldsymbol{\kappa} \rangle = T_0 + T_\sigma i \boldsymbol{\sigma} \cdot \mathbf{n}, \qquad [23]$$

where the functions T_0 and T_σ depend on four (rotation) scalar variables; for example,

$$E_{\text{c.m.}}, \kappa, \kappa', \cos\theta = \hat{\boldsymbol{\kappa}} \cdot \hat{\boldsymbol{\kappa}}',$$

or

$$E_{\text{c.m.}}, E_a, E_b, \theta,$$

or q^2 could replace $\cos\theta$ or θ, where $\mathbf{q} = \boldsymbol{\kappa}' - \boldsymbol{\kappa}$ is the three-momentum transfer (in c.m.). For the on-energy-shell case, Eq. 23 will give the elastic scattering amplitude in Eqs. 8 and 10. The dependence on θ can be expanded in partial waves, as in Eq. 12. The isospin expansion given in Eq. 15 also applies to T_0 and T_σ.

The nonrelativistic forms we have just given are appropriate for Schroedinger mechanics, which we use in much of the book. This is not necessarily a poor approximation for nucleons in the frame of a nucleus, but it is generally not adequate for the mesons, which do move relativistically in a nucleus in many of the cases we discuss. It is simple enough to correct for the relativistic *kinematics* of the mesons by replacing the mass m by the energy $E(k) = (k^2 + m^2)^{1/2}$ throughout (e.g., in Eq. 21). This is often called a semirelativistic approach, and is still appropriate for a Schroedinger dynamical theory. For a fully relativistic dynamical theory, however, there is more to be said regarding off-shell t-matrices. We discuss a few points.

For the fully relativistic case, we define the t-matrix in terms of the four-momentum variables k, p, k', p', and α, β as before for spin-isospin, where $k \equiv (k_0, \mathbf{k})$:

$$\langle k', p', \beta | T | k, p, \alpha \rangle. \qquad [24]$$

The energy variables k_0, k'_0, p_0, p'_0 now appear in the t-matrix, but these variables may be "off-mass-shell," that is, $k^2 \equiv k_0^2 - \mathbf{k}^2 \neq m^2$. We call the t-matrix in Eq. 24 "completely off-mass-shell" (see Fig. 2.2). It should be clear that with energy conservation $(k'_0 + p'_0 = k_0 + p_0)$ there are two more variables in Eq. 24 than in the nonrelativistic form given by Eq. 20. Translational invariance in four-space $(t, \mathbf{x})$ leads to conservation of total four-momentum $P = (P_0, \mathbf{P}) = k + p = k' + p'$. (The energy, but not the mass, is conserved in Eq. 24, thus "off-mass-shell.") This gives a factor in Eq. 24 of

$$\delta^4(k' + p' - k - p) = \delta(k'_0 + p'_0 - k_0 - p_0)\,\delta(\mathbf{k}' + \mathbf{p}' - \mathbf{k} - \mathbf{p}) \qquad [25]$$

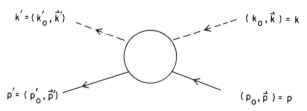

Fig. 2.2 Representation of t-matrix, Eq. 24, with relativistic kinematics. Dashed line refers to meson, solid line to nucleon.

where the energy conservation is that already given in Eq. 1, and the momentum conservation corresponds to Eq. 22.

The Lorentz transformation properties of an on-mass-shell t-matrix are well defined; for transformation from frame 1 to frame 2, one has (e.g., Gol–64)

$$\langle k_2' p_2' \beta | T | k_2 p_2 \alpha \rangle = \left[\frac{\varepsilon_m(k_1)\varepsilon_m(k_1')\varepsilon_M(p_1)\varepsilon_M(p_1')}{\varepsilon_m(k_2)\varepsilon_m(k_2')\varepsilon_M(p_2)\varepsilon_M(p_2')} \right]^{1/2} \langle k_1' p_1' \beta | T | k_1 p_1 \alpha \rangle$$

(on-mass-shell). [26]

This transformation does not automatically apply to the off-mass-shell t-matrix given in Eq. 24, in particular, in the case of a theory that is relativistic, but not covariant, as for example in a potential theory. (See the discussion at the end of Section 5.1b, and, e.g. Hel–76.) However, in completely covariant theories, such as those generated by covariant perturbation theory, or the Bethe-Salpeter equation,* the transformation in Eq. 26 holds for the completely off-mass-shell t-matrices as well. This allows us to express Eq. 24 in terms of an *invariant amplitude*[†]

$$\langle k',p',\beta | T | k,p,\alpha \rangle = -(2\pi)^4 \delta^4(p'+k'-p-k)$$
$$\times \sqrt{\frac{M^2}{4\varepsilon_m(k)\varepsilon_m(k')\varepsilon_M(p)\varepsilon_M(p')}} \, (k',p',\beta | M | k,p,\alpha)$$

[27]

where the amplitude M is independent of Lorentz frame and may therefore be calculated, for example, in the c.m. frame. Since M is a Lorentz scalar (including rotations and parity) it may be expressed in terms of Lorentz scalars. From the four-momenta k,p,k',p', using conservation Eq. 25, one can form six independent scalars: for example,

$$s \equiv P^2 = (k+p)^2 = (k'+p')^2$$
$$t \equiv (k'-k)^2 = (p-p')^2 \equiv q^2 \qquad [28]$$
$$k^2, k'^2, p^2, p'^2.$$

*The Bethe-Salpeter equation is the relativistic analog of the Lippmann-Schwinger integral equation for the t-matrix (for which, see Eqs. 48 and 49, p. 29); it is equivalent to a relativistic Schroedinger equation (see Bro–69).
[†]The definition of T_{ba} hence corresponds to $S_{ba} = \delta_{ba} - iT_{ba}$, in place of Eq. 1. The definition of M follows standard conventions (e.g., Moo–69).

The scalar variable $s = W^2$ where W is the c.m. energy: $W = P_0$ (c.m.) (**P** (c.m.)$\equiv 0$); q is the four-momentum transfer: $t = q^2 = q_0^2 - \mathbf{q}^2$. The last four variables are the off-shell masses of the four lines in Fig. 2.2. Scalars may also involve the spin of the nucleon through the "bilinear covariants" for a Dirac particle (e.g., Bjo–64). This involves a number of combinations of γ, σ (Dirac), and the four-momenta, which we do not write out for the general case. For the on-mass-shell case, the invariant amplitude is written in the form

$$(k',p',\beta|M|k,p,\alpha) = \bar{u}_\beta(p')\,\hat{T}(s,t,k,k')u_\alpha(p), \qquad [29a]$$

$$\hat{T}(s,t,k,k') = A(s,t) + \tfrac{1}{2}\gamma\cdot(k+k')B(s,t) \quad \text{(on-mass-shell)}, \qquad [29b]$$

where $\hat{T}$ is a Dirac (4×4) matrix, and u and $\bar{u}$ are Dirac spinors (e.g., Bjo–64). This form can be reduced to the Pauli-spin-matrix form (Eq. 10) in the c.m., if we write

$$\begin{aligned} f(E,\theta) &= f_1 + f_2\cos\theta \\ g(E,\theta) &= -f_2\sin\theta \end{aligned} \qquad [30a]$$

$$\begin{aligned} f_1 &= \frac{\varepsilon_M(|\mathbf{k}|) + M}{8\pi E}\left[A + (E - M)B\right] \\ f_2 &= \frac{\varepsilon_M(|\mathbf{k}|) - M}{8\pi E}\left[-A + (E + M)B\right] \end{aligned} \qquad [30b]$$

where $E = W = \sqrt{s}$, and $|k|$ is the magnitude of the meson (or nucleon) momentum in the c.m. The standard partial wave analysis given in Eq. 12 and the isospin decomposition given by Eq. 15 may be applied to Eq. 30 (see, e.g., Moo–69). (It is worth noting that the form given in Eq. 29 also applies to the partially off-mass-shell case, where the nucleon is on-mass-shell, $p^2 = p'^2 = M^2$, but the meson is not.)

2.2 Information from πN and KN Scattering

Most of the information we have about the interaction of mesons and nucleons comes from the analysis of scattering experiments. These have been carried out for lab energies up to and above 100 GeV (see *Review of Particle Properties*, Par–78). However, we are here interested in lower energies, $0 < T_m < 1$ GeV, with more emphasis on the first few hundred MeV, which are better studied and of somewhat more interest for nuclear studies. (We use the symbol T_m for meson kinetic energy (lab): $T_m = \varepsilon_m(k) - m$.) Measurements exist in this energy range for total cross sections, differential cross sections for elastic scattering, and polarization for charged mesons on protons. [Scattering information for neutron targets is obtained from meson-deuteron scattering, usually under kinematic conditions where the impulse approximation is expected to apply (see Chapter 3). These data are not always necessary because of isospin invariance (below).] The scattering data provide a determination of the elastic scattering amplitude given in Eqs. 8 and 10. Since we are interested in the strong-interaction part, the Coulomb scattering effects must be removed. This is not an entirely model independent procedure, but it is probably sufficiently accurate for the small charges involved ($\pm e$) (see, e.g., Rop–65 for Coulomb analysis in πp). The interference of the Coulomb and strong-interaction scattering (at small angles for all but the very lowest energies) helps remove some ambiguities in signs of amplitudes. The remaining strong-interaction amplitudes can then be expanded in partial waves (Eq. 12) and isospin (Eq. 15).

2.2a Pion-Nucleon Scattering

For $\pi^\pm p$, elastic and charge-exchange scattering are possible at all energies:

$$\pi^+ + p \rightarrow \pi^+ + p, \qquad\qquad [31a]$$

$$\pi^- + p \rightarrow \pi^- + p, \qquad\qquad [31b]$$

$$\pi^- + p \rightarrow \pi^0 + n. \qquad\qquad [31c]$$

Inelastic channels become significant for $T_\pi \gtrsim 300$ MeV. To the extent that charge invariance holds for $\pi + N$, or equivalently, that isospin is conserved, the three amplitudes (Eqs. 31a–c) are determined by the two isospin amplitudes $f_{1/2}, f_{3/2}$, or $f^{(\pm)}$ (Eq. 16). In fact, the measurement of all three amplitudes given in Eqs. 31a–c provides a test of charge invariance, which appears to hold to a few percent, at least for $T_\pi \sim 100$–300 MeV, where it has been tested (Hen–69).

In Fig. 2.3 we show a summary of measured total cross sections for $\pi^\pm p$ within the energy range $0 < T_\pi \lesssim 2$ GeV (Bar–68). The most evident features of the cross sections are the strong energy-dependence and the strong charge-dependence (difference between $\pi^+ p$ and $\pi^- p$, and therefore between $T = \frac{3}{2}$ and $\frac{1}{2}$). For comparison, we show the total cross section for $p + p$ (Hes–58), which decreases monotonically over the same range of lab energy. At low energy ($T < 40$ MeV) $\sigma_{pp} > \sigma_{\pi p}$. We may conclude that the

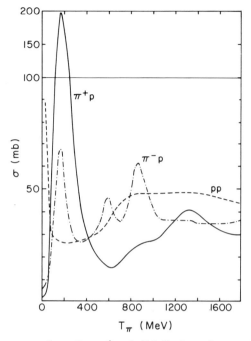

Fig. 2.3 Total cross sections for π^+p (solid line) and π^-p (dot-dashed line) scattering as a function of lab kinetic energy T_π (data from Bar–68). For comparison the total pp cross section (Hes–58) is shown. Note change in vertical scale.

πN interaction is comparable in strength to, but very different in detail from, the nucleon-nucleon interaction.

We learn more by considering differential cross sections with separation into partial waves. This analysis has been done by a number of groups (Rop–65, Bar–68, Alm–72, Car–73). Since the partial wave decomposition is not unique for each energy at which data are available, the usual procedure assumes some regular energy dependence of phase shifts to cut down on the number of possibilities.

As an example, we list in Table 2.1 a set of phase shifts produced by one analysis (Rop–65) for the lower energy range of $0 < T_\pi < 345$ MeV. We list the real phase shifts $\delta_{l\pm}$ (Eq. 13b) for $l = 0$ (s-wave) and $l = 1$ (p-wave). The largest d-wave phase shifts are $|\delta_2| \sim 3$–$9°$ at the highest energy represented. The inelasticity parameters η (Eq. 13a) are almost unity ($\eta \simeq 1$, no inelasticity) in this energy region. A more recent analysis (Car–73) gives similar phase shifts for $90 < T_\pi < 310$ MeV.

Table 2.1 Low Energy Pion-Nucleon Phase Shifts (from Rop-65)[a]

T_π (MeV)	S_{11}	S_{31}	P_{11}	P_{31}	P_{13}	P_{33}
20	4.4	−2.2	−0.43	−0.25	−0.10	1.45
58	6.8	−4.8	−1.5	−1.08	−0.46	8.0
120	8.9	−8.8	−2.2	−2.9	−1.17	32.4
170	10.0	−12.1	−0.32	−4.6	−1.74	72.4
220	11.1	−15.1	4.68	−6.5	−2.23	109.5
310	12.9	−21.8	22.7	−10.8	−2.84	136.8

[a]In degrees, entries have been rounded to $0.1°$ or $0.01°$.

The largest phase shift in this energy range is for the $l = 1$, $J = \frac{3}{2}$, $T = \frac{3}{2}$ wave (denoted P_{33} in the notation $P_{2T,2J}$), where $\delta(P_{33})$ rises through $90°$ for $T_\pi \simeq 195$ MeV. This is associated with a resonance, which is broad and unsymmetrical in energy. It is possible to express the resonant partial wave amplitude (c.m.), for $\eta = 1$, as

$$\kappa f_{33} = \frac{\frac{1}{2}\Gamma(\omega)}{\omega_R - \omega - (i/2)\Gamma(\omega)} \qquad [32a]$$

in terms of a resonant energy ω_R and width $\Gamma(\omega)$, where we use the pion c.m. energy ω. However, the width must be energy dependent. The parametrizations

$$\Gamma(\omega) = \left(\frac{\kappa}{\kappa_R}\right)^3 \Gamma_R \qquad [32b]$$

and

$$\Gamma(\omega) = \frac{\omega_R}{\omega}\left(\frac{\kappa}{\kappa_R}\right)^3 \Gamma_R \qquad [32c]$$

are often used, where κ is the pion c.m. momentum and $\omega = (\kappa^2 + m_\pi^2)^{1/2}$ (the κ^3 corresponding to a p-wave "penetration factor" $\propto \kappa^{2l+1}$). The second form corresponds to the Chew-Low "effective range formula" (Che-56) which can be written

$$\frac{\kappa^3 \cot \delta_{33}}{\omega} = \frac{1}{\lambda}(1 - \omega r) \qquad [33a]$$

or

$$\cot\delta_{33} = \frac{\omega_R - \omega}{\frac{1}{2}\Gamma(\omega)}, \qquad \frac{1}{2}\Gamma(\omega) = \frac{\lambda\kappa^3}{\omega r}, \qquad r = \frac{1}{\omega_R} \qquad [33b]$$

based on a model discussed in Section 2.4. A fit to the phase shift $\delta_{33}(\omega)$ gives $\omega_R \simeq 2.14\ m\ (=300\ \text{MeV})$ and Γ about 110–122 MeV. It is now conventional to label resonances in terms of the total energy $(\pi + N)$ in c.m., $E_{\text{c.m.}} = W = \sqrt{s}$; for the P_{33} resonance, the energy is taken as $E_{\text{c.m.}} = 1232$ MeV, and the label is $\Delta(1232)$ (see, e.g., Par–78). Note that the peak in the total cross section corresponding to the resonance is at $T_\pi \sim 185$ MeV, lower than the energy $(T_\pi \sim 195\ \text{MeV})$ at which $\delta_{33} = 90°$: this follows from the κ^{-2} dependence of the total (i.e., angle integrated) cross section for this partial wave

$$\sigma_{33} = \frac{8\pi}{\kappa^2}\sin^2\delta_{33}.$$

The s-wave phase shifts δ_{2T} are roughly equal in magnitude, and opposite in sign for $T = \frac{1}{2}, \frac{3}{2}$. The scattering lengths (Eq. 14a) are given in Table 2.2. Notably they are small compared to nuclear dimensions $\left(|a_{2T}| \sim 0.1\ m^{-1}\right)$ and are almost pure isovector; that is, from Eq. 16b

$$a^{(+)} = \tfrac{1}{3}(a_1 + 2a_3) \simeq -0.005\ m_\pi^{-1}$$

$$a^{(-)} = \tfrac{1}{3}(a_1 - a_3) \simeq 0.087\ m_\pi^{-1}, \qquad [34]$$

Table 2.2 Pion-Nucleon Scattering Lengths and Volumes

	s-wave scattering lengths a_{2T} (in $10^{-3}m_\pi^{-1}$)	
a_1	170 ± 4	181 ± 8
a_3	-92 ± 2	-89 ± 5
	(Bug–73)	(Sam–72a)

	p-wave scattering volumes $a_{2T,2J}$ (in $10^{-3}m_\pi^{-3}$)	
a_{33}	204 ± 5	204 ± 16
a_{13}	-27 ± 6	-16 ± 5
a_{31}	-43 ± 7	-37 ± 4
a_{11}	-84 ± 10	-54 ± 8
	(Sam–72b)	(Sal–74)

Table 2.3 Lowest Energy Resonances in Pion-Nucleon System (from Par–78)

	$T=\frac{1}{2}$				$T=\frac{3}{2}$		
P_π(MeV/c)	Mass (MeV)	L,J	Γ (MeV)	P_π (MeV/c)	Mass (MeV)	L,J	Γ (MeV)
a	N (939)	$1,\frac{1}{2}$	$\left(\begin{array}{c}\text{nucleon:}\\\text{stable}\end{array}\right)$	300	Δ (1232)	$1,\frac{3}{2}$	115
660	N (1470)	$1,\frac{1}{2}$	200	960	Δ (1650)	$0,\frac{1}{2}$	140
740	N (1520)	$2,\frac{3}{2}$	125	1000	Δ (1670)	$2,\frac{3}{2}$	200^b
760	N (1535)	$0,\frac{1}{2}$	100	1030	Δ (1690)	$1,\frac{3}{2}$	250^b
1000	N (1670)	$2,\frac{5}{2}$	155				
1030	N (1688)	$3,\frac{5}{2}$	140				

^{a}Below threshold. ^{b}Less certain identification.

using the numbers of Bug–73. These numbers are not very well determined, since they involve extrapolation of the low energy data to zero energy.

The p-waves (other than P_{33}) have small phase shifts in the low energy region as shown in Table 2.1. The phases all start negative at low energy, as is shown also by the scattering volumes (Eq. 14b) listed in Table 2.2. Even at $T_\pi \simeq 0$, the P_{33} wave has the strongest interaction, although the P_{11} wave is comparable in strength. With increasing energy, the phase δ_{11} changes sign near $T_\pi = 170$ MeV, and eventually rises through 90°, defining a resonance at $E_{\text{c.m.}} = 1470$ MeV, denoted as N(1470). By convention, the nucleon is labeled as N(939), and $T=\frac{1}{2}$ resonances are denoted N($E_{\text{c.m.}}$), while $T=\frac{3}{2}$ resonances are labeled $\Delta(E_{\text{c.m.}})$, beginning with $\Delta(1232)$. In fact, a number of resonances appear in various partial waves; in Table 2.3 we list some of the lower energy resonances that are most clearly identified.

The threshold for inelastic πN scattering is at $E_{\text{c.m.}} = 1218$ MeV, or $T_\pi \simeq 173$ MeV, where $\pi + \text{N} \rightarrow 2\pi + \text{N}$ becomes energetically possible. However the inelastic channels do not really play an important role for $T_\pi < 400$ MeV. Above these energies, the inelasticity parameter η (Eq. 13a) may be considerably less than unity in some partial wave, indicating the loss of flux from the πN channel to $\pi\pi$N (or other) channels. A useful method of displaying the partial wave parameters δ, η in the presence of inelasticity $\eta < 1$, is the Argand diagram, shown in Fig. 2.4. The partial wave amplitude $f_l(E)$ for given E, multiplied by κ, is represented as a point z within a circle of unit diameter in the complex plane, centered at $z_0 = +i/2$. From

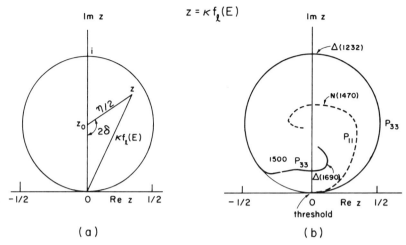

Fig. 2.4 Argand diagrams for partial wave amplitudes as complex functions of energy: (a) the phase variables defined in Eq. 35; (b) the locus of amplitudes for πN scattering in the P_{33} (solid line) and P_{11} (dashed line) partial waves.

Eq. 13a we have

$$\kappa f_l(E) = \frac{1}{2i}\,(\eta \exp 2i\delta - 1), \qquad\qquad [35]$$

from which we find that $|\kappa f_l|$ is the modulus of the point z, $\eta/2$ the distance $|z - z_0|$, and 2δ the angle denoted. A resonant phase, rising through $\delta = 90°$, is represented as a trajectory (in E), crossing the $\mathrm{Re}\,z = 0$ line within the circle at $\delta = 90°$. In Fig. 2.4b we illustrate the trajectories traced by the P_{33} and P_{11} amplitudes, showing the $\Delta(1232)$ (elastic) resonance ($\eta = 1$) and the N(1470) (inelastic) resonance ($\eta \lesssim 0.3$).

Another useful characterization of the πN scattering is in terms of the forward elastic scattering amplitude $f(E, 0°)$. This quantity appears in simple approximations for pion-*nucleus* reactions in later chapters, but we generally need the amplitude defined in the *laboratory* frame (in which the initial target is stationary.) However, the combination $f(E, 0°)/\kappa$ is an invariant (Lorentz), so the lab and c.m. amplitudes are simply related

$$\frac{f(E, 0°)_{\mathrm{lab}}}{k_{\mathrm{lab}}} = \frac{f(E, 0°)_{\mathrm{c.m.}}}{\kappa}. \qquad\qquad [36]$$

If partial wave amplitudes are available, the forward amplitude is constructed from the sum indicated by Eqs. 12 and 13, with $\theta = 0$. In general the amplitude may also be obtained as follows. The optical theorem, which follows from the unitary relation (Eq. 3), relates $\mathrm{Im} f(E, 0°)$ to the total cross section:

$$\frac{4\pi}{k} \mathrm{Im} f(E, 0°) = \sigma_{\mathrm{tot}}(E) \quad \text{(optical theorem).} \quad [37]$$

(Note that Eq. 37 is invariant, as is the total cross section.) The $\mathrm{Re} f(E, 0°)$ is obtained from the analysis of interference of the Coulomb and strong-interaction scattering, at small angles, extrapolating to $\theta = 0°$. The real and imaginary parts of $f(E, \theta)$ are related through forward dispersion relations, of which there have been many applications to πN scattering (Ham–63, Bra–73). (These relations involve integration over E.)

2.2b Kaon-Nucleon Scattering

The K^+N, K^0N systems (labeled KN) have strangeness $S = 1$, while the K^-N, $\overline{K}^0N$ systems (labeled $\overline{K}N$) have $S = -1$. The scattering for these two cases differs greatly, largely because there are a number of baryons or low-mass baryon resonances with $S = -1$, but none with $S = +1$. This is consistent with the quark picture, in which baryons contain only the $S = -1$ strange quark (λ or s), while mesons may carry $\lambda \equiv s$ with $S = -1$ or the anti-quark ($\bar{\lambda} \equiv \bar{s}$) with $S = +1$. The difference can be seen in the total cross sections for K^+p and K^-p shown in Fig. 2.5. The K^-p cross

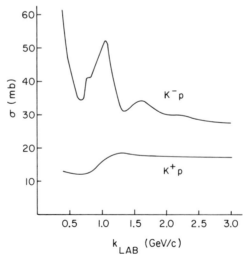

Fig. 2.5 Total cross sections for K^-p and K^+p as a function of lab momentum (data from Par–78).

section for $S = -1$ is considerably larger for low energies and shows more energy structure than that for K^+p ($S = 1$).

Since the K^+N, K^0N systems are the simpler we discuss them first. Scattering experiments are only done with K^+, but both p and n targets are needed to separate the isospin $T = 0, 1$ contributions (Eq. 18). This requires experiments with deuteron targets, as mentioned above. For low energies $T_K < 150$ MeV s-wave scattering dominates, and the scattering is almost entirely $T = 1$. The scattering lengths are as follows for $T = 0, 1$ (Ste–64):

$$a_0 = 0.04 \pm 0.04 \text{ fm}$$

and

$$a_1 = -0.29 \pm 0.02 \text{ fm}, \qquad [38]$$

which, like the πN scattering lengths (Table 2.2) are small in nuclear dimensions. Above $T_K = 150$ MeV p-wave scattering and the $T = 0$ channel become important, but there are no resonances (Mar–75). The threshold for inelastic processes starts at about $T_K = 250$ MeV, with $K^+ + p \rightarrow K^+ + N + \pi$, which is not significant until $T_K > 650$ MeV, at which $K^+ + p \rightarrow K^+ + \Delta$ (1232) is allowed. (For KN scattering, the meson lab momentum k_L is usually quoted: for $T_K \simeq 250$ MeV, $k_L \simeq 556$ MeV/c.)

The K^-N, K^0N systems are much more complicated, as we have indicated. Again, experiments can be done with K^- on p or n (actually, deuteron) targets, to separate the isospin $T = 0, 1$ channels (Eq. 18). In principle, there is sufficient information to separate these amplitudes from the proton-target reactions

$$K^- + p \rightarrow K^- + p \qquad [39a]$$

and

$$\rightarrow \overline{K}^0 + n \qquad [39b]$$

using the isospin amplitudes (Eq. 18). (The threshold for charge exchange (b) is positive, unlike the $\pi^-p \rightarrow \pi^0n$ case, at $T_K = 8.1$ MeV, or $k_L = 90$ MeV/c.) What is totally different about the K^-p reactions is the existence of *inelastic* channels, down to $T = 0$, involving pion plus hyperon,

$$K^- + p \rightarrow \pi^0 + \Lambda \quad (T = 1) \qquad [39c]$$

$$\rightarrow \pi^\pm + \Sigma^\mp \quad (T = 0, 1) \qquad [39d]$$

$$\rightarrow \pi^0 + \Sigma^0 \quad (T = 0), \qquad [39e]$$

as well as the less important three-body channel

$$K^- + p \to \pi + \pi + \Lambda. \qquad [39f]$$

Since the channels given in Eq. 39a–e may all be strongly coupled, the analysis of K^-p scattering must be done in a coupled-channel formalism, in which the t-matrix for Eq. 39 is given as a matrix with indices T and α, where $T = 0, 1$ and $\alpha = \overline{K}N$ and $\pi\Sigma$ for $T = 0$, or $\alpha = \overline{K}N$, $\pi\Sigma$, and $\pi\Lambda$ for $T = 1$. This gives a 2×2 channel matrix for $T = 0$, and 3×3 for $T = 1$. (Actually, the analysis is easier to perform in terms of the "K-matrix," which is a hermitian (in fact, real symmetric) matrix closely related to the t-matrix (see the original paper of Dal–60).) Determination of the K-matrix elements for low energy ($T_K < 80$ MeV) K^-p reactions has been done (Mar–69a) in a zero-range approximation, which involves energy-independent matrix elements and s-wave scattering only—the multichannel equivalent of the scattering-length approximation in one channel. (For higher partial waves, see Kim–71 and Mar–77a.) The equivalent scattering lengths in the $\overline{K}N$ channel are complex (Mar–69a):

$$a_0 = -1.66 + i\,0.69 \text{ fm},$$
$$a_1 = -0.09 + i\,0.54 \text{ fm}. \qquad [40]$$

These indicate stronger interactions for low energy $\overline{K}N$ than for KN.

The coupled system given in Eq. 39a–e (with $S = -1$) has a number of resonant states: those at lower energy are listed in Table 2.4, with the

Table 2.4 Lowest Energy Resonances in $\overline{K}$-Nucleon System (from Par–78)

	$T = 0$				$T = 1$		
P_K (MeV/c)	Mass (MeV)	L, J	Γ (MeV)	P_K (MeV/c)	Mass (MeV)	L, J	Γ (MeV)
a	$\Lambda(1115)$	$1, \frac{1}{2}$	$\left(\begin{array}{c}\text{weak}\\\text{decay}\\\text{only}\end{array}\right)$	a	$\Sigma(1193)$	$1, \frac{1}{2}$	$\left(\begin{array}{c}\text{weak}\\\text{decay}\\\text{only}\end{array}\right)$
a	$\Lambda(1405)$	$0, \frac{1}{2}$	40	a	$\Sigma(1385)$	$1, \frac{3}{2}$	35
389	$\Lambda(1520)$	$2, \frac{3}{2}$	16	720	$\Sigma(1660)$	$1, \frac{1}{2}$	100^b
740	$\Lambda(1670)$	$0, \frac{1}{2}$	40	740	$\Sigma(1670)$	$2, \frac{3}{2}$	50
780	$\Lambda(1690)$	$2, \frac{3}{2}$	60				

^{a}Below $\overline{K}p$ threshold. ^{b}Less certain identification.

conventional label $\Lambda(E_{c.m.})$ for states of $T=0$, and $\Sigma(E_{c.m.})$ for states of $T=1$. (The stable Λ is $\Lambda(1116)$ and Σ^+ is $\Sigma^+(1189)$.) In general, these resonances appear in all channels, that is, not only in $\overline{K}N$, but also for $\pi\Lambda$, $\pi\Sigma$ "scattering," the latter seen in final state reactions, for example, $\overline{K}p\rightarrow\Sigma\pi+X$. Note, however, that there are two resonant states *below* threshold for K^-p (at $E_{c.m.}=1432$ MeV): $\Lambda(1405)$ and $\Sigma(1385)$ (also called Y_0^* and Y_1^*, respectively, where the subscripts indicate the isospin of the resonance). These play an important role in the very low energy scattering of $\overline{K}$ from nuclear systems, as for example, in K^--atoms (see Chapter 5).

2.3 Potential Theory

We first discuss the dynamics of meson-nucleon scattering in the most familiar form of quantum mechanics: the Schroedinger equation with a potential interaction. For simplicity we treat the problem of elastic scattering with only a single channel, for example, πN or KN scattering with specific isospin T, below the threshold of inelastic reactions. We write the Schroedinger equation

$$(E_a-H_0-V)|\psi_a^{(\pm)}\rangle=0 \qquad [41a]$$

where a denotes the quantum numbers of the initial plane wave state given in Eq. 2 of the meson and nucleon, or in c.m. given in Eq. 6. We denote the plane wave state by $|\chi_a\rangle$ or simply $|a\rangle$; it satisfies the noninteracting equation

$$(E_a-H_0)|\chi_a\rangle=0 \qquad [41b]$$

with kinetic energy operator H_0; E_a is the kinetic energy (or asymptotic total energy) given by Eq. 7. The scattering states $\psi_a^{(\pm)}$ satisfy the usual boundary conditions at very large separation distance; that is,

$$\psi_a^{(+)}\rightarrow\chi_a+\text{outgoing spherical waves,}$$
$$\psi_a^{(-)}\rightarrow\chi_a+\text{ingoing spherical waves.} \qquad [42]$$

2.3a Equations for Transition Matrices and Transition Operators

It is now possible to define the half off-energy-shell t-matrices introduced in Eq. 19 in terms of the scattering states and the potential:

$$\langle b|T(E_a)|a\rangle\equiv\langle\chi_b|V|\psi_a^{(+)}\rangle \qquad [43a]$$
$$\langle b|T(E_b)|a\rangle\equiv\langle\psi_b^{(-)}|V|\chi_a\rangle \qquad [43b]$$

where a and b have different energy, that is, $E_a \neq E_b$. It is a standard result of scattering theory (e.g., Gol–64) that the on-energy-shell limits of Eq. 43a–b, which coincide, give the on-energy-shell transition amplitude $\langle b|T|a \rangle$ defined in Eq. 1, through which we defined cross sections (Eq. 5) for scattering, and the elastic scattering amplitude (Eq. 8). The half off-energy-shell amplitudes (Eq. 43) may be considered to contain information on the potential and wave function at all distances, while the on-energy-shell amplitude has only large distance (asymptotic) information: for example, the phase shifts. We may see this, for example, by Fourier transforming Eq. 43a;

$$\langle \mathbf{r}|V|\psi_a^{(+)} \rangle = (2\pi)^{-3/2} \int d\kappa_b e^{i\kappa_b \cdot \mathbf{r}} \langle b|T(E_a)|a \rangle \qquad [44]$$

which shows that the combination $V\psi_a^{(+)}$, evaluated at any position $\mathbf{r}$ (relative coordinate), is given by a linear combination of the t-matrices in Eq. 43a for fixed a, but for all κ_b (and E_b), that is, the half off-energy-shell t-matrices.

We may also obtain a dynamical equation for the t-matrices (Eq. 43) if we first use the Lippmann-Schwinger equation to express the scattering states in Eq. 42 (see, e.g., Gol–64, Chapter 5):

$$|\psi_a^{(\pm)} \rangle = |\chi_a \rangle + G_0^{(\pm)}(E_a)V|\psi_a^{(\pm)} \rangle \qquad [45a]$$

with

$$G_0^{(\pm)}(E) \equiv (E \pm i\eta - H_0)^{-1} \qquad [45b]$$

defining the free Green operator or propagator, where the infinitesimal η is to be taken in the limit $\eta \to 0^+$ for the calculation of scattering states in Eq. 45a or t-matrices in Eq. 43. If we now form the matrix elements of Eq. 43 using Eq. 45a for $|\psi_a^{(+)} \rangle$ and the hermitian conjugate of Eq. 45a for $\langle \psi_b^{(-)}|$, we obtain the following linear integral equations for the t-matrices:

$$\langle b|T(E_a)|a \rangle = \langle \chi_b|V|\chi_a \rangle + \sum_c \langle \chi_b|V|\chi_c \rangle \langle \chi_c|G_0^{(+)}(E_a)|\chi_c \rangle \langle c|T(E_a)|a \rangle,$$

$$[46a]$$

$$\langle b|T(E_b)|a \rangle = \langle \chi_b|V|\chi_a \rangle + \sum_c \langle b|T(E_b)|c \rangle \langle \chi_c|G_0^{(+)}(E_b)|\chi_c \rangle \langle \chi_c|V|\chi_a \rangle,$$

$$[46b]$$

where the sum $\sum_c$ over intermediate (plane wave) states represents the sum and integral $\Sigma_\gamma \int d\kappa_c \cdot (2\pi)^{-3}$. The Green operator G_0 is diagonal in c:

$$\langle \chi_c | G_0^{(+)}(E) | \chi_{c'} \rangle = \delta_{cc'}(E + i\eta - E_c)^{-1}. \qquad [47]$$

Note that only $G_0^{(+)}$ appears in Eq. 46; the $+$ may now be omitted. The integral equations in Eq. 46a–b are equivalent to the Schroedinger equation (Eq. 41), as dynamical representations of the scattering. The use of Eq. 46 is mostly as formal equations, which play a considerable role in later chapters; however, *solution* of the integral equations is often less convenient than direct solution of Eq. 41 in configuration space for a given potential V.

Either integral equation of Eq. 46 may now be used to define the fully off-energy-shell t-matrix in Eq. 19, $\langle b | T(E) | a \rangle$, where E is a parameter unrelated to either a or b:

$$\langle b | T(E) | a \rangle = \langle \chi_b | V | \chi_a \rangle + \sum_c \langle \chi_b | V | \chi_c \rangle \langle \chi_c | G_0(E) | \chi_c \rangle \langle c | T(E) | a \rangle$$

$$= \langle \chi_b | V | \chi_a \rangle + \sum_c \langle b | T(E) | c \rangle \langle \chi_c | G_0(E) | \chi_c \rangle \langle \chi_c | V | \chi_a \rangle. \qquad [48]$$

The definition of the completely off-energy-shell t-matrix as the solution of either expression given in Eq. 48 is equivalent to defining an *operator* (transition) $T(E)$ whose plane wave matrix elements $\langle b | T(E) | a \rangle$ form this t-matrix. The operator satisfies the operator equations equivalent to Eq. 48:

$$T(E) = V + VG_0(E)T(E)$$

$$= V + T(E)G_0(E)V. \qquad [49]$$

It is in this form that we often use the off-energy-shell t-matrices for the treatment of meson-nucleon scattering in a many-particle system in Chapters 3 and 4.

2.3b Potentials

To exhibit some explicit forms for potentials, we first write the Schroedinger equation (Eq. 41) in the c.m. using the c.m. momentum κ:

$$(E_a - E(\kappa))\psi_\alpha^{(\pm)}(\kappa) = \sum_\gamma \int \frac{d\kappa'}{(2\pi)^3} \langle \kappa, \alpha | V | \kappa', \gamma \rangle \psi_\gamma^{(\pm)}(\kappa') \qquad [50]$$

where $E(\kappa)$ is the total c.m. energy for c.m. momentum κ; that is, for the nonrelativistic limit, $E(\kappa) = m + M + \kappa^2/2m^*$. As usual, α, γ are spin-isospin labels. Transforming to position space, we obtain

$$\left(E_a - m - M + \frac{\nabla^2}{2m^*} \right) \psi_\alpha^{(\pm)}(\mathbf{r}) = \sum_\gamma \int d\mathbf{r}' \langle \mathbf{r}, \alpha | V | \mathbf{r}', \gamma \rangle \psi_\gamma^{(\pm)}(\mathbf{r}'),$$

$$[51a]$$

where $\mathbf{r}$, $\mathbf{r}'$ are the relative position variables.

By definition, a *local* potential is diagonal in position:

$$\langle \mathbf{r}, \alpha | V | \mathbf{r}', \gamma \rangle = \delta(\mathbf{r} - \mathbf{r}') V_{\alpha\gamma}(\mathbf{r}); \qquad (\text{local } V). \qquad [52a]$$

Thus Eq. 51a becomes the differential equation

$$\sum_\gamma \left\{ \left(E_a - m - M + \frac{\nabla^2}{2m^*} \right) \delta_{\alpha\gamma} - V_{\alpha\gamma}(\mathbf{r}) \right\} \psi_\gamma^{(\pm)}(\mathbf{r}) = 0. \quad [51b]$$

It is easy to show that the momentum representation of a local potential (Eq. 52a) is a function of the momentum transfer $\mathbf{q} = \kappa' - \kappa$ only:

$$\langle \kappa, \alpha | V | \kappa', \gamma \rangle = \tilde{V}_{\alpha\gamma}(\mathbf{q}) \qquad [52b]$$

where $\tilde{V}(\mathbf{q})$ is the Fourier transform of $V(\mathbf{r})$.

A potential which is not diagonal in position is called *nonlocal*, referring to the need for information about the wave function at $\mathbf{r}'$ in order to determine the interaction at $\mathbf{r}$ (see Eq. 51a). The smallest degree of nonlocality is obtained for potentials that depend on finite powers of the momentum (as an operator κ); in position space, this gives derivatives (to finite order), since $\kappa \to -i\nabla$. For example, the spin-orbit potential

$$V_{\text{s.o.}}(r)(\mathbf{r} \times \kappa) \cdot \boldsymbol{\sigma} \qquad [53]$$

is linear in momentum and is "almost local." The Kisslinger potential (Chapter 5) is second order (in momentum or derivatives) and also "almost local." An example of a highly nonlocal potential that cannot be represented by finite powers of momentum is one that operates only in a single partial wave; that is,

$$\langle \mathbf{r}, \alpha | V | \mathbf{r}', \beta \rangle = v_{\alpha\beta}^l(r, r') \sum_m Y_{lm}(\hat{\mathbf{r}}) Y_{lm}^*(\hat{\mathbf{r}}'). \qquad [54a]$$

Clearly this potential has only diagonal matrix elements in l, m (and independent of m for rotational invariance):

$$\langle rlm\alpha|V|r'l'm'\beta\rangle = v_{\alpha\beta}^{l}(r,r')\delta_{ll'}\delta_{mm'} \qquad [54b]$$

as can be demonstrated by angular integration. The nonlocality enters through the form of the angular part, which serves as a projection operator for the particular l-wave. A particularly simple form of Eq. 54 is given by a *separable* potential, for which the $v(r,r')$ factors are as follows:

$$\langle \mathbf{r},\alpha|V|\mathbf{r}',\beta\rangle = \lambda_{\alpha\beta}^{l}g_{l}(r)g_{l}(r')\sum_{m}Y_{lm}(\hat{\mathbf{r}})Y_{lm}^{*}(\hat{\mathbf{r}}')$$

or

$$\langle rlm\alpha|V|r'lm\beta\rangle = \lambda_{\alpha\beta}^{l}g_{l}(r)g_{l}(r'). \qquad [55a]$$

Alternatively, the spin and orbital angular momentum could be coupled: $\mathbf{l}+\mathbf{s}=\mathbf{J}$, for which a separable form would be given by

$$\langle \mathbf{r},\alpha|V|\mathbf{r}',\beta\rangle = \lambda^{J}g_{J}(r)g_{J}(r')\sum_{M}\langle \hat{\mathbf{r}},\alpha|JM\rangle\langle JM|\hat{\mathbf{r}}',\beta\rangle$$

or

$$\langle r(ls)JM|V|r'(ls)JM\rangle = \lambda^{J}g_{J}(r)g_{J}(r'). \qquad [55b]$$

A potential separable in position, as in Eq. 55, is also separable in momentum:

$$\langle \kappa,\alpha|V|\kappa',\beta\rangle = \lambda^{J}\hat{g}_{J}(\kappa)\hat{g}_{J}(\kappa')\sum_{M}\langle \hat{\kappa},\alpha|JM\rangle\langle JM|\hat{\kappa}',\beta\rangle. \qquad [55c]$$

The fully off-energy-shell t-matrix corresponding to this potential is also separable in the form

$$\langle \kappa|T(E)|\kappa'\rangle = \tau^{J}(E)\hat{g}_{J}(\kappa)\hat{g}_{J}(\kappa')\sum_{M}\langle \hat{\kappa}|JM\rangle\langle JM|\hat{\kappa}'\rangle$$

$$\tau^{J}(E) = \lambda^{J}[1-\lambda^{J}I^{J}(E)]^{-1} \qquad [56]$$

$$I^{J}(E) = \frac{1}{(2\pi)^{3}}\int\frac{l^{2}dl\,\hat{g}_{J}^{2}(l)}{E+i\eta-E(l)}$$

which can be shown by using the forms of Eqs. 55c and 56 in the integral equations of Eq. 48.

2.3c   Effective Potentials

The description of meson-nucleon scattering through a Schroedinger equation (Eq. 41) with a potential interaction is certainly an approximation. As we see in the following section, our understanding of the dynamics of the meson-nucleon interaction requires degrees of freedom not included in the Schroedinger equation, such as the meson field itself, which represents the possibility of exciting many-meson (virtual) states in scattering. Also, there may be other kinds of mesons involved in πN and KN scattering—for example, the vector mesons, ρ, ω, ϕ, which may be exchanged. However, we find that for many purposes the Schroedinger approach may still be used through the introduction of an *effective potential*, which, by construction, includes the effects of the other degrees of freedom of the system that have been left out of the Schroedinger wave functions. This device, which allows the use of Schroedinger methods in situations where many (hidden) degrees of freedom are present, is further developed in Chapter 6.

As a simple example, consider a process in which a π or K meson interacts with a nucleon by exchanging a particle "V," for example, a vector meson. In Fig. 2.6a we illustrate the process by a perturbation (Born) diagram in momentum. Such a process may be considered a contribution to a potential interaction between the π or K and nucleon in the sense that the Born amplitude represented by Fig. 2.6a may be associated with the plane-wave matrix element $\langle \kappa' | \mathcal{V} | \kappa \rangle$ of an *effective* potential $\mathcal{V}$. No variables appear in the potential referring to the exchanged particle "V"; they are integrated out in the evaluation of the Born term. The potential $\mathcal{V}$ is a function of κ, κ', and spins. To the extent that the dependence is only on $\mathbf{q} = \kappa' - \kappa$, the effective potential may be *local* (Eq. 52b): this may be the case for exchange of spinless particles, for example. Consider a different process (Fig. 2.6b) in which a meson and nucleon interact by combining into a single system M, which then dissociates back into meson plus nucleon. For example, M might be a particle, distinct from π, K, or N, that can couple to the scattering channel. The process represented may also be considered as a contribution to a potential interaction between meson and nucleon. If the particle or system M has a single (intrinsic) angular momentum J (or l) it will lead to a nonlocal potential as in Eq. 54 or in the simplest cases, separable potentials as in Eq. 55 with $J(l)$ as the only partial wave.

In the simple examples illustrated in Fig. 2.6, the effective interaction is obtained from a Born matrix element for a scattering process. This amplitude may be completely off-energy-shell in that the energy of the meson-nucleon system E need not equal $E(\kappa)$ or $E(\kappa')$. Thus the effective interaction is *energy dependent*—$\mathcal{V}(E)$, with plane wave matrix elements

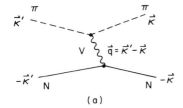

(a)

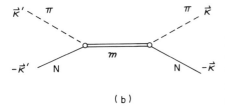

(b)

Fig. 2.6 Diagrams for processes that contribute to the effective potential for πN: (a) exchange of a particle V; (b) formation of a quasi-particle or resonance M.

$\langle \kappa'|\mathcal{V}(E)|\kappa\rangle$. This energy dependence is a general feature of effective potentials that have been derived by suppressing some degrees of freedom, for example, by reducing a many-channel system to a one-channel representation. Some of the consequences of this energy dependence are technical problems: for example, solutions of the effective Schroedinger equation for different E are not orthogonal, and the potential $\mathcal{V}$ will be complex, if any of the coupled channels included are "open," that is, can take flux from the elastic channel. The potential may also be regarded as *nonstatic retarded* in that the E-dependence can be translated into time-delay in the interaction. Many of the ideas for the use of effective interactions follow from the work of Feshbach (Fes–58,62; see also Oku–54). Recent applications of the method of suppressing coupled channels to obtain separable effective interactions are in Lon–74.

Last, it should be pointed out that if processes like those illustrated in Fig. 2.6a and b are included in the effective interaction, the t-matrix or scattering amplitude for meson-nucleon scattering may be obtained from the integral equation given in Eq. 48, for example, in operator form, using Eq. 49:

$$T(E) = \mathcal{V}(E) + \mathcal{V}(E)G_0(E)T(E). \qquad [57\text{a}]$$

If we iterate this equation as a series in $\mathcal{V}(E)$,

$$T(E) = \mathcal{V}(E) + \mathcal{V}(E)G_0(E)\mathcal{V}(E) + \cdots, \qquad [57\text{b}]$$

we see that the scattering process involves repeated exchanges of "V" and repeated formation of M, which we would get by continuing the original Born series. However, not all higher order processes occur in the series given in Eq. 57b: An example is the exchange of two "V" mesons simultaneously (crossed lines) (see Fig. 2.13).

2.4 Pion-Nucleon Dynamics

We now give an elementary treatment of the interaction of pions and nucleons that goes beyond the potential theory picture of the last section. The basic ideas go back to Yukawa, who invented mesons to explain nuclear potentials. (We return to this question in the following section.) The theoretical form is quite similar to that of electrodynamics, and we use some of the tools of quantum field theory but in a somewhat simplified manner (see, e.g., Bjo–64, 65).

2.4a Yukawa Interaction

The pions are represented by a field operator $\vec{\phi}(\mathbf{x})$, which is pseudoscalar in space and isovector in charge (isospin) space. If we let $\alpha = 1, 2, 3$ represent the "cartesian" components in isospin, ϕ_α is a real field (operator). (We work in the Schroedinger picture, where operators are independent of time.) The field operator can be written in terms of annihilation and creation operators for plane-wave states (momentum $\mathbf{k}$, energy ω_k):

$$\phi_\alpha(\mathbf{x}) = \int d\mathbf{k}\, N_k \{ a_\alpha(\mathbf{k}) e^{i\mathbf{k}\cdot\mathbf{x}} + a_\alpha^\dagger(\mathbf{k}) e^{-i\mathbf{k}\cdot\mathbf{x}} \}, \quad N_k = \left[(2\pi)^3 2\omega_k \right]^{-1/2},$$

$$[58]$$

with the conventional normalization for a boson field. The isospin can also be written in "spherical vector" form to correspond to the three charge states π^+, π^0, π^-, using index $v = 1, 0, -1$. Then $a_v^\dagger(\mathbf{k})$ creates a meson of momentum $\mathbf{k}$ and charge v where

$$a_{+1}^\dagger(\mathbf{k}) = -\sqrt{\tfrac{1}{2}} \left[a_1^\dagger(\mathbf{k}) + i a_2^\dagger(\mathbf{k}) \right]$$

$$a_0^\dagger(\mathbf{k}) = a_3^\dagger(\mathbf{k}) \qquad\qquad [59]$$

$$a_{-1}^\dagger(\mathbf{k}) = \sqrt{\tfrac{1}{2}} \left[a_1^\dagger(\mathbf{k}) - i a_2^\dagger(\mathbf{k}) \right]$$

if we use the standard (Condon-Shortley) phases, which allow for conventional vector coupling of isospin. Then the operator $a_v(\mathbf{k}) = (-1)^v (a^\dagger_{-v})^\dagger$ annihilates the meson of momentum $\mathbf{k}$ and charge v.

The interaction of pions with a nucleon is assumed to be linear in the meson field, and local in analogy with electrodynamics. We write the interaction H_I in terms of a Hamiltonian density $H(\mathbf{x})$:

$$H_I = \int d\mathbf{x}\, H(\mathbf{x})$$

$$H(\mathbf{x}) = \vec{j}(\mathbf{x}) \cdot \vec{\phi}(\mathbf{x}) \qquad [60]$$

where $\vec{j}(\mathbf{x})$ is the nucleon current operator at the position $\mathbf{x}$. Since $\vec{\phi}(\mathbf{x})$ is pseudoscalar and isovector, $\vec{j}(\mathbf{x})$ must also be pseudoscalar and isovector, to give a scalar, isoscalar interaction H_I. Therefore $\vec{j}(\mathbf{x})$ also carries an isovector index α: $j_\alpha(\mathbf{x})$, and the dot in Eq. 60 indicates a scalar product in isospin space (i.e., $\vec{j} \cdot \vec{\phi} = \Sigma_\alpha j_\alpha \phi_\alpha$, $\alpha = 1, 2, 3$).

The simplest nonrelativistic current operator with these properties is of the form

$$j_\alpha(\mathbf{x}) = (4\pi)^{1/2} i \frac{f_0}{m_\pi} \tau_\alpha \boldsymbol{\sigma} \cdot \nabla, \qquad [61]$$

where the gradient acts on the field $\phi_\alpha(\mathbf{x})$ only, and τ_α, $\boldsymbol{\sigma}$ are the nucleon (Pauli) isospin and spin operators. The normalization and dimensionless coupling constant f_0 are conventional. If we consider a Dirac nucleon, the simplest operators take the forms

$$j_\alpha(\mathbf{x}) = i g_0 \tau_\alpha \gamma_5 \qquad \text{(PS)}, \qquad [62a]$$

$$j_\alpha(\mathbf{x}) = \frac{g_0}{2M} \tau_\alpha \gamma_5 \gamma \cdot \nabla \qquad \text{(PV)}, \qquad [62b]$$

where the four-gradient ∇ acts on the field ϕ_α. (The time derivative here is actually $\gamma_0 \omega$.) The terms pseudoscalar (PS) and pseudovector (PV) are used to denote the two forms of coupling to the nucleon given in Eq. 62, although both operators in Eq. 62 are actually pseudoscalar, and combine with the pseudoscalar pion field $\phi(x)$ to give a scalar interaction in Eq. 60 (in the πN space). The relativistic forms in Eq. 62 have the proper Lorentz properties of a current. The nonrelativistic current in Eq. 61 is not even

Galilei invariant, and properly belongs to a static theory in which nucleons are fixed (see Section 7.2 on Galilei invariance).

The absorption or emission of a pion by a nucleon are induced by the interaction given by Eq. 60. If we take the matrix element of H_I between plane wave states for absorption of a meson by a nucleon, $(\pi + N \to N)$, we have

$$\langle \mathbf{p}'\beta | H_I | \mathbf{p}\alpha, \mathbf{k}\mu \rangle = \int d\mathbf{x}\, N_k e^{i\mathbf{k}\cdot\mathbf{x}} \langle \mathbf{p}'\beta | j_\mu(\mathbf{x}) | \mathbf{p}\alpha \rangle$$

$$= \int d\mathbf{x}\, N_k e^{i(\mathbf{k}+\mathbf{p}-\mathbf{p}')\cdot\mathbf{x}} \langle \mathbf{p}'\beta | j_\mu(0) | \mathbf{p}\alpha \rangle \qquad [\mathbf{63a}]$$

$$= (2\pi)^3 N_k \delta(\mathbf{p}' - \mathbf{p} - \mathbf{k}) \langle \mathbf{p}'\beta | j_\mu(0) | \mathbf{p}\alpha \rangle$$

for nucleon momentum $\mathbf{p}$, $\mathbf{p}'$, isospin-spin α, β in the initial and final states, respectively, and a pion of momentum $\mathbf{k}$ and charge μ. We have used the transformation $j(\mathbf{x}) = e^{-i\mathbf{p}_{op}\cdot\mathbf{x}} j(0) e^{i\mathbf{p}_{op}\cdot\mathbf{x}}$, where $\mathbf{p}_{op}$ is the nucleon momentum operator. For emission, we take the hermitian conjugate matrix element of Eq. 63a. The two processes are illustrated in Fig. 2.7. Note that the matrix element in Eq. 63a is off-energy-shell, since energy is not conserved: $E(p') \neq E(p) + \omega_k$.

Now there are two interpretations we may give to the matrix element in Eq. 63a. From the point of view of perturbation theory, the plane wave states are unperturbed states of the pion and nucleon, with Eq. 60 as the perturbing interaction. The particles have unrenormalized masses not corresponding to free physical particles. In this case, the matrix element of Eq. 63a is evaluated by straightforward calculation, using Eqs. 58–60. For the static current operator given in Eq. 61, we obtain

$$\langle \mathbf{p}'\beta | j_\mu(0) | \mathbf{p}\alpha \rangle = -(4\pi)^{1/2} \frac{f_0}{m_\pi} \langle \beta | \boldsymbol{\sigma} \cdot \mathbf{k}\, \tau_\mu | \alpha \rangle \qquad [\mathbf{64}]$$

with similar evaluation for the relativistic cases given in Eq. 62. For relativistic calculation, we prefer to calculate amplitudes in a covariant manner, as in Feynman diagrams, rather than in the fixed time manner of Eq. 63a. The results differ only in the use of four-momenta and a consequent conservation of energy (but not of mass); hence the terms

$$\delta^4(p' - p - k) \langle p'\beta | j_\mu(0) | p\alpha \rangle \qquad [\mathbf{63b}]$$

in four-momenta would replace the corresponding three-momenta terms in Eq. 63a.

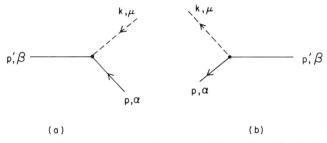

Fig. 2.7 Representation of the matrix elements in Eq. 63 for $\pi N \leftrightarrow N$, showing the kinematic variables.

The second method of treating the transition (absorption) amplitude in Eq. 63a is to use plane wave states of physical pions and nucleons, which propagate freely but with their proper masses. Now the matrix element in Eq. 63a contains terms to all orders in the interaction given by Eq. 58, since the nucleon lines and the π-N vertex are now *renormalized*. The result can be a very complicated function of the plane-wave variables, and is not usually directly calculable from the interaction H_I. However, we can generally characterize the renormalized matrix element in Eq. 63a in terms of a new (renormalized) coupling constant and a *vertex function* or *form factor*, which gives the dependence on the kinematic variables. For the static current operator in Eq. 61, we obtain

$$\langle \mathbf{p}'\beta | j_\mu(0) | \mathbf{p}\alpha \rangle = -(4\pi)^{1/2} \frac{f}{m_\pi} \langle \beta | \boldsymbol{\sigma} \cdot \mathbf{k} \tau_\mu | \alpha \rangle F(\mathbf{k}) \qquad [65]$$

where f is the renormalized coupling constant, and $F(\mathbf{k})$ is the *form factor*; the other factors of Eq. 63a are unchanged. In general, F could depend on $(\mathbf{p}+\mathbf{p}')$ as well as on the momentum transfer $\mathbf{k}$, but it is consistent with the static nucleon model to ignore such dependence. The factor $F(\mathbf{k})$ expresses the probability that a physical nucleon will remain a nucleon (in its ground state) after absorbing a pion of momentum $\mathbf{k}$. Conventionally the normalization is given by setting $F(0) = 1$ (thus fixing f); consequently, $|1 - F(\mathbf{k})|$ is a measure of *inelasticity*, that is, the possibility of reaching excited states of the nucleon. In the static model, $F(\mathbf{k})$ may also be interpreted as the Fourier transform of the density distribution of the physical nucleon ground state:

$$F(\mathbf{k}) = \int d\mathbf{k}\, e^{i\mathbf{k}\cdot\mathbf{x}} \rho(\mathbf{x}). \qquad [66]$$

For the relativistic case, we write the matrix element of Eq. 63a in terms of a *vertex function* Γ:

$$\langle p'\beta|j_\mu(0)|p\alpha\rangle = g\,\Gamma_{\beta\alpha}(p',p), \qquad [67]$$

where the renormalized coupling constant g is determined by fixing Γ at one kinematical point, for which the matrix element of Eq. 67 takes its unperturbed form. At this point $p'^2 = p^2 = M^2$ and $k^2 = m_\pi^2$ (not conserving total mass!). In particular, for the PS case (Eq. 62a),

$$\langle p'\beta|j_\mu(0)|p\alpha\rangle \rightarrow ig\langle p'\beta|\tau_\mu\gamma_5|p\alpha\rangle$$
$$\left(p'^2 = p^2 = M^2, k^2 = m_\pi^2\right). \qquad [68]$$

This particular kinematic point is of further interest, as we show below.

Further, if we evaluate Eq. 68 in the static limit, $|\mathbf{k}|, |\mathbf{p}|, |\mathbf{p}'| \ll M$, we find that Eq. 68 goes over to the static form of Eq. 65 with $F\rightarrow 1$. The magnitudes will agree if the coupling constants are related by

$$f^2 = \frac{g^2}{4\pi}\left(\frac{m_\pi}{2M}\right)^2. \qquad [69]$$

(This limit is also obtained for the PV case in Eq. 62b.) We find that this (dimensionless) quantity, which characterizes the Yukawa coupling of π and N, is measurable.

2.4b Born Terms

In perturbation theory, the lowest order πN scattering process induced by H_I (Eq. 60) is analogous to Compton scattering of photons by charged particles. There are two terms, illustrated in Fig. 2.8, referred to as *Born*

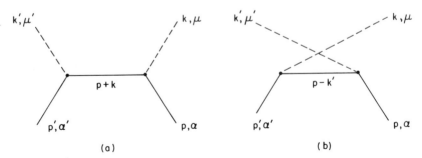

Fig. 2.8 Diagrams for Born terms: (*a*) direct; (*b*) crossed.

terms: (a) *direct* and (b) *crossed*. In the first term, absorption of the pion precedes emission; in the second, the order is reversed. Let us consider these terms, but with the interpretation that the nucleons are to be considered as physical plane-wave states, and the vertices for absorption or emission ($\pi + N \leftrightarrow N$) as renormalized (Eqs. 65 and 67).

Let us calculate the Born amplitudes for a nonrelativistic nucleon, using the static approximation in Eq. 65. We may neglect energy due to nucleon motion, treating the nucleons as fixed (lab frame) except for conservation of momentum at vertices. We write the Born t-matrix (half off-energy-shell, $E_a = \omega_k + M$) as

$$\langle \mathbf{k}', \mathbf{p}', \beta | T(E_a) | \mathbf{k}, \mathbf{p}, \alpha \rangle = \delta(\mathbf{k}' + \mathbf{p}' - \mathbf{k} - \mathbf{p})(2\pi)^3$$

$$\times \left[\langle \mathbf{k}' \beta | T_a(\omega_k) | \mathbf{k} \alpha \rangle + \langle \mathbf{k}' \beta | T_b(\omega_k) | \mathbf{k} \alpha \rangle \right],$$

$$[\,70\,]$$

where we have separated the direct (subscript a) and crossed (subscript b) terms. We calculate in second order, using Eq. 65, assuming $F \to 1$:

$$\langle \mathbf{k}' \beta | T_a(\omega_k) | \mathbf{k} \alpha \rangle = \frac{4\pi f^2}{\omega_k m_\pi^2 \sqrt{4\omega_{k'}\omega_k}} (\boldsymbol{\sigma} \cdot \mathbf{k}')(\boldsymbol{\sigma} \cdot \mathbf{k}) \tau_\beta \tau_\alpha, \qquad [\,71a\,]$$

$$\langle \mathbf{k}' \beta | T_b(\omega_k) | \mathbf{k} \alpha \rangle = \frac{4\pi f^2}{(-\omega_{k'}) m_\pi^2 \sqrt{4\omega_{k'}\omega_k}} (\boldsymbol{\sigma} \cdot \mathbf{k})(\boldsymbol{\sigma} \cdot \mathbf{k}') \tau_\alpha \tau_\beta. \quad [\,71b\,]$$

(We have omitted the pion isospin indices.) Note that the two amplitudes given in Eq. 71 obey a *crossing relation*: $T_a \leftrightarrow T_b$ if we interchange $\omega_k \leftrightarrow -\omega_{k'}$, $\mathbf{k} \leftrightarrow -\mathbf{k}'$, and $\alpha \leftrightarrow \beta$. The entire amplitude in Eq. 70 is invariant under this interchange; it is *crossing symmetric*.

The amplitudes in Eq. 71 may be decomposed into isospin channels and partial waves by simple techniques. First, for isospin we use

$$\tau_\alpha \tau_\beta = \delta_{\alpha\beta} + i\tau_\gamma \varepsilon_{\alpha\beta\gamma} \qquad [\,72a\,]$$

where $\varepsilon_{\alpha\beta\gamma} = \pm 1$ for cyclic (anticyclic) permutations of (1,2,3). We may compare this to the matrix elements of the operator $\vec{I}_\pi \cdot \vec{\tau}$ in the (cartesian) isospin states of the pion (α, β):

$$\langle \beta | \vec{I}_\pi \cdot \vec{\tau} | \alpha \rangle = i\tau_\gamma \varepsilon_{\alpha\beta\gamma}. \qquad [\,72b\,]$$

Hence, for matrix elements,

$$\langle \beta | \tau_\beta \tau_\alpha | \alpha \rangle = \langle \beta | 1 - \vec{I}_\pi \cdot \vec{\tau} | \alpha \rangle$$

$$\langle \beta | \tau_\alpha \tau_\beta | \alpha \rangle = \langle \beta | 1 + \vec{I}_\pi \cdot \vec{\tau} | \alpha \rangle. \qquad [72c]$$

These can be rewritten in terms of Q_T (Eq. 16a):

$$\langle \beta | Q_{1/2} | \alpha \rangle = \tfrac{1}{3} \langle \beta | \tau_\beta \tau_\alpha | \alpha \rangle,$$

$$\langle \beta | Q_{3/2} | \alpha \rangle = \langle \beta | 1 - \tfrac{1}{3} \tau_\beta \tau_\alpha | \alpha \rangle. \qquad [73]$$

From this it is seen that the direct term in Eq. 71a contributes only to the $T = \tfrac{1}{2}$ channel; this is clear, since the intermediate state (Fig. 2.8a) is a nucleon with $T = \tfrac{1}{2}$. The crossed term in Eq. 71b has both $T = \tfrac{1}{2}$ and $\tfrac{3}{2}$.

The spin and angular dependence of Eq. 71 may be decomposed, using the spin analogy of Eq. 72a:

$$(\sigma \cdot \mathbf{k}')(\sigma \cdot \mathbf{k}) = \mathbf{k}' \cdot \mathbf{k} + i \sigma \cdot \mathbf{k}' \times \mathbf{k}. \qquad [74a]$$

These terms clearly correspond to the $l = 1$ terms of the partial wave expansion in Eq. 12. Further, one may construct projection operators for the $l = 1$, $J = \tfrac{1}{2}$ and $\tfrac{3}{2}$ partial waves (following, e.g., Bjo–64, p. 235):

$$\langle \mathbf{k}' | P_{1/2} | \mathbf{k} \rangle = \frac{1}{4\pi k k'} (\sigma \cdot \mathbf{k}')(\sigma \cdot \mathbf{k})$$

$$\langle \mathbf{k}' | P_{3/2} | \mathbf{k} \rangle = \frac{3}{4\pi k k'} \left[\mathbf{k}' \cdot \mathbf{k} - \tfrac{1}{3} (\sigma \cdot \mathbf{k}')(\sigma \cdot \mathbf{k}) \right]. \qquad [74b]$$

Again, the direct term in Eq. 71a has only $J = \tfrac{1}{2}$, corresponding to the nucleon. Combining the isospin and J-projection operators in Eqs. 73 and 74b, we have $P_{2T,2J} = Q_T P_J$, where

$$P_{2T,2J} = \sum_{T_3, M_J} |TT_3, JM_J \rangle \langle TT_3, JM_J|. \qquad [75]$$

We may write the amplitudes in Eq. 71 in terms of partial waves:

$$\langle \mathbf{k}', \beta\mu' | T_s(\omega_k) | \mathbf{k}, \alpha\mu \rangle$$

$$= \frac{2\pi f^2 k k'}{m_\pi^2 \sqrt{\omega_k \omega_{k'}}} \sum_{TJ} \frac{C_{2T,2J}^s}{d_s} \frac{4\pi}{3} \langle \mathbf{k}', \beta\mu' | P_{2T,2J} | \mathbf{k}, \alpha\mu \rangle. \qquad [76]$$

where $s = a$ or b and

$$d_a = \omega_k, \quad C_{11}^a = 9, \quad C_{13}^a = C_{31}^a = C_{33}^a = 0,$$
$$d_b = -\omega_{k'}, \quad C_{11}^b = 1, \quad C_{13}^b = C_{31}^b = -2, \quad C_{33}^b = 4.$$

(We have now included all πN spin, isospin labels: α, μ, β, μ'.)

For the on-energy-shell Born t-matrix, we combine $T_a + T_b$ with $\omega_{k'} = \omega_k$ to obtain

$$\langle \mathbf{k}', \beta\mu' | T | \mathbf{k}, \alpha\mu \rangle = \sum_{TJ} T_{2T,2J}(\omega_k) \langle \mathbf{k}', \beta\mu' | P_{2T,2J} | \mathbf{k}, \alpha\mu \rangle,$$

$$\frac{T_{2T,2J}(\omega_k)}{4\pi} = \frac{2\pi f^2 k^2}{3 m_\pi^2 \omega_k^2} C_{2T,2J} \qquad [77\text{a}]$$

where $C = C^a - C^b$: $C_{11} = 8$, $C_{13} = C_{31} = 2$, $C_{33} = -4$. The partial wave amplitudes (Eq. 12) are given by

$$f_{2T,2J}(\omega_k) = -\frac{\omega_k}{2\pi} \frac{T_{2T,2J}(\omega_k)}{4\pi} = -\frac{f^2 k^2}{3 m_\pi^2 \omega_k} C_{2T,2J}, \qquad [77\text{b}]$$

where the 4π in Eq. 77 corresponds to the chosen norm of $P_{2T,2J}$.

The relativistic calculation of the Born terms, based on a covariant formulation and the relativistic currents (Eq. 62), proceeds on somewhat different lines and is not given here (see, e.g., Bjo–64). There are few differences at low energy for the p-wave terms: aside from kinematic corrections for the recoil of the nucleon the same results in Eq. 77 obtain for the on-energy-shell amplitudes with Eq. 69 relating f and g. (However, the off-mass-shell Feynman amplitudes differ from those of Eq. 76.) The relativistic Born terms also include s-wave (and $l > 1$) πN scattering unlike the static model, which gives only p-wave scattering. These extra waves are generated by nucleon motion in the crossed term, and by the coupling of the pseudoscalar and pseudovector currents (Eq. 62) to *antinucleon* intermediate states.

The forward-scattering Born amplitudes have singularities as a function of ω_k. This singularity is a pole at $\omega_k = 0$ in the static case, which becomes two poles at $\omega_k = \pm(m_\pi^2/2M)$ in the relativistic case (for lab frame). (These are not, of course, physically possible values of ω_k, which equals $\sqrt{m_\pi^2 + k^2}$ on-mass-shell.) The value of the amplitudes at the pole defines the value of f^2 or g^2. The contribution of the pole term is extracted from the evaluation of forward dispersion relations for $\pi^{\pm}p$ scattering, which makes it possible to obtain a (reasonably) model independent evaluation of the coupling constants f or g (Eq. 69). A recent analysis of experimental data (Bug–73) gives

$$f^2 = 0.079 \pm 0.001. \qquad [78]$$

Having determined the coupling constant, we may return to the Born amplitudes for p-wave scattering in the static approximation (Eq. 77) and see what connection, if any, there is with the experimentally determined partial wave amplitudes. Presumably the Born approximation should be best at low energy: it will certainly not produce resonant behavior and will not even remain unitary at high energy. It is interesting to compare the scattering volumes (Eq. 14b)

$$a_{2T,2J} = -\frac{1}{3}\frac{f^2}{m_\pi^3}C_{2T,2J} \qquad \text{(static approximation)} \qquad [79]$$

with the experimental results (Table 2.2). One finds that the signs of Eq. 79 and the rough order-of-magnitude set by Eq. 78 are in agreement with the tabulated values, but the relative sizes of the different $a_{2T,2J}$ are wrong. For predictions of the s-waves, one must use the relativistic Born terms (e.g., Bjo–64, Kal–64), and one finds that the Born amplitudes agree poorly with the scattering lengths (Eq. 14a) of Table 2.2, for example, and that the predictions for PS and for PV differ in magnitude, although they lead to the same Born amplitudes for p-wave. Clearly the Born terms alone do not provide a theory of πN scattering.

2.4c Chew-Low Theory

Chew and Low (Che–56) introduced a theoretical approach to πN scattering which goes well beyond the Born approximation, but which makes use of the qualitative connection between the Born terms and experiment, at very low energies, for the p-waves (Eq. 79). We review this approach, since it provides a framework for dealing with the strong interactions of the πN system and has had some success in giving a theory of p-wave scattering for $T_\pi \lesssim 300$ MeV, that is, where elastic πN scattering dominates (see Section 2.2). Specifically, this approach leads to a resonance in the 3, 3 partial wave, and not in the other p-waves in this energy domain. We develop the theory of Chew and Low in the *static model* to which it was originally applied; this gives the simplest formulation (see the review article of Wic–55). This should be considered only an outline: modern developments have much extended the methods and have removed some of the more severe approximations used.

The basic method of the Chew-Low formulation is to use only physical states of the system of pions interacting with a nucleon, and thereby to avoid some of the difficulties of perturbation theory with quantum fields. This is what we have already done in interpreting the vertices of Figs. 2.7 and 2.8 in terms of renormalized coupling constants and form-factors or vertex functions (Eqs. 65 and 67). We therefore expect that the Born terms (Eqs. 70 and 71) will appear in this approach.

We work in a Schroedinger formalism, with a Hamiltonian for one (fixed) nucleon plus a pion field interacting with the nucleon through the Yukawa interaction H_I (Eq. 60). We write

$$H = H_0 + V$$

and

$$V = H_I - \delta M.$$

[**80**]

The ground state of this Hamiltonian is the nucleon, with state vector ψ_0 and mass (or energy) M,

$$H\psi_0 = M\psi_0.$$

[**81**]

We have defined H_0 to include the mass renormalization δM, which is therefore subtracted from H_I. Starting with Eq. 60 for H_I and Eq. 58 for $\phi_\alpha(x)$, we perform the x-integration, and obtain H_I in the form

$$H_I = \int d\mathbf{k} N_k \left[j(\mathbf{k})a(\mathbf{k}) + j^\dagger(\mathbf{k})a^\dagger(\mathbf{k}) \right],$$

[**82**]

where

$$j(\mathbf{k}) = \int e^{i\mathbf{k}\cdot\mathbf{x}} j(\mathbf{x})\,d\mathbf{x}, \qquad j^\dagger(\mathbf{k}) = \int e^{-i\mathbf{k}\cdot\mathbf{x}} j(\mathbf{x})\,d\mathbf{x} = j(-\mathbf{k}),$$

and spin and isospin indices are suppressed. For the static model, we assume the current operator to be of the form of Eq. 61; suppressing momentum conservation, we have $j(\mathbf{k}) = -(4\pi)^{1/2}(f_0/m_\pi)\tau_\alpha\boldsymbol{\sigma}\cdot\mathbf{k}$.

For πN scattering, we want scattering eigenstates $\psi_\mathbf{k}^{(+)}$ of H, which can be separated asymptotically as in Eq. 42, into a plane wave χ_a for a meson of momentum $\mathbf{k}$, and the fixed nucleon (target), and outgoing spherical waves. For total energy below threshold for producing additional mesons (e.g., $\pi + N \rightarrow \pi + \pi + N$) the outgoing waves are those for elastically scattered pions; above that threshold, inelastic waves will appear. The plane wave state may be constructed in terms of the nucleon state ψ_0 by creating a plane wave meson with the creation operator $a^\dagger(\mathbf{k})$:

$$\chi_a = a^\dagger(\mathbf{k})\psi_0$$
$$\psi_{\mathbf{k}}^{(+)} = a^\dagger(\mathbf{k})\psi_0 + \phi_{\mathbf{k}}^{(+)} \qquad [83]$$

where $\phi_{\mathbf{k}}^{(+)}$ represents the (outgoing) scattered wave. We calculate the t-matrix (half off-energy-shell) for πN scattering from the formula*:

$$\langle b|T(E_a)|a\rangle = \lim_{\eta \to 0^+} \langle \chi_b|H - M - \omega_{k'}|\psi_{\mathbf{k}}^{(+)}\rangle \qquad [84]$$

where $|\chi_b\rangle = a^\dagger(\mathbf{k}')|\psi_0\rangle$ is a plane wave state with the final asymptotic momentum $\mathbf{k}'$. Commuting $a(\mathbf{k}')$ with H, we may write

$$\langle \mathbf{k}'|T(\omega_k)|\mathbf{k}\rangle = (2\pi)^3\langle b|T(E_a)|a\rangle = (2\pi)^3 N_{k'}\langle \psi_0|j^\dagger(\mathbf{k}')|\psi_{\mathbf{k}}^{(+)}\rangle$$
$$[85]$$

where we have relabeled the t-matrix in the pion momenta $\mathbf{k}, \mathbf{k}'$ and energy ω_k. We have used two of the following commutators:

$$[V, a(\mathbf{k})] = -j^\dagger(\mathbf{k})N_k, \qquad [V, a^\dagger(\mathbf{k})] = j(\mathbf{k})N_k,$$
$$[H_0, a(\mathbf{k})] = -\omega_k a(\mathbf{k}), \qquad [H_0, a^\dagger(\mathbf{k})] = \omega_k a^\dagger(\mathbf{k}). \qquad [86]$$

These follow from Eq. 82 with normal boson commutation relations $[a(\mathbf{k}'), a^\dagger(\mathbf{k})] = \delta(\mathbf{k}' - \mathbf{k})$; H_0 includes the kinetic energy of free pions. (Note that χ_a is normed $\langle \chi_b|\chi_a\rangle = (2\pi)^{-3}\langle \mathbf{k}'|\mathbf{k}\rangle = \delta(\mathbf{k}' - \mathbf{k})$.)

To evaluate Eq. 85, we first write the Schroedinger equation for the scattering state given in Eq. 83 in the form

$$(H - M - \omega_k)a^\dagger(\mathbf{k})|\psi_0\rangle = (\omega_k + M - H)|\phi_{\mathbf{k}}^{(+)}\rangle. \qquad [87]$$

Using Eqs. 81 and 86 to evaluate the left-hand side of Eq. 87, we obtain

$$N_k j(\mathbf{k})|\psi_0\rangle = (\omega_k + M - H)|\phi_{\mathbf{k}}^{(+)}\rangle \qquad [88a]$$

which can be inverted to solve for $|\phi_{\mathbf{k}}^{(+)}\rangle$:

$$|\phi_{\mathbf{k}}^{(+)}\rangle = N_k(\omega_k + i\eta + M - H)^{-1}j(\mathbf{k})|\psi_0\rangle, \qquad [88b]$$

---

*This formula is a generalization of Eq. 43a, and is discussed in Gol–64, Chapter 5. This result allows us to proceed directly to Eq. 85, which was obtained by an alternate but equivalent method in Wic–55, Che–56. See also Eqs. 6.22, 6.23 below, and footnote, p.278.

where $+i\eta$ gives outgoing waves. The scattered wave given in Eq. 83, with Eq. 88b, may be inserted in Eq. 85 to yield

$$(2\pi)^{-3}\langle \mathbf{k}'|T(\omega_k)|\mathbf{k}\rangle = N_{k'}\langle \psi_0|j^\dagger(\mathbf{k}')a^\dagger(\mathbf{k})|\psi_0\rangle$$
$$+ N_{k'}N_k\langle \psi_0|j^\dagger(\mathbf{k}')(\omega_k + M - H + i\eta)^{-1}j(\mathbf{k})|\psi_0\rangle.$$
$$[89]$$

To evaluate the first term of Eq. 89, we first find $\langle \psi_0|a^\dagger(\mathbf{k})$, using the commutators in Eq. 86 as follows:

$$\langle \psi_0|a^\dagger(\mathbf{k})H = \langle \psi_0|a^\dagger(\mathbf{k})(M - \omega_k) - \langle \psi_0|j(\mathbf{k})N_k,$$
$$\langle \psi_0|a^\dagger(\mathbf{k}) = -N_k\langle \psi_0|j(\mathbf{k})(\omega_k - M + H)^{-1}.$$
$$[90]$$

Then

$$N_{k'}\langle \psi_0|j^\dagger(\mathbf{k}')a^\dagger(\mathbf{k})|\psi_0\rangle = N_{k'}\langle \psi_0|a^\dagger(\mathbf{k})j^\dagger(\mathbf{k}')|\psi_0\rangle$$
$$= -N_{k'}N_k\langle \psi_0|j(\mathbf{k})(\omega_k - M + H)^{-1}j^\dagger(\mathbf{k}')|\psi_0\rangle,$$
$$[91]$$

and the t-matrix in Eq. 89 may be written in the form

$$\langle \mathbf{k}'|T(\omega_k)|\mathbf{k}\rangle = (2\pi)^3 N_{k'}N_k\big[\langle \psi_0|j^\dagger(\mathbf{k}')(\omega_k + M - H + i\eta)^{-1}j(\mathbf{k})|\psi_0\rangle$$
$$- \langle \psi_0|j(\mathbf{k})(\omega_k - M + H)^{-1}j^\dagger(\mathbf{k}')|\psi_0\rangle\big].$$
$$[92]$$

This form exhibits the *crossing symmetry* of the t-matrix: the two terms of Eq. 92 interchange, if $\mathbf{k}\leftrightarrow-\mathbf{k}'$ and $\omega_k\leftrightarrow-\omega_k$ (and $\alpha\leftrightarrow\beta$, had we kept the labels). (We assume that $j(\mathbf{k})$ has a definite parity: $j(-\mathbf{k}) = \pm j(\mathbf{k})$, as in Eq. 61. Also, we must include an otherwise extraneous $+i\eta$ in Eq. 91, to complete the symmetry.) Note that this differs from the crossing relation for the Born terms in Eq. 71, for which $\omega_k\leftrightarrow-\omega_k'$, although the distinction disappears on-energy-shell. The contribution of the "crossed" term to the scattering comes from mesons already in the nucleon ground state, which may be emitted $[j^\dagger(\mathbf{k}')]$ before the incoming pion is absorbed $[j(\mathbf{k})]$. Equation 92 is the dynamical equation we use for the πN t-matrix. Except for the suppression of nucleon motion (static model), the equation is exact for a Yukawa interaction (Eqs. 60 and 82). The t-matrix may now be defined completely off-energy-shell by replacing ω_k in Eq. 92 by a free energy parameter ω. Because of the ω-dependence of the potential term in Eq. 91, this off-energy-shell extension differs from that given earlier in Eq. 48, based on the Lippmann–Schwinger equation.

We may put the dynamical equation given in Eq. 92 in a form suitable for approximation and calculation by introducing a complete set of states $\{n\}$ of the Hamiltonian H into the expression and expanding:

$$\langle \mathbf{k}' | T(\omega) | \mathbf{k} \rangle = (2\pi)^3 N_{k'} N_k \sum_n \left\{ \frac{\langle \psi_0 | j^\dagger(\mathbf{k}') | n \rangle \langle n | j(\mathbf{k}) | \psi_0 \rangle}{\omega + M - E_n} - \frac{\langle \psi_0 | j(\mathbf{k}) | n \rangle \langle n | j^\dagger(\mathbf{k}') | \psi_0 \rangle}{\omega - M + E_n} \right\} \qquad [93]$$

where we have given the completely off-energy-shell amplitude. This equation is often called the Low equation (Low–55). The right-hand side of Eq. 93 is a sum (or integral) of quantities that are expressed in terms of matrix elements of the nucleon current operator, between physical states of the system, that is, the nucleon (ground) state ψ_0, and the states $\{n\}$ of one (physical) nucleon plus any number of pions. Useful approximations may be obtained by limiting the sum to a few particular states, whose contribution might be expected to dominate the series (Eq. 93).

If we specifically include in Eq. 93 only the intermediate states of one nucleon and no pions (summing over the nucleon spin and isospin indices γ in the intermediate state), we obtain

$$\langle \mathbf{k}' | T^{(0)}(\omega) | \mathbf{k} \rangle = (2\pi)^3 \frac{N_{k'} N_k}{\omega} \sum_\gamma \left\{ \langle \psi_0 | j^\dagger(\mathbf{k}') | \psi_0 \gamma \rangle \langle \psi_0 \gamma | j(\mathbf{k}) | \psi_0 \rangle - \langle \psi_0 | j(\mathbf{k}) | \psi_0 \gamma \rangle \langle \psi_0 \gamma | j^\dagger(\mathbf{k}') | \psi_0 \rangle \right\} \qquad [94]$$

where $E_n = M$ for the nucleon. This expression clearly gives the Born amplitudes illustrated in Fig. 2.8: the first and second terms on the right-hand side of Eq. (94) are the direct (a) and crossed (b) terms. For the static model we take the current matrix elements to be of the form of Eq. 65; hence Eq. 94 may be evaluated explicitly:

$$\langle \mathbf{k}' \beta | T^{(0)}(\omega) | \mathbf{k} \alpha \rangle = \frac{4\pi f^2 (2\pi)^3 N_{k'} N_k F(\mathbf{k}') F(\mathbf{k})}{\omega m_\pi^2} \left\{ (\boldsymbol{\sigma} \cdot \mathbf{k}')(\boldsymbol{\sigma} \cdot \mathbf{k}) \tau_\beta \tau_\alpha - (\boldsymbol{\sigma} \cdot \mathbf{k})(\boldsymbol{\sigma} \cdot \mathbf{k}') \tau_\alpha \tau_\beta \right\} \quad \text{(static model, nucleon term).}$$

$$[95]$$

Note that this is *almost* the same result as we obtained in Eq. 71, using second order perturbation theory for the Born terms, with the renormalized coupling constant f. Since we have

$$N_k = \frac{1}{\sqrt{(2\pi)^3 2\omega_k}},$$

Eq. 95 agrees with Eq. 71 *except* for the form factors in the numerator of Eq. 95, which were dropped in Eq. 71, and the denominator ω in Eq. 95 versus ω_k and $\omega_{k'}$ in Eq. 71a, b. (The half on-energy-shell t-matrix corresponding to Eq. 71 is obtained by setting $\omega = \omega_k$ in Eq. 95; this differs from the perturbation result in Eq. 71 in the crossed term.) The separation of Eq. 95 into isospin and partial wave amplitudes proceeds exactly as in Eqs. 72–76, leading to the (fully off-energy-shell) matrix for the T, J partial wave:

$$\left\langle k' \left| T^{(0)}_{2T,2J}(\omega) \right| k \right\rangle = \frac{2\pi f^2 k' k}{3 m_\pi^2 \omega} \cdot \frac{F(k')F(k)}{(\omega_{k'}\omega_k)^{1/2}} C_{2T,2J}$$

(static model, nucleon term) **[96]**

with the coefficients C defined in Eq. 77. This expression goes over into the on-energy-shell expression for the Born amplitude (Eq. 77a), for $F(k) \to 1$, and $\omega = \omega_{k'} = \omega_k$. Note that this amplitude is separable in the sense defined in Eqs. 55 and 56. (This is a consequence, in part, of the static assumption.)

We may obtain a more complete dynamical equation from Eq. 93 by extending the set of states $\{n\}$ to include scattering states of one pion and one nucleon, as well as the one nucleon states. To get the standard form of the Chew-Low one meson approximation (Che–56, Wic–55) we take the states to be scattering states $\psi_l^{(-)}$ with *ingoing* scattered wave (compare Eq. 83) with energy $\omega_l + M$. Inserting these states into Eq. 93, and adding the nucleon term in Eq. 94, we obtain

$$\langle \mathbf{k}' | T(\omega) | \mathbf{k} \rangle = \langle \mathbf{k}' | T^{(0)}(\omega) | \mathbf{k} \rangle$$

$$+ (2\pi)^3 N_{k'} N_k \sum \int dl \left\{ \frac{\langle \psi_0 | j^\dagger(\mathbf{k}') | \psi_l^{(-)} \rangle \langle \psi_l^{(-)} | j(\mathbf{k}) | \psi_0 \rangle}{\omega - \omega_l} \right.$$

$$\left. - \frac{\langle \psi_0 | j(\mathbf{k}) | \psi_l^{(-)} \rangle \langle \psi_l^{(-)} | j^\dagger(\mathbf{k}') | \psi_0 \rangle}{\omega + \omega_l} \right\}, \qquad \textbf{[97]}$$

where we have not explicitly given the spin-isospin labels, which would be summed in the intermediate state. The matrix elements in the second term of Eq. 97 may be evaluated using the steps leading to Eq. 85 to obtain

$$\langle \mathbf{k}' | T(\omega_k) | \mathbf{k} \rangle = (2\pi)^3 \langle \psi_{\mathbf{k}'}^{(-)} | j(\mathbf{k}) | \psi_0 \rangle N_k$$
$$= (2\pi)^3 \langle \psi_0 | j^\dagger(\mathbf{k}) | \psi_{\mathbf{k}'}^{(-)} \rangle^* N_k. \qquad [98]$$

Since the static current operator in Eq. 65 is hermitian (for "cartesian" isospin), we may replace $j^\dagger$ by j in evaluating Eq. 97, in terms of the t-matrices given in Eq. 98:

$$\langle \mathbf{k}' | T(\omega) | \mathbf{k} \rangle = \langle \mathbf{k}' | T^{(0)}(\omega) | \mathbf{k} \rangle$$
$$+ \int \frac{d\mathbf{l}}{(2\pi)^3} \left\{ \frac{\langle \mathbf{k}' | T(\omega_l) | \mathbf{l} \rangle^* \langle \mathbf{l} | T(\omega_l) | \mathbf{k} \rangle}{\omega - \omega_l} - \frac{\langle \mathbf{k} | T(\omega_l) | \mathbf{l} \rangle^* \langle \mathbf{l} | T(\omega_l) | \mathbf{k}' \rangle}{\omega + \omega_l} \right\}$$

(static model, one meson approximation). $\qquad [99]$

This is the Chew-Low equation in the one meson approximation, often called just the Chew-Low approximation. It is an integral equation that gives the fully off-energy-shell t-matrix, or the half off-energy-shell t-matrix, in terms of integrals over *products* of two half off-energy-shell t-matrices:—a nonlinear equation. It provides a dynamical equation for πN scattering, analogous to the Lippmann-Schwinger equation given in Eqs. 46 and 47 for potential scattering. In Eq. 99 the original interaction in Eq. 80 does not appear explicitly; the inhomogeneous term in the integral equation is the one nucleon (Born) term $T^{(0)}$, which does, of course, depend on the interaction. What has been left out of the approximation is the effect of states of more than one meson interacting with the physical nucleon; these are the *inelastic* channels in πN scattering. The energies of these states are $E_n \geqslant M + 2m_\pi$, and therefore do not contribute to the *singularities* of Eq. 93, as long as ω is restricted to lie below the threshold for π production $(\omega < 2m_\pi)$. The effect of including these states on the integral equation given in Eq. 99 for the πN elastic scattering t-matrix would be an addition to the inhomogeneous term $T^{(0)}$, with no new singularity for $\omega < 2m_\pi$, as mentioned. To some extent, this effect may be simulated by allowing the form factor $F(k)$ of $T^{(0)}$ to be an adjustable function, to be determined in part by the experimental πN elastic scattering data.

The structure of the Chew-Low approximation (Eq. 99) may be clarified by the diagrammatic representation of Fig. 2.9. Here (*a*) and (*b*) are the

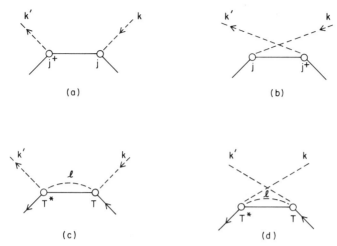

Fig. 2.9 Diagrammatic representation of the Chew-Low one-meson approximation, given in Eq. 99.

two terms of Eq. 95, the Born terms, direct and crossed, with a nucleon intermediate state. The one-meson-and-one-nucleon terms of Eqs. 97 and 99 are shown in (c) and (d), direct and crossed terms, respectively. The circles in (a) and (b) represent the current vertices $(j, j^\dagger)$, and in (c) and (d), the t-matrices (T, T^*).

The Chew-Low equations given in Eq. 99 may be considerably simplified by partial wave decomposition of the t-matrices. The only nonzero terms are in partial waves for which the one-nucleon term $T^{(0)}$ contributes; these are the p-waves with $TJ = (\frac{1}{2}, \frac{1}{2}), (\frac{1}{2}, \frac{3}{2}), (\frac{3}{2}, \frac{1}{2}), (\frac{3}{2}, \frac{3}{2})$ as in Eqs. 76 and 96. (This restriction is a feature of the *static approximation*; nucleon recoil mixes other partial waves, for example, s-waves, as mentioned earlier.) The crossed (last) term of Eq. 99 is responsible for coupling the equations for different TJ: although the TJ for a pion-nucleon scattering is conserved by each t-matrix, the t-matrices that enter the crossed term may carry $T'J' \neq TJ$. This is because the crossing of the pion lines in Fig. 2.9d represents a recoupling of angular momentum at each vertex (t-matrix). Note that the initial and final *momenta* k, k' have interchanged roles in (d); the initial and final angular momenta are similarly interchanged, requiring a recoupling (see, e.g., Kal–64 or Sch–61 for details). The result is that after recoupling the partial wave integral equations form a coupled set of four equations; if the crossing term is neglected, the four TJ equations are decoupled.

A further simplification follows from the separable form of the nucleon term $T^{(0)}$ in the variables k, k', and ω for each partial wave TJ (Eq. 96).

(This feature also follows from the static approximation.) As in the potential scattering case (Eq. 56), the separability of the inhomogeneous term leads to the separability of the off-energy-shell *t*-matrix for each *TJ*:

$$(k'|T_{2T,2J}(\omega)|k) = - \frac{2\pi k' kF(k')F(k)}{(\omega_{k'}\omega_k)^{1/2}} h_{2T,2J}(\omega), \qquad [100]$$

as can be verified by substitution in Eq. 99. The recoupling fixes the coefficients in the equations for the $h_{2T,2J}$. It is conventional to replace the labels $2T$, $2J$ by a single label α, where $\alpha = 1$ for P_{11}, $\alpha = 2$ for *either* P_{13} or P_{31} (both enter symmetrically, since $C_{13} = C_{31}$ in Eq. 96) and $\alpha = 3$ for P_{33}. Then the equations for $h_\alpha(\omega)$ may be put in the form

$$h_\alpha(\omega) = \frac{\lambda_\alpha}{\omega} - \int_0^\infty \frac{l^2\,dl}{\pi} \frac{l^2 F^2(l)}{\omega_l} \left\{ \frac{|h_\alpha(\omega_l)|^2}{\omega - \omega_l + i\eta} + \sum_\beta A_{\alpha\beta} \frac{|h_\beta(\omega_l)|^2}{\omega + \omega_l} \right\}$$

$$[101]$$

with

$$\lambda_\alpha = \frac{2f^2}{3m_\pi^2} \begin{Bmatrix} -4 \\ -1 \\ 2 \end{Bmatrix} \qquad \text{for} \quad \alpha = \begin{Bmatrix} 1 \\ 2 \\ 3 \end{Bmatrix},$$

and the "crossing matrix" $A_{\alpha\beta}$ given by Che–56 and Kal–64:

$$\frac{1}{9} \begin{bmatrix} 1 & -8 & 16 \\ -2 & 7 & 4 \\ 4 & 4 & 1 \end{bmatrix}.$$

(We have inserted the $i\eta$ in Eq. 101 for the on-energy-shell case.) The solutions to the coupled *nonlinear* integral equations (Eq. 101) depend on the coupling constant f^2 and the form factor $F(l)$. Because of the nonlinearity, the properties of solutions may be quite complicated and are not uniquely determined without further conditions. We do not discuss these questions or the methods of solution here; the interested reader is referred to the growing literature (Che–56, Sal–57, Cas–56). It is interesting that if the crossing term is neglected, Eq. 101 has a solution in closed form:

$$h_\alpha(\omega) = \frac{\lambda_\alpha}{\omega} \left[1 + \frac{\lambda_\alpha \omega}{\pi} \int_0^\infty l^2\,dl \frac{F^2(l)l^2}{\omega_l^3(\omega - \omega_l + i\eta)} \right]^{-1} \qquad [102]$$

which is reminiscent of the solution given in Eq. 56 for a separable potential. In fact, an equivalent *energy dependent* potential can be found, which produces the same *t*-matrix as Eq. 102 (see Dov–74).

Now it is consistent with the static assumptions that the form factor $F(l)$ should be a slowly varying function of l for low momentum, presumably decreasing above some "cutoff" momentum of the order of Mc. The model presumably is only valid for momenta well below the cutoff. (Note that the integrals in the static model diverge unless there is some cutoff imposed: This is a limitation of the model.) With the assumed behavior of $F(l)$, the integral in Eq. 102 becomes approximately constant as a function of ω, with *negative* real part. Now, if the sign of λ_α is negative, the denominator in square brackets in Eq. 102 grows in magnitude with increasing ω. Since the factor λ_α/ω is simply the single nucleon contribution to h_α (Eqs. 96 and 100), the result is a decreasing modification of the single nucleon term (Eq. 96) as ω grows. This is the prediction for the P_{11}, P_{13}, and P_{31} partial waves, for which λ_α is negative ($\alpha = 1, 2$) in Eq. 101.

For the P_{33} partial wave, λ_α is positive ($\alpha = 3$), and the behavior of Eq. 102 is totally different. Now it is possible for the real part of the denominator of Eq. 102 to vanish for some value of ω, depending on the value of λ_3, or equivalently, of f^2, as well as of the integral itself (and therefore, of the cutoff parameters.) At this point, the *t*-matrix is purely imaginary. If we construct the partial wave amplitude $f_{33}(\omega_k)$ from Eqs. 77b, 100 and 102 on-energy-shell, we find that it has a resonant form, as in Eq. 32a, with an "effective range" from which we may write

$$\frac{k^3 \cot\delta_{33}}{\omega_k} = \frac{1}{\lambda_3 F^2(k)}(1 - \omega_k r_3). \qquad [103]$$

This is similar to the formula given in Eq. 33a with the nonconstant $\lambda = \lambda_3 F^2(k)$, which gives additional momentum dependence through $F^2(k)$ to the width given in Eq. 33b. The prediction of the *possibility* of a resonance in the P_{33} channel, and not in the other partial waves, is the major success of the Chew-Low static approximation we have just discussed. The quantitative solution of the model requires fixing the form factors and coupling constant and solving the full equation given in Eq. 101, including crossing. As long as the effective range approximation given by Eq. 103 remains valid, for small momentum, the coupling constant can be extracted from it by extrapolating the phase shift data to $\omega_k = 0$. This procedure leads to a value of

$$f^2 = 0.087 \qquad [104]$$

which is comparable to, but somewhat higher than, the value extracted from forward dispersion relations (Eq. 78).

It is appropriate at this point to say a word more about form factors in πN scattering. The function $F(l)$ we have just discussed serves as a convergence or cutoff factor in the integrand of Eq. 101. Because of the separability of the t-matrix in Eq. 100 in the Chew–Low model, the same form factor gives the off-shell behavior, through the term $F(k')F(k)$, which is the same for all four p-waves. Thus the "range" of the t-matrix in all channels is connected through Eq. 101 or Eq. 103, to the position, width, and possibly the shape of the 3,3 resonance. In terms of a function of the form $F(k)=(1+k^2\alpha^{-2})^{-1}$, a range parameter of $\alpha\sim1.4$ $Mc\sim10$ $m_\pi c$ as adopted in Che–56 and Sal–57 is consistent with the resonance parameters, as we have mentioned above. However, if one drops the crossing term in Eq. 101 and uses Eq. 102 instead to fit data, the parameter is found to be smaller, for example, $\alpha\sim3.7$ $m_\pi c$ (Lay–61). This is similar to the range found by Lon–74 for a separable potential fit to phase shifts in the 3,3 channel (if treated nonrelativistically). The ranges already depend on the form of the theory used. In addition, one can include the inelasticity of the πN scattering above the 3,3 energy to improve the fit to phase shifts at higher energies (Baj–75, Ern–78); this also affects the range parameters. A good summary of the present state of this question is given in Ern–78: they obtain $\alpha\sim5.5$ $m_\pi c$ (see also Tho–78).

As a dynamical model of πN scattering, for energies below the inelastic threshold, the Chew–Low theory provides a fairly successful picture. It predicts from the Yukawa interaction taken in a static limit that p-wave scattering will dominate, that the P_{33} wave may resonate, but the other partial p-waves will have small negative phase shifts of the same order as predicted by the Born approximation (e.g., 79). The theory clearly omits s-waves, which do matter in this energy domain, and $l \geqslant 2$ partial waves. Even the best versions are not quantitatively successful with the small p-waves or with the P_{33} wave *above* the resonance. The coupling to inelastic channels is not properly treated. Still, the Chew–Low model appears to provide the proper mechanism for generating the P_{33} resonance from the Yukawa interaction. Even with improved versions of the πN dynamics, which we discuss in the following subsection, the Chew–Low model, suitably generalized, still underlies the behavior of the P_{33} wave.

If we *assume* the existence of a resonant state in the πN scattering, it is possible to approach the Chew–Low theory from another point of view (see, e.g., Bro–75). Returning to the Low equation (Eq. 93), we choose the set of intermediate states to be the nucleon states ψ_0, as in Eq. 94, and the resonant states Δ, which represent an unstable particle, the $\Delta(1232)$, with $TJ=(\tfrac{3}{2},\tfrac{3}{2})$ which decays into $\pi+$N in a P_{33} wave. The mass (energy) and width of the Δ is given by, for example, the effective range formula in Eqs. 33 and 103 with $M_\Delta=M+\omega_R$ (neglecting nucleon motion). Then the

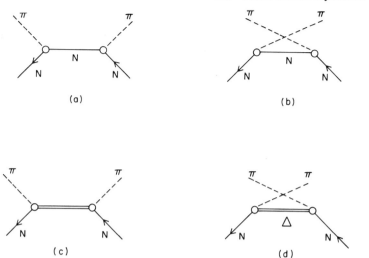

Fig. 2.10 Representation of Chew-Low theory using the nucleon and the $\Delta(1232)$ as intermediate states.

t-matrix in Eq. 93 may be represented as in Fig. 2.10. The diagrams (a) and (b) are the nucleon or Born terms, as before. The new diagrams (c) and (d) represent the Δ as the intermediate state: direct and crossed terms. The calculation of the new terms is similar to that of the nucleon term in Eq. 94, except that now we require the amplitudes for $\pi+N \rightarrow \Delta$: $\langle \Delta | j(\mathbf{k}) | \psi_0 \rangle$ for the numerator. If the Δ were stable, the denominators in Eq. 93 would be simply $\omega \pm (M - M_\Delta)$. The simplest way to include the instability of the Δ is to add the width $M_\Delta \rightarrow M_\Delta - i\frac{1}{2}\Gamma$; a more complete development would involve integration of the Δ term over the spectrum of the Δ. In this version, the knowledge of the $\pi+N \rightarrow \Delta$ amplitudes replaces the need to solve the integral equation in Eq. 99.

Returning to the deficiencies of the Chew-Low model, we note that what makes it tractable and simple also requires the omission of physical effects that may be important for a quantitative description of πN scattering. The static assumption decouples partial waves, but it violates the Lorentz invariance of the theory and leads to ambiguities, for example, in the treatment of the c.m. versus the lab frame. The decoupling of partial waves may not affect (strongly) the P_{33} wave, but the P_{33} wave could affect the small partial waves, through the neglected coupling. The treatment of higher energy intermediate states is not adequate. To some extent these deficiencies are corrected in a different version of the theory developed from partial wave dispersion relations solved by the N/D method. This approach, while possibly quite appropriate to the πN problem, does not provide us with general methods for π-nucleus scattering, so we do not pursue it further here. (See, however, Section 3.4b for a dispersion treatment of πd scattering.) The interested reader is referred to Ham–67, Moo–69, and Bra–73.

Other omissions from the Chew-Low theory concern processes not induced by the Yukawa interaction. We discuss these next.

2.4d Exchange Diagrams, Summary of πN Dynamics

One of the dynamical processes that must be included in a theory of πN scattering is the interaction between pions. We illustrate one such possibility in Fig. 2.11a, where the circle represents a scattering interaction between two pions. (Such interactions are not given by the static Yukawa interaction, although they become possible if excitation of nucleon-antinucleon pairs from the vacuum is allowed.) Now, if we characterize the $\pi\pi$ states by partial waves TJ, we find that there are resonant states at various energies in different partial waves. We illustrate the resonant interaction of two pions involved in πN scattering in Fig. 2.11b. In Fig. 2.11c we reinterpret this process as the *exchange* of the $\pi\pi$ resonant state (or unstable particle) between the π and the N. (In a sense, the crossed Chew-Low diagrams of Fig. 2.10b and d may be considered to represent the exchange of a N or Δ between the π and N; in this case there is also an additional "exchange" of the external pion and nucleon at each vertex.)

The resonant exchange amplitude for the process illustrated in Fig. 2.11c can be written as a Born amplitude proportional to

$$\langle \mathbf{k'}\mathbf{p'}|T_x|\mathbf{kp}\rangle \propto \frac{\langle \mathbf{k'}|J_x(\mathbf{q})|\mathbf{k}\rangle\langle \mathbf{p'}|G_x^\dagger(\mathbf{q})|\mathbf{p}\rangle}{q^2 + m_x^2} \qquad [105]$$

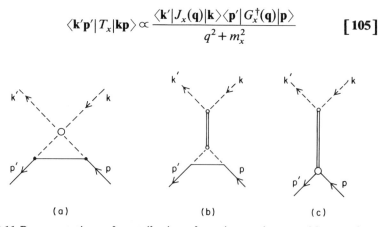

(a)   (b)   (c)

Fig. 2.11 Representations of contribution of π-π interactions to πN scattering: (a) π-π scattering; (b) π-π resonance; (c) π-π resonance as an exchange process between π and N, as in Eq. 105.

where m_x and $\mathbf{q}$ are the mass and three-momentum of the exchanged state, treated here as a particle x. The vertex operators (currents) $G_x^\dagger, J_x$ describe the coupling of the exchanged particle (x) to the nucleon and to the pion, respectively: the vertex functions (matrix elements) will, in general, depend on the momenta and spins (isospins). For very low momenta (e.g., in the πN c.m.) the vertices normally become independent of momenta, or depend only on low powers:—for example, p^l where l is a small integer (spin). In the simplest case, J_x and $G_x^\dagger$ become momentum independent spin-isospin matrices, characterized by coupling constants $g_{x\pi\pi}$ and g_{xNN}:

$$T_x \propto \frac{g_{x\pi\pi} g_{xNN} \left[J_x G_x^\dagger \right]}{q^2 + m_x^2}.$$ [106]

Following the discussion of local potentials (Section 2.3b), an exchange term of the form of Eq. 106, depending only on momentum transfer q, leads to a local potential in position as in Eq. 52. This has a Yukawa potential form

$$V_x(r) \propto g_{x\pi\pi} g_{xNN} \left[J_x G_x^\dagger \right] \frac{e^{-m_x r}}{r}.$$ [107]

The range of the potential is defined by the inverse mass m_x^{-1} $(= \hbar c / m_x =$ Compton wave length) of the exchanged particle. Since the exchanged particles are unstable with a spread of mass, the potential is more like a superposition of potentials (Eq. 107) of differing ranges. The lighter the mass of the exchanged particle, the longer the range of the interaction.

The best known state of the $\pi\pi$ system which contributes to the interaction is the ρ meson: a resonant state, with $m_\rho = 770$ MeV, $\Gamma \simeq 150$ MeV, $T = 1, J^\pi = 1^-$, that is, a vector meson which couples strongly to the π and N. Since it carries isospin $T = 1$, it couples to the π as $\vec{J}_\rho \propto \vec{I}_\pi$ and to the N as $\vec{G}_\rho^\dagger \propto \vec{\tau}$ (isovector coupling); thus the Born term (Eq. 106) is isovector coupled:

$$T_\rho \propto \vec{I}_\pi \cdot \vec{\tau}.$$ [108]

That is, ρ-exchange contributes to isovector πN scattering. It is believed that the ρ-exchange dominates the low energy πN s-wave scattering, giving the isovector scattering length $a^{(-)}$ (Eq. 34). This is the basic idea behind the use of current algebra or Partially Conserved Axial Current (PCAC) for the theory of the πN scattering lengths (see, e.g., Wei–66).

There is also strong $\pi\pi$ scattering in the $T = 0$, $J^\pi = 0^+$ partial wave. Although there is no clear resonance, there is a strongly interacting wave, centered at about 1300 MeV, and very broad in energy (Par–78). Whether properly or not, in view of the large width, this wave is sometimes described as a particle, ε (also, "σ" meson), which couples to the π and N. Since it is isoscalar ($T=0$), it contributes to isoscalar πN scattering. However, the isoscalar πN s-wave scattering at low energy is very weak; the scattering length is $a^{(+)} = -0.005\ m_\pi^{-1}$ (Eq. 34). It is believed that this comes from a destructive interference of the effect of ε-exchange, with other contributions, for example, the s-wave part of the *relativistic* Born amplitude for the Yukawa coupling (see the end of Section 2.4b). At higher energies, the ε-exchange may contribute comparably to the ρ.

To summarize our picture of the dynamics of πN scattering for energies below strong inelasticity, we find that the Chew-Low static model gives us some understanding of the p-wave amplitudes, based on Yukawa coupling of the π and N. The resonance in the P_{33} wave is perhaps best understood. The interactions in the p-wave are highly nonlocal (and separable in the static model). Other physical features, such as s-wave scattering, may involve exchange of ρ or other mesons between the π and N. These exchanges may be adequately described by local interactions of moderately short range ($m_x^{-1} \sim 0.3$ fm).

A complete quantitative theory, based on Hamiltonian or Lagrangian quantum mechanics, that combines all these features has not been given. However, there has been an attempt to include all of this physics in a compact theoretical approach using partial wave dispersion relations.* (See Ham–67 and Bra–73 for an account of these methods.) The dynamics in this method are given in terms of the analytic structure of partial wave amplitudes, that is, on-mass-shell amplitudes, that are continued as functions of complex values of their kinematic variables. The analytic structure enters in terms of poles (Born terms) involving exchanges of particles (as well as cut singularities, from this or other sources). The physics enters, therefore, in terms of the selection of particles to be exchanged and the strengths (residues) of the poles—that is, coupling constants. With a modest number of exchanged particles ($N, \Delta, \rho, \varepsilon$) a fairly good picture of πN scattering in the s, p, d, and f partial waves emerges for scattering energies at which elastic scattering dominates. This tends to justify the picture we have assumed for πN scattering, even if details of the static Chew-Low theory, for example, are incorrect. Unfortunately, it is not clear how to use the dispersion methods for multiple scattering, which is the major problem of this book; thus we resort to the approximate static model, or ρ-exchange when required.

We should briefly mention why we have not devoted a section to kaon-nucleon dynamics: Although some of the same ideas presented here

---

*An alternative approach is given by the so-called hard meson current algebra; see, for example, Nat–73a.

may apply in that case, less has been done with the low energy scattering theory than in the πN case. This is due in part to the increased complexity involved in channel coupling ($KN \leftrightarrow \pi Y$). See, however, Bra–69, who reviews particularly the application of dispersion theory to KN.

2.5 Nucleon-Nucleon Interaction

We conclude this chapter with a brief account of the role of mesons in the interaction between nucleons. This topic may seem at first glance not directly relevant to the subject of this book, which is mostly about scattering reactions of mesons on nuclei. However, according to our present understanding, it is the exchange of mesons among nucleons that provides the interaction energy to hold nuclei together. It is true that our understanding of nuclear structure has developed with little reference to the meson degrees of freedom in the nucleus. Generally, nuclear models have been based on Schroedinger mechanics, in which nucleons interact through potentials, whatever the origin of those potentials may be. We also use the Schroedinger equation in much of this present work to describe multiple scattering of mesons from nuclear targets. It is important, throughout this approach, that we remember that mesons are involved in the nuclear interactions as well as in the scattering processes, even though the Schroedinger methods carry us a long way toward handling our theoretical problems. We return to the question of dealing with all the meson degrees of freedom in a satisfactory way in Chapter 6.

The idea that meson exchange underlies the interaction between nucleons is as old as the theory of mesons itself; Yukawa introduced the idea of a meson (i.e., a boson lighter than the nucleons) to play the role in nuclear forces that the photon (virtual) plays in the Coulomb interaction between charged particles (Yuk–35). The pion, discovered in 1947, turns out to be the lightest meson that can play that role.

Consider the lowest order diagram for the exchange of a meson between two nucleons in a scattering process, which we illustrate in Fig. 2.12. If the meson is taken to be a pion, and the coupling to nucleons to be the Yukawa static interaction of Eq. 61, then the scattering amplitude corresponding to the figure may be calculated (second order Born term (renormalized); see, e.g., Bjo–64 and Sak–67)

$$\langle \mathbf{p}_1', \mathbf{p}_2' | T | \mathbf{p}_1, \mathbf{p}_2 \rangle = (2\pi)^3 \delta(\mathbf{p}_1' + \mathbf{p}_2' - \mathbf{p}_1 - \mathbf{p}_2) \langle \mathbf{p}' | T | \mathbf{p} \rangle$$

$$\langle \mathbf{p}' | T | \mathbf{p} \rangle = -4\pi \frac{f^2}{m_\pi^2} (\vec{\tau}_1 \cdot \vec{\tau}_2) \frac{(\boldsymbol{\sigma}_1 \cdot \mathbf{q})(\boldsymbol{\sigma}_2 \cdot \mathbf{q})}{\omega_q^2} , \qquad [109]$$

where $\mathbf{p}, \mathbf{p}'$ are the relative momenta of the two nucleons ($2\mathbf{p} = \mathbf{p}_2 - \mathbf{p}_1$) and

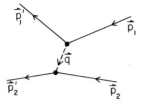

Fig. 2.12 One-meson exchange contribution to nucleon-nucleon interaction.

q is 'the momentum transfer ($\mathbf{p}' - \mathbf{p}$) carried by the pion in the NN c.m. frame. (We have calculated the on-energy-shell t-matrix in Eq. 109.)

Now let us consider the amplitude (Eq. 109) as an *effective potential* between the nucleons in the sense discussed above in Section 2.3c for Fig. 2.6a. This is called the one-pion-exchange (OPE) potential. (For the exchange of other mesons, the general name one-boson-exchange (OBE) potential is used.) The potential should actually be taken from the fully off-energy-shell t-matrix corresponding to Eq. 109, which has an additional energy in the denominator. However, for low energy nuclear physics, we can neglect the energy dependence ($\Delta E \ll m_\pi < \omega_q$) and use Eq. 109 as a *static interaction*. In position space, the potential is "almost" local in the sense of Eq. 52; we take the Fourier transform of Eq. 109 to obtain

$$V(r) = \frac{f^2}{m_\pi^2} (\vec{\tau}_1 \cdot \vec{\tau}_2)(\boldsymbol{\sigma}_1 \cdot \nabla)(\boldsymbol{\sigma}_2 \cdot \nabla) \frac{e^{-m_\pi r}}{r}, \qquad [110]$$

where r is the relative NN distance. This is the standard OPE potential, with the Yukawa form $\exp(-m_\pi r)/r$ whose range is given by $m_\pi^{-1} \simeq$ 1.4 fm. The spin-gradient terms lead to a spin-dependent central potential plus a tensor potential term (see, e.g., Bro–76b).

The strength of the potential is controlled by the πN coupling constant f^2 (and the pion mass m_π). Now, since the pion is the lightest meson that can be exchanged between nucleons, it provides the longest range potential. The long range part of the potential is most effective, relative to shorter range parts, in partial waves of high angular momentum. It has been shown from the study of the phase shifts for NN scattering in higher partial waves (for $E_{lab} < 300$ MeV, $1 < l \leqslant 5$) that the OPE form is consistent for the long range potential with $f^2 \lesssim 0.07 \pm 0.01$ (see Mor–60). This is in fair agreement with the values quoted from πN scattering analysis: $f^2 = 0.079$ in Eq. 78; $f^2 = 0.087$ in Eq. 104.

With the discovery of the existence of other mesons that couple strongly to nucleons, the attempt has been made to include in the NN potential a number of OBE contributions, corresponding to the lightest of these mesons (see Bro–76b). These usually include the ρ and ε (or σ) "mesons"

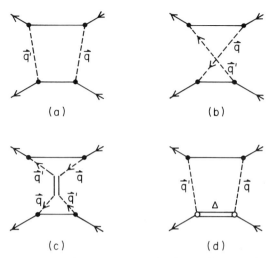

Fig. 2.13 Two-meson exchange processes in the nucleon-nucleon interaction.

mentioned above in Section 2.4d as exchange particles in πN scattering*. The vector meson ω couples strongly to the nucleon but not to the π, and is also included ($m_\omega = 783$ MeV, $T = 0, J = 1^-$). The ρ, ω, and ε then combine with the π to give the middle-range nuclear potential ($0.3 < r < 1.4$ fm). With a number of adjustable parameters in the various coupling constants, as well as the "mass" of the ill-defined ε, some success has been obtained in fitting NN phase-shifts with such an OBE potential. However, there is no reason to think the story ends here.

What may happen once may happen twice: if one meson may be exchanged between two nucleons, why not two (or more)? Consider the two-meson exchange diagrams of Fig. 2.13. The process (a) may certainly contribute to the amplitude for NN scattering. However, if we interpret the OBE process of Fig. 2.12 as a potential $V(q)$ to be used in a Schroedinger equation, or equivalently, in the Lippmann-Schwinger equation given as Eq. 49, the t-matrix we obtain (e.g., from Eq. 49) will contain a term like that of Fig. 2.13a, to second order in $V(q)$. So (a) is not a process which contributes to our *effective potential*, although it will contribute to the scattering through the OBE potential. However, diagram b, in which two mesons are exchanged simultaneously, is not generated by Eq. 49 from the OBE potential; (b) is a "true" two-meson contribution to the effective potential for NN scattering.

*See Table 1.1.

In Fig. 2.13c we show the two exchanged mesons in a resonant state: —e.g., two pions forming a ρ-meson (virtual). This is similar to the process of Fig. 2.11b in πN scattering, which we interpreted as ρ-exchange between the π and N. If we interpret 2.13c similarly as ρ-exchange between the two nucleons, we find it has already been included in the OBE diagram of Fig. 2.12 (with appropriate form factors at the ρNN vertices). Therefore (c) is not to be counted as a two-pion contribution to V; it has already been included in V(OBE).

Last, we show a two-meson exchange process in which the nucleon is excited to a resonant state, for example, the $\Delta(1232)$. This is also a "new" diagram, not generated by the OBE potential. However, we note that if we consider the Chew-Low model for the Δ-resonance in πN scattering, as illustrated in Fig. 2.9, the crossed-Born term (Fig. 2.9b), which contributes to the P_{33} resonance, is already included in the diagram of Fig. 2.13b. Therefore we cannot simply add the terms in Fig. 2.13b and d. However, if we adopt the πN model shown in Fig. 2.10, in which the Δ is a state independent of the crossed-Born term in Fig. 2.10b, we would come to the conclusion that both terms of Fig. 2.13b and d should contribute. The models are different.

We should note that the calculation of two-meson exchange is complicated, and we refer the interested reader to Bro–76b for further details and for references to the research literature. The form of the effective potential generated from two-meson exchange is not as simple as for the OBE case. One expects the order of the range to be given by the inverse sum of the masses exchanged, so that two pions should give the longest range part, which could be longer than the range of the vector mesons in OBE, since $(2m_\pi)^{-1} > m_\rho^{-1}$.

As usual, we have slighted the role of the K-mesons as exchange particles. Because of strangeness conservation, they cannot contribute to the OBE potential for NN. However, they would contribute to the interaction of hyperons (e.g., Λ, Σ) with nucleons and therefore play a role in hypernuclei. (Pion exchange also enters in the hyperon-nucleon interaction.) They could contribute to NN scattering through, for example, the process illustrated by Fig. 2.13b, but the range should be short [$(2m_K)^{-1} < 0.2$ fm].

References

A good general treatment of quantum scattering theory is given in Gol–64, both as it pertains to two-body systems as in this chapter, or for scattering from complex targets as in the rest of this book. For aspects of the pion-nucleon problem, one may consult Moo–69, Bra–73, and Sha–69. The kaon-nucleon problem is discussed by Bra–69. Much useful information on parameters of elementary particles and their resonances may be found in Par–78.

For an introduction to field theoretic methods, see Bjo–64,65, Sak–67, and Sch–61. Chew-Low theory (Che–56) is given a clear pedagogic treatment in Wic–55 and in Kal–64. The dispersion approach is presented in Ham–67 and Bra–73. The two-nucleon interaction is the subject of a recent book by Brown and Jackson (Bro–76b).

3 Mesons and Deuterons — Three-Body Systems

3.1 Introduction to Multiple Scattering

The deuteron is the lightest complex nucleus, and, in some senses, the simplest to treat dynamically. For most purposes it may be considered as a two-body system consisting of a proton and neutron interacting through a potential. Other degrees of freedom may generally be neglected for interactions not involving too high energy. This relative simplicity makes the deuteron an attractive target for studying meson interactions. At the same time, the internal degrees of freedom of the deuteron are sufficient to allow meson-deuteron reactions to exhibit many of the features also exhibited by heavier nuclear targets. Thus we may use the deuteron as an elementary test case for the study of meson interactions with nuclei in general. (The greater diffuseness of the deuteron relative to heavier nuclei somewhat limits this in detail.) However, special theoretical techniques are available for treating the dynamics of three-body systems, such as a meson scattering from a deuteron, which deserve separate discussion. We see explicitly certain aspects of the general πA scattering problem, which remain less explicit in Chapter 4. For this reason we discuss the meson-deuteron system in this chapter, before introducing the more general study of meson-nuclear interactions.

Let us consider the reactions induced by π and K mesons on deuterons. We consider charged mesons, which have sufficiently long lifetimes and which can be easily focused or collimated to make proper beams. Information on neutral mesons can usually be obtained by isobaric invariance, using the fact that deuterons have isospin $T = 0$. Thus πd reactions have total isospin $T = 1$, and Kd have $T = \frac{1}{2}$. However, K^+d and K^-d are distinguished by different strangeness: $S = +1$ and -1, respectively. We list reactions accessible at lower energies only (less than a few hundred MeV).

62

For $\pi^+ d$, there are three types of scattering reactions:

Elastic	$\pi^\pm + d \rightarrow \pi^\pm + d$	[1a]
Inelastic	$\pi^\pm + d \rightarrow \pi^\pm + n + p$	[1b]
Charge exchange	$\pi^\pm + d \rightarrow \pi^0 + \left\{ \begin{matrix} pp \\ nn \end{matrix} \right\}.$	[1c]

Since the deuteron is the only bound state of two nucleons, the inelastic and charge exchange reactions lead to continuum, or deuteron breakup states. There is no possibility for the deuteron of double charge exchange (π^+, π^-) without new particles being produced as well.

Pions may be absorbed with or without the emission of electromagnetic (EM) radiations:

Absorption	$\pi^\pm + d \rightarrow \left\{ \begin{matrix} pp \\ nn \end{matrix} \right\}$	[1d]
Radiative absorption	$\pi^\pm + d \rightarrow \left\{ \begin{matrix} pp \\ nn \end{matrix} \right\} + \gamma.$	[1e]

The reactions listed all involve strong interactions, with possible Coulomb interaction as well; the last necessarily involves EM interactions. We ignore weak interactions.

For K^+ induced reactions, we have only scattering:

Elastic	$K^+ + d \rightarrow K^+ + d$	[2a]
Inelastic	$K^+ + d \rightarrow K^+ + n + p$	[2b]
Charge exchange	$K^+ + d \rightarrow K^0 + p + p.$	[2c]

No strong interaction absorption conserving strangeness can occur, since no stable $S = 1$ baryon exists.

For K^- induced reactions, the analogous scattering reactions are as follows:

Elastic	$K^- + d \rightarrow K^- + d$	[3a]
Inelastic	$K^- + d \rightarrow K^- + n + p$	[3b]
Charge exchange	$K^- + d \rightarrow \overline{K}^0 + n + n,$	[3c]

also strangeness exchange scattering, with the following possibilities

$$K^- + d \rightarrow \quad \begin{matrix} \pi^0 + \Lambda + n \\ \pi^0 + \Sigma^0 + n \\ \pi^\pm + \Sigma^\mp + n \\ \pi^- + \Sigma^0 + p \\ \pi^0 + \Sigma^- + p \end{matrix} \qquad [3d]$$

(which may involve charge exchange as well).

Strong absorption of K^- can occur, leading to the following hyperon-nucleon final states:

$$K^- + d \rightarrow \Lambda + n$$
$$\Sigma^0 + n \qquad\qquad [3e]$$
$$\Sigma^- + p.$$

Radiative absorption of K^- by deuterons is also possible.

3.1a Dynamics: Elementary Multiple Scattering Theory

For a first approach to the treatment of meson-deuteron dynamics, we consider only elastic scattering. We assume that mesons interact with nucleons, and nucleons with each other, through two-body potentials. The elastic scattering of a meson by a deuteron can then be represented by a Schroedinger equation of the form

$$\left[\varepsilon(k) + E_d - h_0 - H_d - V\right]\psi_{k,d}^{(+)} = 0 \qquad [4]$$

for a scattered meson of initial momentum k, where $\varepsilon(k)$ is the kinetic energy of the meson and h_0 is the meson kinetic energy operator. The deuteron Hamiltonian is H_d, and the deuteron ground state energy is E_d, which is negative ($E_d \simeq -2.22$ MeV). The meson-deuteron interaction V can be written as a sum over the two target nucleons

$$V = \sum_{i=1,2} V_i \qquad [5]$$

where V_i is the meson-nucleon interaction. The elastic scattering amplitude is then given by (compare Eqs. 2.19 and 2.43a)

$$\langle k', d | T(\varepsilon(k)) | k, d \rangle = \langle k', d | V | \psi_{k,d}^{(+)} \rangle \qquad [6]$$

where $\langle k', d |$ is the plane wave state of a meson of (final) momentum k', and the deuteron in its ground state, while $\psi_{k,d}^{(+)}$ is the full scattering state of Eq. 4. As in Section 2.3 we may obtain a linear operator equation* for the transition operator $T(\varepsilon)$ in the form of

$$T(\varepsilon) = V + V G_N(\varepsilon) T \qquad [7a]$$

---

*Eq. 7 is later rewritten in the form of Eq. 38.

with the Green operator

$$G_N(\varepsilon) = [\varepsilon + E_d - h_0 - H_d + i\eta]^{-1}. \qquad [7b]$$

Now, although Eq. 7 is a compact dynamical equation for the operator which gives us the elastic scattering amplitude, it is not a particularly tractable equation in terms of methods of solution. We defer further discussion of solution of the scattering equations to the following section. Here we consider approximation schemes for calculating the amplitudes.

The basic approach we consider here is multiple scattering theory. The underlying idea is that the scattering on the deuteron can be expressed in terms of a series of two-body collisions involving the meson with each nucleon separately. Formally, this can be done exactly in what is called a multiple scattering expansion. The approximations follow from the idea that the nuclear dynamics plays a less important role in the scattering than does the interaction of the meson with each nucleon separately. This allows us to treat the meson-nucleon scatterings in the multiple scattering series as approximately free, but with unusual (off-energy and momentum-shell) kinematics (see Section 2.1). The dynamics of the target between scatterings may also be simplified in this approximation. The domains within which this central idea might be expected to work are those in which the interaction energies of the meson with each nucleon exceed the scale of the nuclear interaction in the deuteron—for example, for high-energy mesons or with short-range meson-nucleon interactions. We come back to this question in Section 3.3 and also in Chapter 4, where we give a more general treatment of multiple scattering theory.

Let us express these ideas in terms of Eq. 7. First we expand T as a doubly infinite series in the potentials V_1 and V_2 (suppressing the explicit dependence on ε):

$$T = V_1 + V_2 + V_1 G_N V_1 + V_2 G_N V_2 + V_1 G_N V_2 + V_2 G_N V_1 + \cdots \quad [8]$$

(see also the discussion of Fig. 3.3 at the beginning of Section 3.2b). We regroup the infinite series involving V_1 or V_2 alone, defining $(i = 1,2)$

$$t_i = V_i + V_i G_N V_i + V_i G_N V_i G_N V_i + \cdots \qquad [9a]$$

which can be formally summed to give the linear equation

$$t_i = V_i + V_i G_N t_i. \qquad [9b]$$

If we continue to group terms in Eq. 8 by summing every series involving repetitions of $V_1, V_1 + V_1 G_N V_1 + \cdots$, and similarly for V_2, we obtain the new series

$$T = t_1 + t_2 + t_1 G_N t_2 + t_2 G_N t_1 + t_1 G_N t_2 G_N t_1 + \cdots \qquad [10]$$

in which the operators replace the interactions V_i, and in which successive interactions in any one term are on different nucleons. This is a multiple scattering series expansion (formally exact) usually called the Watson series. It is treated more fully, for general targets, in Section 4.2. The t-matrix t_i involves the meson interaction with one target nucleon only; however, the nucleon-nucleon interaction enters through H_d in G_N. It is here that we want to invoke an approximation that allows t_i to be obtained from a free meson-nucleon collision; this involves some form of *impulse approximation*, which we discuss further.

The nuclear dynamics still appear in the series given in Eq. 10 through the Green operators, or propagators G_N; we need some approximation here as well. Then we still have to sum the multiple scattering series to obtain an approximate evaluation of the elastic scattering amplitude Eq. 6.

3.1b Elementary Methods

In the *impulse approximation*, t_i of Eq. 9b is replaced by the free meson-nucleon t-operator

$$t_i(\varepsilon) \simeq t_i^{free}(E) \qquad [11a]$$

$$t_i^{free}(E) = V_i + V_i G_0^{(+)}(E) t_i^{free}(E) \qquad [11b]$$

with the free Green operator $G_0^{(+)}(E)$ (see Eq. 2.45b). This t-operator is by assumption independent of the dynamics of the nuclear target, although kinematical effects of the target appear in the evaluation of matrix elements of Eq. 11b. For example, the matrix element of Eq. 6 of the first term of the series in Eq. 10 is the expectation value of t_1^{free} in the deuteron ground state:

$$\langle \mathbf{k}', d | t_1^{free}(E) | \mathbf{k}, d \rangle = \int \psi_d^*(\mathbf{p}') \langle \mathbf{k}', \mathbf{p}' | t_1(E) | \mathbf{k}, \mathbf{p} \rangle \psi_d(\mathbf{p}) \, d\mathbf{p}' \, d\mathbf{p}. \qquad [12]$$

The momentum representation of the meson-nucleon scattering amplitude on the right-hand side of Eq. 12 was introduced in Eq. 2.20. We have not kept explicitly the spin-isospin state labels on the t-matrix or on the deuteron wave functions. (We use the relative momentum variable $\mathbf{p} = \frac{1}{2}(\mathbf{p}_1 - \mathbf{p}_2)$ which in the deuteron c.m. becomes $\mathbf{p} = \mathbf{p}_1 = -\mathbf{p}_2$. Thus for the matrix element of t_2^{free}, we need $\psi_d(-\mathbf{p}), \psi_d^*(-\mathbf{p}')$.) The t-matrix under the integral in Eq. 12 is off-energy-shell, since, although $|\mathbf{k}'| = |\mathbf{k}|$ for elastic meson-deuteron scattering, $|\mathbf{p}'| \neq |\mathbf{p}|$ in general; in fact $\mathbf{p}' + \mathbf{k}' = \mathbf{p} + \mathbf{k}$ by momentum conservation (see Eq. 2.22). The energy parameter E in Eq. 12 is not determined by the approximation in Eq. 11a; the choice may depend on the treatment of the entire series of Eq. 10. For example, if one ignores the motion of the nucleons in evaluating the t-matrix (except for momentum conservation)

$$\langle \mathbf{k}', \mathbf{p}' | t(E) | \mathbf{k}, \mathbf{p} \rangle \simeq \langle \mathbf{k}' | t(E) | \mathbf{k} \rangle (2\pi)^3 \delta(\mathbf{p}' + \mathbf{k}' - \mathbf{p} - \mathbf{k}), \qquad [13]$$

then $E = \varepsilon(k)$, the kinetic energy of the meson, is a consistent choice. (This is a fixed scatterer assumption.)

The sum of the first two terms of Eq. 10 ($T \simeq t_1 + t_2$) gives the *single scattering* approximation, which may be valid for weak scattering or for dilute systems. (This works for electron scattering.) With the additional approximation in Eq. 13 and ignoring spin-dependence of the scattering, we get the simple expression

$$\langle \mathbf{k}', d | T | \mathbf{k}, d \rangle \simeq \{ \langle \mathbf{k}' | t_1 | \mathbf{k} \rangle + \langle \mathbf{k}' | t_2 | \mathbf{k} \rangle \} F_d(\mathbf{q}) \qquad [14]$$

where $\mathbf{q} = \mathbf{k}' - \mathbf{k}$ is the momentum transfer, and $F_d(\mathbf{q})$ is the elastic form factor of the deuteron

$$F_d(\mathbf{q}) = \int \psi_d^*(\mathbf{p} - \tfrac{1}{2}\mathbf{q}) \psi_d(\mathbf{p}) \, d\mathbf{p}$$

$$= \int e^{-i\mathbf{q}\cdot\mathbf{r}/2} |\psi_d(\mathbf{r})|^2 \, d\mathbf{r}. \qquad [15]$$

Figure 3.1 illustrates the form factor of the deuteron, obtained from the analysis of electron scattering (Eli–69). For spin dependent scattering we also require form-factorlike terms with $\langle \sigma_1 + \sigma_2 \rangle$ in the deuteron ground state.

To evaluate terms beyond the single scattering approximation, we must deal with the spectrum of states of the deuteron that appear if we expand the Green function G_N in eigenstates:

$$G_N = \sum_n \frac{|n\rangle\langle n|}{\varepsilon(k) + i\eta - h_0 - \Delta E_n} \qquad [16]$$

where $\Delta E_n = E_n - E_d$ is the excitation energy of the two-nucleon state n. The only approximation we discuss here is *closure*, with ΔE_n set $\simeq 0$, which may be a valid assumption for high energies $E(k) > \Delta E_n$. (See further

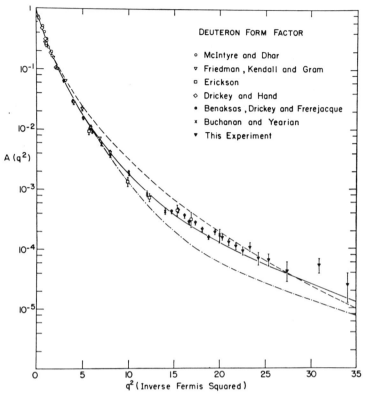

Fig. 3.1 Form factor for the deuteron, Eq. 15, obtained from electron scattering (from Eli–69).

discussion in Section 4.3.) Then

$$G_N \simeq \left[\varepsilon(k) + i\eta - h_0 \right]^{-1} \qquad [17]$$

is independent of nucleon motion as in Eq. 13. We may evaluate, for example, the third term of Eq. 10 (double scattering) using both Eqs. 13 and 17, again ignoring spin dependence:

$$\langle \mathbf{k}', \mathbf{d} | t_1 G_N t_2 | \mathbf{k}, \mathbf{d} \rangle \simeq \int \frac{d\mathbf{l}}{(2\pi)^3} \frac{\langle \mathbf{k}' | t_1 [\varepsilon(k)] | \mathbf{l} \rangle \langle \mathbf{l} | t_2 [\varepsilon(k)] | \mathbf{k} \rangle F_d(\mathbf{k}' + \mathbf{k} - 2\mathbf{l})}{\varepsilon(k) - \varepsilon(l) + i\eta} .$$

$$[18]$$

This expression, although calculable, is not particularly simple, and more important, does not directly lead to a method of evaluating the entire series in Eq. 10 without more simplifying assumptions.

One such assumption, again expected to apply at high scattering energies, is that only $\varepsilon(l) \simeq \varepsilon(k)$ contribute to Eq. 18, so that we can make a linear approximation (see also Chapter 4, Eqs. 147 and 148) to the denominator

$$\varepsilon(k) - \varepsilon(l) + i\eta \simeq \mathbf{v} \cdot (\mathbf{k} - \mathbf{l}) + i\eta$$
$$= v(k - l_{\parallel}) + i\eta \qquad [19a]$$

dropping higher order terms in $(k - l_{\parallel})$, where $\mathbf{v}$ is the velocity

$$\mathbf{v} \equiv \nabla_k \varepsilon(k) \qquad [19b]$$

that is parallel to $\mathbf{k}$. We denote the projection of $\mathbf{l}$ parallel to $\mathbf{k}$ or $\mathbf{v}$ by $l_{\parallel}$; the perpendicular projections are $\mathbf{l}_{\perp}$. Note that Eq. 19 implies that forward scattering dominates. We then calculate Eq. 18 by integration of $l_{\parallel}$ along $(-\infty, \infty)$, using a contour that closes in the upper half plane* and drastically assuming that only the pole due to the denominator Eq. 19a contributes. We obtain

$$-\frac{1}{2} \frac{2\pi i}{v(2\pi)^3} \int d^2 \mathbf{l}_{\perp} \langle \mathbf{k}' | t_1 | \mathbf{l} \rangle \langle \mathbf{l} | t_2 | \mathbf{k} \rangle F_d(\mathbf{k}'_{\perp} + \mathbf{k}_{\perp} - 2\mathbf{l}_{\perp}). \qquad [20a]$$

If we further assume that the t-matrices depend only on the differences $(\mathbf{k}' - \mathbf{l}), (\mathbf{l} - \mathbf{k})$—that is, momentum transfers (and on $\varepsilon(k)$, which we have suppressed)—then it is convenient to rewrite Eq. 20a with changed variables:

$$\mathbf{q}' = \tfrac{1}{2}(\mathbf{k} + \mathbf{k}' - 2\mathbf{l})_{\perp}, \qquad \mathbf{q} = (\mathbf{k}' - \mathbf{k}):$$

$$\langle t_1 G_N t_2 \rangle = -\frac{i}{2v(2\pi)^2} \int d^2 \mathbf{q}' \, t_1\left(\tfrac{1}{2}\mathbf{q} + \mathbf{q}'\right) t_2\left(\tfrac{1}{2}\mathbf{q} - \mathbf{q}'\right) F_d(2\mathbf{q}').$$

$$[20b]$$

With these approximations it is possible to give a compact expression for the elastic scattering t-matrix through double scattering, that is, the first four terms of Eq. 10. We combine Eqs. 14 and 20b, multiplying the latter

---

*The form factor requires that we close the contour in the upper half plane.

by 2 to include the term $t_2 G_N t_1$, which contributes identically, to obtain

$$\langle \mathbf{k}', \mathrm{d} | T | \mathbf{k}, \mathrm{d} \rangle \simeq \{ t_1(\mathbf{q}) + t_2(\mathbf{q}) \} F_\mathrm{d}(\mathbf{q})$$

$$- \frac{i}{(2\pi)^2 v} \int d^2 \mathbf{q}' t_1(\tfrac{1}{2}\mathbf{q} + \mathbf{q}') t_2(\tfrac{1}{2}\mathbf{q} - \mathbf{q}') F_\mathrm{d}(2\mathbf{q}')$$

(single and double scattering). [21]

If we replace t-matrices by scattering amplitudes, $f(\mathbf{q}) = -(\varepsilon/2\pi) t(\mathbf{q})$, we obtain Glauber's formula for the elastic scattering amplitude for a high energy particle (meson) on a deuteron* (Gla–55, 59, 67, 70):

$$f_\mathrm{d}(\mathbf{q}) = \{ f_\mathrm{p}(\mathbf{q}) + f_\mathrm{n}(\mathbf{q}) \} F_\mathrm{d}(\mathbf{q})$$

$$+ \frac{i}{2\pi k} \int d^2 \mathbf{q}' f_\mathrm{p}(\tfrac{1}{2}\mathbf{q} + \mathbf{q}') f_\mathrm{n}(\tfrac{1}{2}\mathbf{q} - \mathbf{q}') F_\mathrm{d}(2\mathbf{q}') \quad \text{(Glauber)},$$

[22]

where f_p and f_n are the scattering amplitudes for a meson on a proton and neutron, respectively. Glauber's result is derived differently, in a theory with fixed nucleons (see Section 4.4) using, however, an eikonal approximation equivalent to the linearization in Eq. 19. In Glauber's theory, there are no further multiple scattering terms; Harrington (Har–69) has shown that this is consistent with the Watson expansion (which we have followed here) under the eikonal and fixed nucleon approximations, and for a local potential interaction (see also Section 4.4).

The formula in Eq. 22 gives a useful method of estimating the importance of double scattering (see also Gla–55, 67). We assume that $f(\mathbf{q})$ may be replaced by $f(0)$ (for f_p or f_n) since forward scattering ($\mathbf{q} = 0$) is favored by the form factors. Then, for forward scattering on the *deuteron*, Eq. 22 becomes, using $F_\mathrm{d}(0) \equiv 1$,

$$f_\mathrm{d}(0) = f_\mathrm{p}(0) + f_\mathrm{n}(0) + \frac{i}{2\pi k} f_\mathrm{p}(0) f_\mathrm{n}(0) \int d^2 \mathbf{q}' F_\mathrm{d}(2\mathbf{q}'). \quad [23]$$

Using Eq. 15, we can evaluate the integral in terms of the deuteron density:

$$\int d^2 \mathbf{q}' F_\mathrm{d}(2\mathbf{q}') = \int d^2 \mathbf{q}' \int d\mathbf{r} \, e^{-i\mathbf{q}' \cdot \mathbf{r}} |\psi(\mathbf{r})|^2 = (2\pi)^2 \int_{-\infty}^{\infty} dz \, |\psi(z)|^2,$$

[24]

*We use $k = \varepsilon v$. Note that our $F_\mathrm{d}(\mathbf{q}) = S(\tfrac{1}{2}\mathbf{q})$ in Gla–67.

which can also be expressed (for spherically symmetric $|\psi|^2$) as

$$(2\pi)4\pi \int_0^\infty dr |\psi(r)|^2 \equiv 2\pi \langle r^{-2} \rangle_d \qquad [25]$$

in terms of the mean inverse-square-radius of the deuteron. Thus the ratio of the contribution of double to single scattering in Eq. 23 is

$$\frac{\text{double}}{\text{single}} = \frac{f_p(0)f_n(0)}{f_p(0)+f_n(0)} \frac{i}{k} \langle r^{-2} \rangle_d. \qquad [26]$$

It is interesting to estimate this ratio for π^+d scattering. At the 3,3 πN resonance ($T_\pi \sim 195$ MeV lab), we have

$$k \sim 1.8 m_\pi \qquad \langle r^{-2} \rangle \sim 0.5 m_\pi^2$$

$$f_p(0) \sim 1.4 i m_\pi^{-1} \qquad f_n(0) = \tfrac{1}{3} f_p(0) \qquad [27]$$

and the ratio is ~ -0.1. For higher energy, the ratio tends to decrease as $f(0)$ decreases and with k^{-1}. Therefore, for this energy range, we expect forward πd elastic scattering to be dominated by the single scattering terms. At larger angles, the double (or higher order) scattering plays a more important role.

The sequence of approximations and simplifications we have introduced, beginning with Eq. 11, have served to illustrate how, in simplest terms, meson-deuteron elastic scattering can be calculated in terms of free meson-nucleon t-matrices and the form factor of the deuteron. The approximations, which ignore nucleon motion and nucleon-nucleon interaction, may be valid in some domains, but are in general too restrictive. In the following section we use a more general theory that does not have these limitations. In Chapter 4, where we develop the Watson multiple scattering theory more fully, we also investigate these approximations.

Rather than make restrictive approximations in evaluating the Watson multiple scattering series, it is also possible to incorporate the assumption that the nucleons may be treated as fixed directly into the formulation of the scattering theory. This approach, called the *fixed scatterer approximation*, is attractive because it is somewhat simpler to develop, and to calculate. This formulation and its relation to the Watson theory are discussed in Chapter 4, Section 4.1 and 4.3. Recent developments with application to meson-deuteron scattering have been given by Gibbs (Gib–71).

3.2 Faddeev Equations for Three-Body Systems

We have thus far discussed meson-deuteron scattering from the point of view of a multiple-scattering approach based primarily on an assumption that the nucleons remain fixed during the scatterings; we now relax that assumption. It is the special virtue of the three-body problem, say π-d scattering for specificity, that it embodies the complexities of scattering on a many-particle system—in particular the appearance of off-energy-shell πN amplitudes and multiple scatterings in the presence of the dynamics of the scattering system—while still permitting complete, exact solution for potential scattering. (Though in practice, as we shall see, the numerical solution generally requires a further assumption of separable πN potentials.) This makes the exact solution of the three-body meson-nucleus scattering problem extremely interesting for the entire study of nuclear physics with mesons, since it provides us with a testing ground in which we can evaluate explicitly the effects of various approximations that must be invoked for other systems in a much less controlled fashion. Furthermore, to the degree that we can obtain a reliable description of meson-deuteron scattering in a potential theory, we can explore deviations from experiment in their relationship to field theory effects of coupling to channels other than those with the meson and two nucleons. We can also generalize meson-deuteron scattering to an artificial three-body problem involving the meson, an active nucleon, and a nuclear core to which the nucleon is bound in order to study detailed features, especially kinematical, of general meson-nucleus scattering.

3.2a General Features of the Three-Body Problem

What makes the quantum mechanical three-body problem difficult to solve? In order to explore this question briefly and in order to introduce notation for later use, let us first review the standard approach to the two-body problem. We take as our point of departure the Lippmann-Schwinger equation written in momentum space:

$$\langle \mathbf{k}_1' \mathbf{k}_2' | t(E) | \mathbf{k}_1 \mathbf{k}_2 \rangle = \langle \mathbf{k}_1' \mathbf{k}_2' | V | \mathbf{k}_1 \mathbf{k}_2 \rangle$$

$$+ \int \langle \mathbf{k}_1' \mathbf{k}_2' | V | \mathbf{k}_1'' \mathbf{k}_2'' \rangle \frac{1}{E - (k_1''^2/2m_1) - (k_2''^2/2m_2) + i\eta}$$

$$\times \langle \mathbf{k}_1'' \mathbf{k}_2'' | t(E) | \mathbf{k}_1 \mathbf{k}_2 \rangle \frac{d\mathbf{k}_1''}{(2\pi)^3} \frac{d\mathbf{k}_2''}{(2\pi)^3} \qquad [28]$$

for the scattering of two particles of masses m_1 and m_2 with initial momenta $\mathbf{k}_1, \mathbf{k}_2$ and final momenta $\mathbf{k}_1', \mathbf{k}_2'$ and a potential V acting between them. In order to solve this problem we effectively reduce it to a one-body situation by introducing relative and total momenta:

$$\mathbf{p} \equiv \frac{1}{m_1 + m_2}(m_2\mathbf{k}_1 - m_1\mathbf{k}_2), \qquad \mathbf{P} \equiv \mathbf{k}_1 + \mathbf{k}_2,$$

$$\mathbf{p}' \equiv \frac{1}{m_1 + m_2}(m_2\mathbf{k}_1' - m_1\mathbf{k}_2'), \qquad \mathbf{P}' \equiv \mathbf{k}_1' + \mathbf{k}_2', \qquad [29]$$

where we note that the relative momenta $\mathbf{p}$ and $\mathbf{p}'$ are Galilei invariant, being the difference of two velocities, $\mathbf{p} = m_1 m_2 (m_1 + m_2)^{-1}(\mathbf{v}_1 - \mathbf{v}_2)$, and that the Jacobian of the transformation in Eq. 29 is unity. Furthermore, the total energy of the two-particle system,

$$\mathcal{E} = \frac{k_1^2}{2m_1} + \frac{k_2^2}{2m_2} = \frac{p^2}{2m_{\text{red}}} + \frac{P^2}{2m_{\text{tot}}}, \qquad m_{\text{red}} \equiv \frac{m_1 m_2}{m_1 + m_2}, \qquad m_{\text{tot}} = m_1 + m_2,$$

$$[30]$$

is made up of a part referring to the internal energy involving the relative momentum and reduced mass, m_{red}, and a part which is the energy of motion of the full system with total momentum and total mass m_{tot}. The energy E at which the amplitude t is evaluated in Eq. 28 can, of course, be different from $\mathcal{E}$, since in general we may wish to study a situation that is off-energy-shell. Indeed, for the scattering problem we require transition amplitudes with $k_1''^2/2m_1 + k_2''^2/2m_2 \neq E$ for the intermediate state integration of Eq. 28, but we can restrict ourselves to an initial energy such that $\mathcal{E} \equiv k_1^2/2m_1 + k_2^2/2m_2 = E$, since this is what enters for calculating the scattering cross section (which also has final energy $k_1'^2/2m_1 + k_2'^2/2m_2 = E$; it is only on the way to solving for t that we require off-energy-shell values). With this restriction, we refer to the amplitude as half off-energy-shell, as defined in Eq. 2.19.

When the relative and total momenta of Eq. 29 are inserted into the Lippmann-Schwinger equation (Eq. 28) we obtain

$$\langle \mathbf{p}'\mathbf{P}'|t(E)|\mathbf{pP}\rangle = \langle \mathbf{p}'\mathbf{P}'|V|\mathbf{pP}\rangle$$

$$+ \int \langle \mathbf{p}'\mathbf{P}'|V|\mathbf{p}''\mathbf{P}''\rangle \frac{1}{E - (p''^2/2m_{\text{red}}) - (P''^2/2m_{\text{tot}}) + i\eta}$$

$$\times \langle \mathbf{p}''\mathbf{P}''|t(E)|\mathbf{pP}\rangle \frac{d\mathbf{p}''}{(2\pi)^3} \frac{d\mathbf{P}''}{(2\pi)^3}. \qquad [31]$$

Translational invariance for the potential V—for example, for a local potential, a function in configuration space of the difference of the position coordinates $\mathbf{r}_1 - \mathbf{r}_2$ only—requires

$$\langle \mathbf{p'P'}|V|\mathbf{pP}\rangle = (2\pi)^3 \delta(\mathbf{P'} - \mathbf{P})\langle \mathbf{p'}|V|\mathbf{p}\rangle, \qquad [32]$$

where the delta function embodies the conservation of the total momentum in the interaction and the accompanying matrix element is a function of the Galilei invariant quantities $\mathbf{p}$ and $\mathbf{p'}$ only. One might initially suppose that the mathematical pathology of the delta function of Eq. 32 would create problems in solving Eq. 31, but this is in fact not the case, since we can anticipate overall total momentum conservation in the problem and make the ansatz

$$\langle \mathbf{p'P'}|t(E)|\mathbf{pP}\rangle = (2\pi)^3 \delta(\mathbf{P'} - \mathbf{P})\langle \mathbf{p'}|t(E)|\mathbf{p}\rangle, \qquad [33]$$

whence

$$\langle \mathbf{p'}|t(E)|\mathbf{p}\rangle = \langle \mathbf{p'}|V|\mathbf{p}\rangle + \int \langle \mathbf{p'}|V|\mathbf{p''}\rangle \frac{1}{E - \left(p''^2/2m_{\text{red}}\right) - \left(P^2/2m_{\text{tot}}\right) + i\eta}$$

$$\times \langle \mathbf{p''}|t(E)|\mathbf{p}\rangle \frac{d\mathbf{p''}}{(2\pi)^3}. \qquad [34]$$

The total system motion is then completely eliminated from the problem except for fixing the energy available to the internal motion as $E - (P^2/2m_{\text{tot}})$, and here it is convenient to work in the center-of-mass system for which the total momentum vanishes, $\mathbf{P} = 0$, so that E refers to internal energy only. The integral equation in Eq. 34 is very satisfactory from the mathematical viewpoint. It has no general pathologies and can be solved for a broad class of potentials—indeed for all those local potentials, for example, that are square integrable in configuration space, that is, for which $\int V^2(\mathbf{r})\,d\mathbf{r}$ is finite (Wei–64). (A problem that might appear to be of concern, namely, the encountering of a bound state from which special difficulties might be anticipated, since it cannot be dealt with in a perturbation expansion, is no worry because for nonrelativistic two-particle scattering states $E > 0$ it is impossible to produce a bound state $E < 0$ while conserving energy and momentum.)

The situation for a three-particle problem* starts off in parallel to the two-particle case. We again have a Lippmann-Schwinger equation, which

---

*General background can be found in Tho–77.

we take preliminarily in the form

$$\langle k'_1 k'_2 k'_3 | T(E) | k_1 k_2 k_3 \rangle = \langle k'_1 k'_2 k'_3 | V | k_1 k_2 k_3 \rangle$$

$$+ \int \langle k'_1 k'_2 k'_3 | V | k''_1 k''_2 k''_3 \rangle$$

$$\times \frac{1}{E - (k''^2_1/2m_1) - (k''^2_2/2m_2) - (k''^2_3/2m_3) + i\eta}$$

$$\times \langle k''_1 k''_2 k''_3 | T(E) | k_1 k_2 k_3 \rangle \frac{dk''_1}{(2\pi)^3} \frac{dk''_2}{(2\pi)^3} \frac{dk''_3}{(2\pi)^3},$$

$$[35]$$

where now the interaction is made up of the three pairwise interactions

$$V = V_1 + V_2 + V_3, \qquad V_i \equiv V(\mathbf{r}_j, \mathbf{r}_k), \qquad i,j,k \text{ cyclic} \qquad [36]$$

in a standard cyclic notation such that V_1 refers to the interaction between the pair $(2,3)$, and so forth. The energy E is the sum of meson kinetic energy and deuteron ground state energy, that is, $E = \varepsilon + E_d$ as in Eq. 7b. Now the matrix element of the interaction involves

$$\langle k'_1 k'_2 k'_3 | V | k_1 k_2 k_3 \rangle = (2\pi)^3 \delta(\mathbf{k}'_1 - \mathbf{k}_1) \langle k'_2 k'_3 | V_1 | k_2 k_3 \rangle$$

$$+ (2\pi)^3 \delta(\mathbf{k}'_2 - \mathbf{k}_2) \langle k'_1 k'_3 | V_2 | k_1 k_3 \rangle$$

$$+ (2\pi)^3 \delta(\mathbf{k}'_3 - \mathbf{k}_3) \langle k'_1 k'_2 | V_3 | k_1 k_2 \rangle. \qquad [37]$$

The delta functions now express the fact (see Fig. 3.2a) that the interaction V_1 involves only particles 2 and 3 so that particle 1 propagates during the interaction with no change in its momentum and so forth. Moreover, there is no trivial way to remove this delta function as there was in the two-body case, since the possibility for successive interactions among the various pairs (Fig. 3.2b) means that no such momentum conservation $\mathbf{k}'_1 = \mathbf{k}_1$ for particle 1 exists in the full amplitude. These delta functions then generate a substantive problem in solving the integral equation (Eq. 35) because after we factor out the delta function of overall momentum conservation $(2\pi)^3 \delta(\mathbf{k}'_1 + \mathbf{k}'_2 + \mathbf{k}'_3 - \mathbf{k}_1 - \mathbf{k}_2 - \mathbf{k}_3)$ we are left with yet another delta function in the (V part of the) kernel of the integral equation. This prevents the kernel from being square integrable or compact and therefore precludes a guarantee that we can solve the integral equation (Wei–64).

In addition to the delta function pathology we have just noted, there is another difficulty in trying to use Eqs. 35 and 36 as they stand. This is

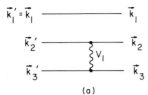

(a)

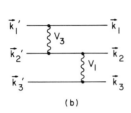

(b)

Fig. 3.2 Diagrams that appear in the consideration of systems of three interacting particles. (*a*) A case in which one particle propagates through without scattering, which expresses itself in the appearance of a delta function factor in its momentum labels, $\delta_{\mathbf{k}_1'\mathbf{k}_1} \to (2\pi)^3 \delta(\mathbf{k}_1' - \mathbf{k}_1)$. In (*b*), particle 1 interacts and no such factor enters. The notation is such that V_i is the interaction between particles j and k, $\{i,j,k\} = \{1,2,3\}$.

because for many interactions V_i that may enter in interesting physical problems there may occur strong, short-range forces that will lead to large or divergent matrix elements of V_i and thus preclude the use of any perturbative approach. As we see in the course of dealing with the delta-function pathology, we regroup our equations so that the possibly problematical interactions V_i are replaced with relatively well-behaved scattering amplitudes t_i, thus giving us an added bonus. In fact, as we see in Chapter 4 when we deal with πA scattering for $A > 2$, the delta-function difficulty does not arise because we cast the problem into a form such that the scattering diagrams are always connected through the action of the nuclear Hamiltonian. In that situation, a major motivation for using an approach similar to that for the three-body problem is precisely the elimination of V_i in favor of t_i.

3.2b Dynamical Equations for Scattering Amplitudes

The solution to this problem was set forth by Faddeev (Fad–61a, b, 63, 65; see also Lov–64a, b, Wei–64), and consists in a rearrangement of the infinite series implied by the Lippmann-Schwinger three-body equations such that the troublesome terms are dealt with completely before the full equation is solved. Schematically we can see how this works by dividing the three-body amplitude of Eq. 35 into three parts $T = T_1 + T_2 + T_3$, depending on whether the last (leftmost) interaction took place between 2 and 3 (pair 1, amplitude T_1), 1 and 3 (T_2) or 1 and 2 (T_3). Then, as in Fig. 3.3, we can resum all those terms in T_1, that is, in which the last interaction is between 2 and 3. Some of these will involve no other interactions and

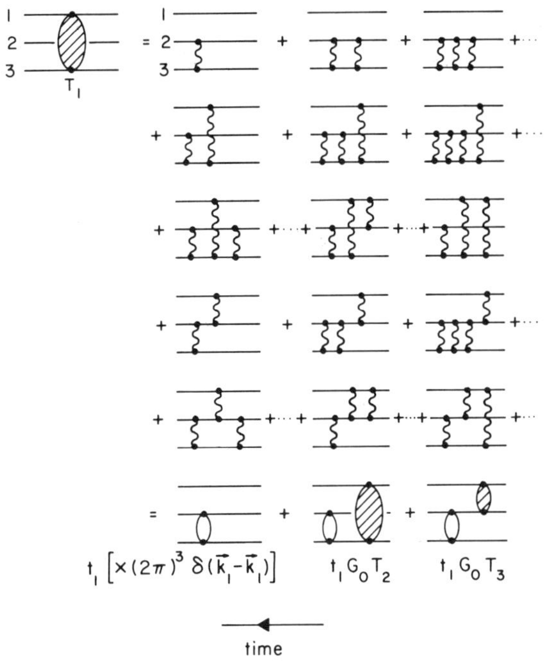

Fig. 3.3 Sample diagrams for three interacting particles. The diagrams are to be read from right to left. The full amplitude T is expressed as a sum over three subamplitudes, defined according to which pair $jk \to i$, $\{i,j,k\} = \{1,2,3\}$, interacted last; in this case the example shown is for T_1 in which the pair 23 scatters last. The integral equation for T_1 then involves in a natural way the amplitude t_1 for two-particle scattering by 23, the three-particle propagation G_0, and the *other* subpartitions T_2 and T_3 of the full amplitude.

will sum to give t_1, the full two-particle amplitude for the 2, 3 system (times a momentum-preserving delta-function for particle 1, of course). Others will have various possible interactions between 2 and 3 preceded by an interaction between another pair, 1 and 3 or 1 and 2, and that preceded by any possible sequence of interactions; these diagrams will thus have t_1 at the left and either T_2 or T_3 to the right of it, separated by a three-particle propagator. Although the delta function that existed in the original Lippmann-Schwinger equation persists, it will now be connected to other interacting pairs when the coupled equations for T_1, T_2, and T_3 are iterated. Faddeev showed that the kernel of the resulting coupled integral equations after *five* iterations (producing t^5 terms) is compact, so a unique solution to the equation is guaranteed.

To construct and explore the coupled integral equations with greater precision, we consider the scattering of particle* 2 (the meson, for us) on a bound state of 1 and 3 (two nucleons bound to form a deuteron). The relevant integral equation is

$$T = (V_1 + V_3) + (V_1 + V_3) \frac{1}{E - K - V_2 + i\eta} T, \qquad [38]$$

where $V_1 + V_3$ expresses the interaction of the projectile with each nucleon in the target, and the full dynamics of the target are carried in the propagator

$$G^{(2)} \equiv \frac{1}{E - K - V_2 + i\eta}. \qquad [39]$$

Eqs. 38 and 39 are merely a rewritten version of Eqs. 5, 7a, and 7b. Thus we have envisioned a Hamiltonian for the three-particle system

$$H = K + V = (K_1 + K_2 + K_3) + (V_1 + V_2 + V_3), \qquad [40]$$

with kinetic energy operators K_i, $i = 1, 2, 3$ for the three particles and interactions V_i, $i = 1, 2, 3$, defined cyclically as before. This Hamiltonian is to be broken up as

$$H = H_2 + H_2' = (K + V_2) + (V_1 + V_3) \qquad [41]$$

to include the dynamics of system 2 in the unperturbed (presumably solved) part H_2, whence its appearance in the propagator G_2.

In a problem involving meson scattering on a more complex nucleus ($A > 2$) than the deuteron, the parallel procedure would lead to a sum over the A interactions of the meson with the A nucleons in H', but with the full nuclear dynamics in H_0 and thus in the propagator. There is then no hope of dealing with the full nuclear dynamics exactly, and we must resort to various approximation schemes to handle the nuclear physics while proceeding with the multiple-scattering aspect of the problem. These two features are then treated in a very unsymmetric way.[†] In the $A = 2$ case, however, the problem represented by $H_2 = K + V_2$ is the two-nucleon problem (in the presence of a noninteracting meson) and can be solved

---

*This notation differs slightly from that of the previous section in that the nucleons are now labeled 1 and 3 (not 1, 2), and the (previously) unlabeled meson is now called 2. This has become fairly conventional for the three-body formulation.

†These methods form the main subject matter of this book and are treated at length in the following chapters.

exactly. For our purpose, this is most conveniently done by noting

$$G^{(2)} = \frac{1}{E - K - V_2 + i\eta} = \frac{1}{E - K + i\eta} + \frac{1}{E - K - V_2 + i\eta} V_2 \frac{1}{E - K + i\eta},$$

[42]

as can be verified by multiplying through with $E - K - V_2 + i\eta$ from the left. Then

$$G^{(2)} = G_0 + G^{(2)} V_2 G_0,$$

[43]

where G_0 is the unperturbed three-body propagator

$$G_0 \equiv \frac{1}{E - K + i\eta}.$$

[44]

Solving Eq. 43 formally, we have

$$G^{(2)} = G_0 (1 - V_2 G_0)^{-1},$$

[45]

and inserting this in the right-hand side of Eq. 43, we obtain

$$G^{(2)} = G_0 + G_0 (1 - V_2 G_0)^{-1} V_2 G_0 = G_0 + G_0 t_2 G_0,$$

[46]

where

$$t_2 = (1 - V_2 G_0)^{-1} V_2$$

[47]

is the two-body t-matrix for scattering of 1 and 3. Equation 47 can be put in the form of the two-body Lippmann-Schwinger equation, given as Eq. 28, by multiplying from the left with $1 - V_2 G_0$ to get

$$t_2 = V_2 + V_2 G_0 t_2.$$

[48]

Now for the desired amplitude of Eq. 38 we write

$$T = T_1 + T_3,$$

[49]

with

$$T_i = V_i + V_i G^{(2)} T, \qquad i = 1, 3.$$

[50]

Using Eq. 46 we obtain

$$T_i = V_i + V_i G_0 T + V_i G_0 t_2 G_0 T$$
$$= V_i + V_i G_0 (T_i + T_j) + V_i G_0 t_2 G_0 T, \qquad [51]$$
$$i = 1,3; \qquad j = 1,3; \qquad j \neq i;$$

Eqs. 49 and 51 parallel our previous Eq. 8, although there the nuclear dynamics are retained in the propagator G_N, while here we deal with them more explicitly and in a manner more symmetrical with πN dynamics. We now subtract $V_i G_0 T_i$ from both sides of the equation to get

$$(1 - V_i G_0) T_i = V_i + V_i G_0 T_j + V_i G_0 T_2, \qquad [52]$$

with

$$T_2 \equiv t_2 G_0 T \equiv t_2 G_0 (T_1 + T_3), \qquad [53]$$

and divide in Eq. 52 by $(1 - V_i G_0)$, again using Eq. 47 to identify the two-body amplitude for the ith system;

$$T_i = t_i + t_i G_0 T_j + t_i G_0 T_2, \qquad i = 1,3; \qquad j = 1,3; \qquad j \neq i. \quad [54]$$

Making Eqs. 53 and 54 more explicit, we arrive at the Faddeev equations for this problem,

$$T_1 = t_1 + t_1 G_0 T_2 + t_1 G_0 T_3,$$
$$T_2 = \qquad t_2 G_0 T_1 + t_2 G_0 T_3,$$
$$T_3 = t_3 + t_3 G_0 T_1 + t_3 G_0 T_2, \qquad [55]$$

a set of coupled integral equations (parallel in content to Eq. 10) whose higher iterations, as we have noted, lead to a compact kernel and hence to a unique solution. Note that there is no inhomogeneous or driving term for T_2 since particles 1 and 3 are bound and do not have a corresponding asymptotic scattering component. The coupled structure of Eq. 55 assures that the relevant diagrams are "linked" (see Figs. 3.2 and 3.3), so no dangerous δ-functions persist in the kernels of the integral equations, and, as noted above, the latter may eventually be proved to be compact (Fad–65).

The equations given as Eq. 55 and the formal manipulations leading to them, are just the equivalent of the schematic approach based on Fig. 3.3. They are very close in their appearance to the Watson series that we derive in the next chapter, but differ in that they embody a piece involving T_2 and t_2 which is just the solution of the np problem, here treated on an equivalent footing with the scattering in the πn and πp systems. We again stress that a further virtue of the Faddeev equations is that they contain the amplitudes t_i rather than the potentials V_i. These may be more directly accessible experimentally, and—more important—will already have summed over large matrix elements that may arise from V if it involves very strong interactions, as in the case of a short-range hard-core repulsion or the like. The t-matrix in such a situation will have well-behaved elements, unlike V, thus making Eq. 55 especially convenient.

3.2c Choice of Variables; Partial Wave Analysis

We choose our kinematic variables by generalizing the scheme which led to the two-body Eq. 29, again using a cyclic notation in which i refers to the single particle i and to the remaining two-particle subsystem j and k (i,j,k cyclic). The total momentum is

$$\mathbf{P} = \mathbf{k}_1 + \mathbf{k}_2 + \mathbf{k}_3, \qquad [56]$$

the internal relative momentum of the j,k subsystem is

$$\mathbf{p}_i = \frac{1}{m_j + m_k}(m_k \mathbf{k}_j - m_j \mathbf{k}_k), \qquad [57]$$

and the relative momentum between this two-particle subsystem and particle i is

$$\mathbf{q}_i \equiv \frac{1}{m_i + m_j + m_k}\left[m_i(\mathbf{k}_j + \mathbf{k}_k) - (m_j + m_k)\mathbf{k}_i \right]. \qquad [58]$$

The Jacobian of this transformation is unity, and since $\mathbf{p}_i \propto \mathbf{v}_j - \mathbf{v}_k$ and $\mathbf{q}_i \propto m_j(\mathbf{v}_j - \mathbf{v}_i) + m_k(\mathbf{v}_k - \mathbf{v}_i)$ both these variables are Galilei invariant. The three-particle kinetic energy is then given by the generalization of Eq. 30,

$$\frac{k_1^2}{2m_1} + \frac{k_2^2}{2m_2} + \frac{k_3^2}{2m_3} = \frac{p_i^2}{2\eta_i} + \frac{q_i^2}{2\nu_i} + \frac{P^2}{2M}, \qquad [59]$$

where the right-hand side contains three terms: (1) the j,k internal energy, with its relative momentum $\mathbf{p}_i$ and reduced mass

$$\eta_i = \frac{m_j m_k}{m_j + m_k}, \qquad [60]$$

(2) the internal energy of that aggregate with particle i involving that

internal momentum $\mathbf{q}_i$ and the corresponding reduced mass

$$\nu_i = \frac{m_i(m_j + m_k)}{m_i + m_j + m_k}, \qquad [61]$$

and (3) the total system kinetic energy derived from the motion of the total mass

$$M = m_i + m_j + m_k = m_1 + m_2 + m_3 \qquad [62]$$

with momentum $\mathbf{P}$. This total momentum $\mathbf{P}$ will, of course, be conserved. The relative momenta $\mathbf{p}_i$ and $\mathbf{q}_i$, $i = 1,2,3$ may be chosen three different ways, the various choices naturally being redundant. Indeed, a pair $\{\mathbf{p}_i, \mathbf{q}_i\}$ and $\{\mathbf{p}_j, \mathbf{q}_j\}$ are related through

$$\begin{pmatrix} \mathbf{p}_j \\ \mathbf{q}_j \end{pmatrix} = \begin{bmatrix} -\dfrac{\eta_j}{m_k} & \dfrac{\eta_j}{\nu_i} \\ -1 & -\dfrac{\eta_i}{m_k} \end{bmatrix} \begin{pmatrix} \mathbf{p}_i \\ \mathbf{q}_i \end{pmatrix}, \qquad i,j,k \text{ cyclic.} \qquad [63]$$

This mapping from one set of variables to another is especially important because the dynamics of each two-particle subsystem i will be described most naturally in terms of its own variables $\mathbf{p}_i$ and $\mathbf{q}_i$, since then $\mathbf{p}_i$ represents *its* internal relative momentum and $\mathbf{q}_i$ is an extraneous, preserved momentum relating to the motion of a noninteracting third particle from the perspective of that ith two body subsystem. Thus we have

$$\langle \mathbf{P}'\mathbf{p}_i'\mathbf{q}_i' | V_i | \mathbf{P}\mathbf{p}_i\mathbf{q}_i \rangle = (2\pi)^6 \delta(\mathbf{P}' - \mathbf{P}) \delta(\mathbf{q}_i' - \mathbf{q}_i) \langle \mathbf{p}_i' | V_i | \mathbf{p}_i \rangle \qquad [64]$$

and

$$\langle \mathbf{P}'\mathbf{p}_i'\mathbf{q}_i' | t_i | \mathbf{P}\mathbf{p}_i\mathbf{q}_i \rangle = (2\pi)^6 \delta(\mathbf{P}' - \mathbf{P}) \delta(\mathbf{q}_i' - \mathbf{q}_i) \langle \mathbf{p}_i' | t_i | \mathbf{p}_i \rangle. \qquad [65]$$

Taking the first of the Faddeev equations given in Eq. 55 as an example, and removing the delta function $(2\pi)^3 \delta(\mathbf{P}' - \mathbf{P})$ for total momentum conservation in parallel to Eqs. 31–34 we have, in the three-particle center-of-mass system,

$$\langle \mathbf{p}_1'\mathbf{q}_1' | T_1(E) | \phi_0 \rangle = \langle \mathbf{p}_1'\mathbf{q}_1' | t_1 \left(E - \frac{q_1'^2}{2\nu_1} \right) | \phi_0 \rangle$$

$$+ \sum_{j=2,3} \int (2\pi)^3 \delta(\mathbf{q}_1' - \mathbf{q}_1'') \langle \mathbf{p}_1' | t_1 \left(E - \frac{q_1'^2}{2\nu_1} \right) | \mathbf{p}_1'' \rangle$$

$$\times \frac{1}{E - (p_1''^2/2\eta_1) - (q_1''^2/2\nu_1) + i\eta} \langle \mathbf{p}_1''\mathbf{q}_1'' | \mathbf{p}_j''\mathbf{q}_j'' \rangle$$

$$\times \langle \mathbf{p}_j''\mathbf{q}_j'' | T_j | \phi_0 \rangle \frac{d\mathbf{p}_1''}{(2\pi)^3} \frac{d\mathbf{q}_1''}{(2\pi)^3} \frac{d\mathbf{p}_j''}{(2\pi)^3} \frac{d\mathbf{q}_j''}{(2\pi)^3}, \qquad [66]$$

where ϕ_0 is the initial state (meson and deuteron, in our case) and we have inserted two complete sets of states defined for the variables $\{\mathbf{p}_1'', \mathbf{q}_1''\}$ and $\{\mathbf{p}_j'', \mathbf{q}_j''\}$ in order to have each amplitude taken with respect to its own natural variables. The amplitude T_1 of Eq. 66, which pertains to the situation in which the last interaction is that of the meson with a nucleon, exhibits in very explicit fashion how the meson-nucleon scattering is to be averaged over the motion of the "active" nucleon, as it is influenced by the force between it and the other nucleon (t_2 in T_2—and in T_3). The energy argument of t_1 at which the scattering takes place is influenced by the energy carried by the other nucleon, which determines the energy left for the "1" πN system; the intermediate momenta for t_1 are weighted by the other dynamic features of the problem through T_j, $j = 2, 3$. For three-body systems, all the kinematics and dynamics enter explicitly and are treated exactly (within the confines of a nonrelativistic, potential theory and so on), while, as we see, for $A > 2$ these features can be treated only in approximate ways.

In order to reduce the (alarming!) number of continuous variables that appear in Eq. 66 we carry out a partial wave analysis (see Ahm-65, Bal-69, Sti-70a, Slo-72, Bol-74, and Man-76, for example) including the spin and isospin degrees of freedom of the participating particles, which up until now we have suppressed. Then

$$|\mathbf{p}_i \mathbf{q}_i; S_i, S_i^z; s_i, s_i^z; T_i, T_i^z; \tau_i, \tau_i^z\rangle = \sum_{L_i J_i, l_i j_i, JM, TM_T} |p_i, q_i;$$

$$\times \left[(L_i S_i) J_i, (l_i s_i) j_i \right] JM; (T_i t_i) TM_T\rangle$$

$$\times \left\{ \left[Y_{L_i}(\hat{\mathbf{p}}_i) \otimes \chi_{S_i} \right]_{J_i} \otimes \left[Y_{l_i}(\hat{\mathbf{q}}_i) \otimes \chi_{s_i} \right]_{j_i} \right\}_{JM}$$

$$\times \left\{ \phi_{T_i} \otimes \phi_{\tau_i} \right\}_{TM_T}, \qquad [67]$$

where S_i is the total spin of the pair i and S_i^z is its magnetic projection, s_i and s_i^z are correspondingly the spin and projection of the ith particle. The isospins T_i, T_i^z and τ_i, τ_i^z are defined in parallel for the pair and particle. In the coupled representation, the relative orbital angular momentum L_i of the pair $i = (j, k)$ is coupled to the pair spin to produce the total angular momentum of the pair J_i. The relative orbital angular momentum l_i between the ith particle and the pair is coupled to the spin of that particle to yield the resultant j_i. Then J_i and j_i are coupled to give the total angular momentum J and its projection M, the isospin being treated in parallel fashion.

In such a basis, and with α_i referring to the discrete quantum numbers as on the right-hand side of Eq. 67, the two-particle amplitudes have matrix elements

$$\langle p_i' q_i'; \alpha_i' | t_i | p_i, q_i; \alpha_i \rangle = \frac{(2\pi)^3}{q_i^2} \delta(q_i' - q_i) \delta_{\alpha_i' \alpha_i} t_i^{L_i S_i J_i}\left(p_i', p_i; E - \frac{q_i^2}{2\nu_i} \right).$$

$$[68]$$

The overlap brackets between different natural variables $\{\mathbf{p}_i, \mathbf{q}_i\}$ and $\{\mathbf{p}_j, \mathbf{q}_j\}$ in the coupled representation $\langle p_i, q_i; \alpha_i | p_j, q_j; \alpha_j \rangle$ can be evaluated (Man–76) using Eq. 63 in terms of kinematical factors, angular momentum recoupling coefficients, and the independent angles between the vector variables. They also carry an energy-conserving delta function $\delta[p_i^2/2\eta_i + q_i^2/2\nu_i - p_j^2/2\eta_j - q_j^2/2\nu_j]$. When the partial wave decomposition is introduced into Eq. 66, we obtain

$$\langle p_1', q_1'; \alpha_1' | T_1(E) | \phi_0 \rangle = \langle p_1', q_1'; \alpha_1' | t_1\left(E - \frac{q_1'^2}{2\nu_1} \right) | \phi_0 \rangle$$

$$+ \sum_{j=2,3} \sum_{\alpha_1'' \alpha_j''} \int \frac{p_1''^2 dp_1''}{(2\pi)^3} \frac{q_1''^2 dq_1''}{(2\pi)^3} \frac{p_j''^2 dp_j''}{(2\pi)^3} \frac{q_j''^2 dq_j''}{(2\pi)^3}$$

$$\times \langle p_1', q_1'; \alpha_1' | t_1\left(E - \frac{q_1'^2}{2\nu_1} \right) | p_1'', q_1''; \alpha_1'' \rangle$$

$$\times \frac{1}{E - (p_1''^2/2\eta_1) - (q_1''^2/2\nu_1) + i\eta}$$

$$\times \langle p_1'', q_1''; \alpha_1'' | p_j'', q_j''; \alpha_j'' \rangle$$

$$\times \langle p_j'', q_j''; \alpha_j'' | T_j | \phi_0 \rangle. \qquad [69]$$

The four integrations here are reduced to two by virtue of Eq. 68 and the energy-conserving delta function in the $\langle i | j \rangle$ overlap bracket. Equation 69 is thus part of a two-variable coupled integral equation, and as such represents the central result in our present approach. This two-variable equation is just barely within the domain of soluble problems on modern computers. To make its solution more accessible, it is common to use a sort of a "model" of the full Faddeev treatment in the form of a separable interaction that eliminates one variable in the equations. The amplitude in Eq. (68) is then taken as (cf. Eq. 2.56)

$$t_i^{L_i S_i J_i}(p_i', p_i; W) = g_i(p_i')\tau_i(W)g_i(p_i), \qquad [70]$$

where the factors $g_i(p_i)$ must be obtained by fitting a separable potential to the relevant two-particle data for each channel, and, from the Lippmann-

Schwinger equation with energy-independent interaction,

$$\tau_i^{-1}(W) = 1 - \int \frac{g_i^2(p_i)}{W - (p_i^2/2\eta_i) + i\eta} \frac{p_i^2 \, dp_i}{(2\pi)^3}. \qquad [71]$$

Then we can trivially eliminate dependence on the unwanted p_1 variable, separability basically allowing us to convert the three-body problem into a two-body one for the relative motion of the particle with respect to the pair. The required ansatz is

$$\langle p_1', q_1'; \alpha_1' | T_1(E) | \phi_0 \rangle = g_1(p_1')\langle q_1'; \alpha_1' | T_1(E) | \phi_0 \rangle, \qquad [72]$$

and the coupled one-variable integral equations for $\langle q_i'; \alpha_i' | T_i(E) | \phi_0 \rangle$ become

$$\langle q_i'; \alpha_i' | T_i(E) | \phi_0 \rangle = \langle q_i'; \alpha_i' | t_i \left(E - \frac{q_i'^2}{2\nu_i} \right) | \phi_0 \rangle$$

$$+ \sum_{\alpha_j} \int K_{ij}^{\alpha_i'\alpha_j}(q_i', q_j) \langle q_j; \alpha_j | T_j(E) | \phi_0 \rangle \, dq_j.$$

$$[73]$$

This can be solved—tediously—by fairly standard techniques. The kernel K in Eq. 73 is a function of two variables q_i' and q_j given by an integral over p_j having a structure that is a generalization of the integral in Eq. 71.

3.2d Relation to the Problem for More Complex Targets

In broad outline, it is therefore clear that the numerical analysis to be dealt with for a solution of the three-body problem involves the coupling of a moderate number of integral equations (for T_1, T_2, T_3, and the various relevant channels) with kernels having two continuous variables and an integral over a third. When these continuous variables are made discrete for purposes of computation, we thus need three separate meshes with, optimistically, ~ 20 points on each, or $\sim 20^3 \cdot n$ if n coupled amplitudes are involved. Both the required storage and computer time for such a problem press the limits of computers rather severely, even after we have invoked separability. We stress these mundane details because they make clear that even once a four-body, five-body,... formalism is created (Yak–67, Kar–74) there is little hope at present of using it successfully in a practical calculation. (Indeed, it will require as input the solution of the three-body, four-body,... problem.) Thus even for $\pi - {}^3$He scattering we are forced to use an approach that relinquishes any pretense of exactitude in favor of a practicable approximation scheme.

As a consequence, the fully defined kinematics of Eqs. 66 and 69, in which the energies and momenta for the elementary πN amplitudes are unambiguous, are the subject of approximative statements in πA scattering with $A > 2$. For example, since we are not able to solve the A nucleon nuclear-structure problem there on the same footing as the scattering problem, we use the equivalent of G_2 of Eq. 39 as the basic propagator in Eq. 38 for T in an approximate way. The energy at which to evaluate the "free" amplitude $t_{\pi N}$ for use in T may then be determined by the value for this parameter which optimizes such approximations. In this regard the deuteron case, which we can solve exactly by the Faddeev equations, gives us somewhat less guidance than we might like, since the deuteron is much more weakly bound and diffuse than for the $A > 2$ systems of interest to us.

3.3 Applications of the Faddeev Equations

The techniques of the Faddeev equations have been applied to calculations of πd and Kd scattering. Before we discuss specific examples, we should consider briefly what such applications can teach us. The Faddeev integral equations, as derived in the previous section, are equivalent to a three-body Schroedinger equation. Solution of the integral equations gives therefore a complete dynamical description within the Schroedinger equation. In terms of multiple scattering, all orders are calculated exactly, including interactions among the target particles. Practical considerations limit this generality, however. To limit the complexity to solving coupled integral equations in one variable, we have restricted the interactions to those that are separable or that lead to separable t-matrices, as in Eq. 70. Further restrictions of the number of partial waves or of channels are often made to limit the size of the computation. So most Faddeev calculations must be considered model calculations: the equations are solved "exactly," but for somewhat simplified interactions.

In many cases, these limitations are not too severe, and it is well worth that price to have a complete solution. In some physical domains as, for example, near resonances, the restriction to separable interactions and truncation of channels do not seriously distort the physical description, and we may expect to make comparison with experiment. But even when the model falls short of a complete dynamical representation, the calculations have another use—that of providing a method of testing the assumptions and approximations made in other methods of treating scattering, namely, multiple scattering approximations. Thus we may study the impulse approximation, static or fixed scatterer approximations, rates of convergence of multiple scattering series, and so on. This kind of test is very important for study of general targets $(A > 2)$ in multiple scattering theory, for which, as we have noted above, there is no practical equivalent to the Faddeev theory.

It is worth remarking that there are also limitations on the completeness of the Faddeev equations, not based simply on our ability to calculate. The most important are the nonrelativistic formulation, and the restriction to three-body channels (leaving out, e.g., $\pi d \rightarrow NN$ as part of πd scattering). The possibility of overcoming these difficulties is discussed later (Sections 3.4 and 3.5).

3.3a Calculation of Low Energy πd and Kd Scattering

Let us consider the simplest assumptions first. We take all the two-body t-matrices to be separable and interacting in S-wave only. This limitation is appropriate, if at all, for low energies. Thus the two-body partial waves are as follows:

$$\pi N \quad L=0, J=\tfrac{1}{2}, T=\tfrac{1}{2} \text{ and } \tfrac{3}{2}$$
$$KN \quad L=0, J=\tfrac{1}{2}, T=0 \text{ and } 1 \qquad [74]$$
$$NN \quad L=0, J=1, T=0.$$

The NN channel has the quantum numbers of the deuteron. Because of the restriction to S-wave meson-nucleon scattering, there is no nucleon spin-flip, and the NN $(L=0, J=0, T=1)$ channel is not coupled to the deuteron.

Separability of the t-matrices leads to a coupled-integral form of the Faddeev equations, such as that given in Eq. 73. We want the t-matrices to be derivable from separable potentials, as in Eq. 2.55. These may be specified for each channel in Eq. 74 by giving the strength λ and the functional form $g(p)$. Simple analytic forms have often been used to simplify the calculation of the kernel K of Eq. 73 in which they occur. For example, the form

$$g(p) = (p^2 + \beta^2)^{-1} \qquad [75]$$

introduced by Yamaguchi is a popular expression carrying one parameter β which may be considered to be an inverse range.

The integral equations given in Eq. 73 involve (matrix) coupling of the discrete quantum numbers α, which were originally specified in Eq. 67. The total quantum numbers (J, T, M, M_T) are conserved. With the restrictions given in Eq. 74, only the sums over isospin T_i remain, so for each J, T

partial wave, Eq. 73 gives two coupled equations for $\langle q_j; \alpha_j | T_j(E) | \phi_0 \rangle$ in the variable q_j. (The three coupled equations $i = 1, 2, 3$ reduce to two, using the isospin symmetry $T = 0$ for the deuteron.) Numerical solution of the equations requires careful handling of singularities in the kernels K_{ij}. The scattering amplitudes are obtained from Eq. 49 by integration over the appropriate asymptotic final state—for example, for elastic scattering, over the deuteron wave function

$$\langle q_2' | T(E) | q_2 \rangle = \int d\mathbf{p}_2' \psi_d^*(\mathbf{p}_2') \langle \mathbf{p}_2' q_2' | T_1 + T_3 | \phi_0 \rangle \qquad [76]$$

where $\mathbf{q}_2, \mathbf{q}_2'$ are the initial and final meson momenta in the three-body c.m. and $\psi_d(\mathbf{p}_2)$ is the deuteron wave function. (We have suppressed the sum over isospin labels α_i.)

The number of partial waves (l_2, J) that contribute significantly depends on the scattering energy $E = \mathcal{E}(q_2) + E_d$ (or momentum q_2). We would expect that $l_2(\text{max}) \simeq q_2 d$, where d is of the order of the deuteron size, that is, 2–4 fm. This means that $l_2(\text{max}) \simeq q_2/(50–100 \text{ MeV}/c)$; for example, $l_2(\text{max}) \simeq 3$ to 6 for $q_2 = 300 \text{ MeV}/c$. For the case (1) of S-wave two-body interactions, the two-nucleon spin ($S = 1$ for deuteron) remains uncoupled, so partial wave amplitudes depend only on l_2. In summary, we have an idea of the size of the computational problem for the Faddeev amplitudes for meson deuteron scattering with $q \leqslant 300 \text{ MeV}/c$—up to six pairs of coupled integral equations (singular) in one momentum variable q.

The problem becomes rapidly more complex if we consider more complicated interactions than those included in Eq. 74. If we consider two-body interactions in higher $(L > 0)$ partial waves, the number of coupled equations immediately increases, with the number of allowed values of α (see Eqs. 67 and 73). Yet even at low energies, the πN p-wave interaction cannot be neglected. Similarly, the tensor NN interaction couples $^3S_1 - {}^3D_1$ states in the deuteron, and in the intermediate states included in the Faddeev equations. As soon as $L \neq 0$ interactions enter, spin flip and spin coupling are induced, and many more partial waves (l_2 and J) enter—for example, the P-states of the NN system.

In principle, one could extend the Faddeev equations to include other three-body channels that might be important, such as

$$K^- d \rightarrow \pi Y N$$
$$\pi d \rightarrow \rho n p, \qquad [77]$$

but again the price is the increase in the number of coupled channels. The difficulty of numerical solution rises rapidly with this number. Incorporation of relativity, or coupling to channels with other than three bodies, leads to further complication. Therefore, most practical Faddeev calculations must leave out some elements of the physics, particularly a wealth of inelastic features.

3.3b Examples

1. πd *scattering length.*

As our first specific case, we consider pion-deuteron scattering at zero energy. The scattering length is defined as the elastic scattering amplitude $a_{\pi d} = f_{\pi d}(T_\pi = 0)$. The amplitude is complex, because the absorption reaction $\pi d \rightarrow 2N$ is exothermic, and contributes an imaginary amplitude even for $T_\pi = 0$. (There is also a small imaginary contribution of charge exchange: $\pi^- d \rightarrow \pi^0 nn$, which is usually ignored.) The real part is measured for $\pi^- d$ from the energy shift of the $2P - 1S$ decay line in the π-mesic atom of deuterium (see Chapter 5). Recent measurements may be combined to give (Bai–74)

$$a_{\pi d} \simeq (-0.05 + 0.004i) \, m_\pi^{-1}$$
$$= (-0.07 + 0.006i) \, \text{fm} \qquad [78]$$

with considerable ($\gtrsim 30\%$) uncertainty.

For $T_\pi = 0$, the scattering is in the $l_2 = 0, J = 1, T = 1$ partial wave only. Faddeev equations have been solved, with separable interactions in S-waves only (Pet–73 and Myh–74), and also, including πN p-waves and the $^3S_1 - {}^3D_1$ NN tensor interaction (Afn–74). There is great uncertainty in the input data, since there is considerable freedom in fitting the separable potentials to the low energy πN phase shifts, which are not well known down to $T_\pi \sim 0$. Therefore, the numerical results range from $-0.07 \leqslant a_{\pi d} \lesssim -0.03$ fm, depending particularly on the chosen values of the πN scattering lengths (see Table 2.2). (The calculated $a_{\pi d}$ for ordinary Faddeev theory is real, since absorption and charge exchange are not included; see, however, Afn–74 and below.)

In order to facilitate the discussion of the Faddeev results, it is useful to turn to multiple-scattering theory and employ a simple estimate. If we ignore isospin for the moment, we may take over a result due to Brueckner, who used a fixed scatter assumption, with p-wave meson-nucleon scattering. In the zero energy limit, his result (Bru–53) has the form

$$f_{\pi d}(T_\pi = 0) = \left\langle \frac{a_1 + a_2 + (2a_1 a_2 / r)}{1 - (a_1 a_2 / r^2)} \right\rangle \qquad [79]$$

where the expectation value is over the deuteron ground state, with r the relative distance of the nucleons. This simple form sums all orders of multiple scattering for this simple static case (with short range interactions). Inclusion of isospin, which allows for multiple charge-exchange πN scattering, leads to a more complicated form (see Kol–73a). However, to second order (in the scattering lengths) the amplitude is given by

$$f_{\pi d}(T_\pi = 0) = 2a^{(+)}c_\pi + 2\left[(a^{(+)})^2 - 2(a^{(-)})^2\right]\langle \frac{1}{r}\rangle c_\pi^2 + \cdots \quad [80]$$

where we use the isoscalar and isovector scattering lengths of Eq. 2.34, $a = a^{(+)} - a^{(-)}\vec{I}_\pi \cdot \vec{\tau}$, and $c_\pi = (1 + m_\pi/M)(1 + m_\pi/M_d)^{-1}$ corrects for reduced masses. The higher-order terms are of order $c_\pi^3 a^3 \langle r^{-2}\rangle$ as in Eq. 79. The expansion given in Eq. 80 would be expected to converge for small values of the ratio $\langle a/r\rangle$ (although the inverse powers of r are taken within the expectation value as in Eq. 79).

The empirical fact is that the πN scattering lengths are small on the scale of $\langle 1/r\rangle^{-1} \sim 2$ fm for the deuteron; we have from Eq. 2.34 that $a^{(+)} = 0.007$ fm and $a^{(-)} = 0.12$ fm. Therefore the series given in Eq. 80 should converge rapidly, for reasonably regular deuteron wave functions. However, since the isoscalar length $a^{(+)}$ is small compared to the isovector length $a^{(-)}$, the second term in Eq. 80 may be as big as the first: double (isovector) scattering is as important as single (isoscalar) scattering. With the values of $a^{(+)}$ given above, Eq. 80 gives (Kol–73a)

$$a_{\pi d} = (-0.015 - 0.025 + \cdots) \text{ fm} = -0.040 \text{ fm} \quad [81]$$

where we give the values of the first two terms separately. The rest of the series contributes $+0.003$ fm in this case, which is indeed a correction of order $c_\pi^3 a^3 \langle r^{-2}\rangle$ to the first two orders. Clearly the multiple-scattering series converges rapidly in this static, s-wave approximation. The sum $(-0.037$ fm$)$ is similar to the calculated result $(-0.0453$ fm$)$ of the Faddeev equations, for s-wave scattering, based on the same values of the scattering lengths (Pet–73). The $(\sim 20\%)$ difference between the static and Faddeev results may be understood as due to the finite range of the πN interaction, the energy-dependence of the πN t-matrix, and to the recoil (or excitation) of the nucleons, all of which contribute in the Faddeev theory, but not in Eq. 80. These effects were estimated (Kol–73a) as corrections to the static series in Eq. 80, to be -0.006 fm (although individual corrections may be larger), which then brings the multiple-scattering (s-wave) result $(-0.037 - 0.006 = -0.043$ fm$)$ closer to the Faddeev result. In this particular case, the multiple-scattering series gives a good approximation, even in low (second) orders to the "exact" theory with the same interaction model.

However, even for zero-energy πd scattering, the p-wave πN amplitudes may be expected to contribute, once nucleon motion is taken into account. The simple scattering contribution is estimated (Kol–73a) to be about $+0.006$ fm, but the net effect on the series is smaller because of destructive interferences with the s-wave scattering in higher orders. This small effect has also been obtained in a Faddeev calculation that included the πN P_{33} channel (Afn–74). (As noted earlier, the inclusion of p-wave scattering considerably enlarges the Faddeev calculation, as does the inclusion of the deuteron D-state, which is also in Afn–74.)

The Faddeev (and multiple-scattering) results are consistent with the experimental value expressed in Eq. 78, given the uncertainty of the latter. As noted, however, there are uncertainties as well in our knowledge of $a_{\pi N}$, or more important, of the low energy πN interaction. In addition, the results quoted above do not include the effect of π-absorption, which contributes both a real and an imaginary part to $a_{\pi d}$. Estimates of this effect (Afn–74 and Miz–77) give a contribution of $(-0.007 - -0.010$ fm$)$ to Re $a_{\pi d}$, with Im $a_{\pi d}$ from $(0.006 - 0.014$ fm$)$. (Absorption is discussed in Section 3.5)

2. K$^+$d *scattering.*

The earliest applications of Faddeev theory to meson-deuteron scattering were calculations by Hetherington and Schick of Kd scattering at low energies ($q_K < 300$ MeV$/c$) (Het–65a,b,66). The K$^+$d case is somewhat simpler, since only scattering channels are involved (see Section 3.1). We discuss this case first.

For K$^+$N scattering, the assumption that only s-waves contribute is a very good approximation at low energy (see Section 2.2). The $T=1$ channel (e.g., K$^+$p) is characterized by a scattering length, given by Eq. 2.38, of $a_1 = -0.29$ fm; the $T=0$ length is less well known, but is believed to be much smaller: $|a_0| \ll |a_1|$. The range of interaction is also believed to be small, that is, $\lesssim 0.5$ fm. If we use the static multiple-scattering model of Brueckner (Eq. 79) corrected for isospin and reduced masses, we have, through second order,

$$f_{K^+d}(T_K = 0) = 2a^{(+)}C_K + 2\left[(a^{(+)})^2 - \tfrac{3}{4}(a^{(-)})^2\right]\langle\tfrac{1}{r}\rangle C_K^2 + \cdots$$

$$[82]$$

where

$$a^{(+)} = \tfrac{1}{4}a_0 + \tfrac{3}{4}a_1, \; a^{(-)} = \tfrac{1}{2}a_0 - \tfrac{1}{2}a_1, \; C_K = \frac{(1 + m_K/M)}{(1 + m_K/M_d)}.$$

If we simply ignore the contribution of a_0 to Eq. 82, we have

$$a_{K^+d} \simeq \frac{3}{2}a_1 C_K + \frac{3}{4}a_1^2 C_K^2 \langle \frac{1}{r} \rangle + O\left(a_1^3 C_K^3 \langle \frac{1}{r^2} \rangle\right). \qquad [83]$$

Since $|a_1| \langle r^{-1} \rangle_d \sim 0.15$, this should be a rapidly convergent series, and well approximated by the first (single scattering) term

$$a_{K^+d} \simeq \frac{3}{2}a_1 C_K = -0.54 \text{ fm} \qquad \text{(single scattering).} \qquad [84]$$

For low energies, this approximation leads to a forward elastic differential cross section for K^+d of the form

$$\frac{d\sigma}{d\Omega}(0°) \simeq \frac{(a_{K^+d})^2}{1 + a_1^2 q_K^2} = \frac{2.8 \text{ mb}}{1 + a_1^2 q_K^2}. \qquad [85]$$

The simple static approximations given in Eqs. 83–85 are not a bad representation of the elastic scattering cross sections, calculated with similar input ($a_0 = 0.04$ fm) (Het–65b). (Only $l = 0$ separable two-body interactions were included.) At the lowest momentum ($q_K = 110$ MeV/c) the forward elastic differential cross section from the Faddeev calculation is $d\sigma/d\Omega(0°) \sim 2.25$ mb/sr. Using only the single scattering term, the result is ~ 2.75 mb/sr, in agreement with the impulse result given in Eq. 85. The reduction from that is partly accounted for in the static approximations by the -10% contribution of the double-scattering term (positive) in Eq. 83, which reduces the cross section by $\sim 20\%$. However, in the Faddeev calculation for this momentum, the reduction due to double scattering is smaller ($\sim -10\%$) with another -10% coming from higher order scattering. In detail, it turns out that the leading higher-order term is triple scattering: $KN - NN - KN$, which is of second order in a_{KN}, and which contributes largely to Im f_{Kd}. This is a correction from unitarity to the static approximation; although the effect on the elastic cross section is small, it makes a major contribution to the inelastic processes and therefore to the total cross section for K^+d. This effect decreases with increasing energy.

Another result of the Faddeev calculations is that the corrections to the single scattering approximation (double- and higher-order scattering) are largest for the $l_2 = 0$ partial wave (K^+d). Since the single scattering amplitude is forward peaked, due to the deuteron form factor (Eq. 15), the deviation from single-scattering grows with increasing angle. This feature is expected in all multiple-scattering theories.

The elastic K^+d cross section has not been (well) measured below $q \sim 640$ MeV/c, but the total cross section is measured down to 366 MeV/c

where $\sigma = 21.4$ mb. This is somewhat higher than the Faddeev result for 230 MeV/c, which is $\sigma = 16.9$ mb.

3. K⁻d scattering.

The K⁻d reactions are somewhat more complicated than the K⁺d reactions, because of the coupling to πYN channels (Eq. 77), as discussed in Section 3.1. Since these channels also involve three particles, they could in principle be treated along with the original three-body channels (K⁻ + N + N) in an expanded set of Faddeev equations. The differences of masses between the channels (e.g., $m_K + m_N - m_\pi - m_\Lambda = 177$ MeV) that are not so small create problems in the nonrelativistic approach. In addition to the π + Y + N channels, K⁻d can also lead to two-body channels Y + N, through strong interaction absorption of the K⁻.

The early work of Hetherington and Schick ignored the hyperon channels and simply dealt with the KNN three-body problem, through Faddeev equations (Het–65a, 66). As for K⁺d, s-wave separable potentials were introduced to represent the two-body interactions; these were adjusted to assumed values (as of 1965) of the K⁻N scattering lengths, for $T = 0, 1$. For example, Kim's analysis gives $a_0 = (-1.674 + i0.722)$ fm, $a_1 = (-0.003 + i0.688)$ fm. The imaginary parts represent coupling to the πY channels in K⁻N scattering.

The scattering lengths in this case are of the same order of magnitude as the deuteron size $\langle r^{-1} \rangle^{-1}$ so that we do not expect an expansion in the scattering lengths, like Eq. 82, to converge (rapidly) for the low energy K⁻d scattering amplitude. This is indeed borne out by the numerical results of the Faddeev equation calculations: the K⁻d elastic scattering amplitude is poorly represented by the first few multiple scattering terms (obtained by iterating the Faddeev integral equations) for $100 < q_K < 300$ MeV/c, although the representation is better for the higher than for the lower momentum. As in the K⁺d calculations, a large correction comes from the NN interaction in excited states: this denotes, presumably, a failure of the static approximation, which should not be surprising at the low energies considered. [A complete static formula similar to Eq. 79 has been derived and evaluated by Chand (Cha–63).] Also, as in the K⁺d calculations, the major contributions of multiple scattering, beyond the single-scattering term, are to the $l_2 = 0$ partial wave of the K⁻d amplitude. This is consistent with the notion that higher-order multiple scattering involves relatively short distances in the deuteron ($a\langle r^{-1} \rangle$ large), which contribute most to $l_2 = 0$.

The Faddeev calculations for K⁻d solve the three-body scattering problem for the kinematical domain in which simpler multiple scattering approximations would fail. However, as we noted earlier, there is still more to the physics of this problem than can be treated by a three-body potential model for $K^- + N + N$. One problem involves coupling to the three-body hyperon channels $\pi + Y + N$. In principle this could be accommodated in an enlarged set of coupled equations like those of Faddeev; this has not been done. The two-body channels $(Y + N)$ have to be handled differently; we consider this in the next section.

However, there is a different problem that is related to the coupling to the hyperon channels, which we discuss now. This arises from the resonances in the $Y + \pi$ system, which show up as resonances below threshold for $K^- N$ (i.e., at negative kinetic energy in the $K^- N$ channel, but positive kinetic energy in the $Y\pi$ channel; see Section 2.2). The effects of these resonances were not included by Hetherington and Schick, since their potentials were designed to fit the $K^- N$ scattering data (i.e., at positive kinetic energy). Yet the resonances may affect $K^- d$ scattering, as may be seen from the following simple consideration. Although, strictly speaking, the channel coupling does not allow us to use the Faddeev equations from potential theory, we ignore the hyperon channels and assume that we still have Faddeev equations of the form of Eq. 73.

Let us assume that the $K^- N$ t-matrix has a simple resonant energy dependence

$$t_i(\mathcal{E}_i) \propto \frac{1}{\mathcal{E}_i - \mathcal{E}_R + (i/2)\Gamma_R} \qquad [86]$$

where $\mathcal{E}_R$ and Γ_R are the position and width of the resonance, and $\mathcal{E}_i$ the $K^- N$ kinetic energy in the two-body c.m. By our convention, $i = 1$ or 3 for $K^- N$. In the Faddeev equations, $\mathcal{E}_i$ is evaluated at $\mathcal{E}_i = E - q_i^2/2\nu_i$ (see, e.g., Eq. 66 and the discussion between Eqs. 56–62) where E is the inital energy of the $K^- d$ system in the three-body c.m. This can be written in terms of q_2, the momentum of the K^-, and E_d, the ground state energy of the deuteron, as

$$E = \frac{q_2^2}{2\nu_2} + E_d. \qquad [87]$$

Thus the two-body resonance given in Eq. 86 enters at a three-body energy

$$E = \mathcal{E}_R + \frac{q_i^2}{2\nu_i} = \frac{q_2^2}{2\nu_2} + E_d, \qquad i = 1,3. \qquad [88]$$

Even if the resonance is below threshold in the two-body system, $\mathcal{E}_R < 0$, it will contribute to the K^-d system with positive kinetic energy $(q_2^2/2\nu_2)$ because of the possibility of sufficiently large q_i^2 to satisfy Eq. 88. We recall that q_i is the momentum of the spectator nucleon i. The average in the deuteron ground state $\langle q_i^2/2\nu_i \rangle$ is 5–10 MeV. The K^-N resonance closest to threshold is the $\Lambda(1405)$, with $\mathcal{E}_R = -27$ MeV, $\Gamma \simeq 40$ MeV, and $\Sigma(1385)$ lies 20 MeV further below. Therefore we expect that the strong energy dependence of these resonances may affect the low energy K^-d scattering in ways not apparent in the short-range potential models discussed above. A calculation for $q_3 = 0$ has been performed (Myh–73a), which illustrates the sensitivity of the K^-d scattering length to the $\Lambda(1405)$ resonance. This effect also has bearing on the level shifts of K-mesic atoms, as is discussed in Section 5.3.

4. πd *scattering for* $50 \leqslant T_\pi \leqslant 250$ MeV.

The application of Faddeev calculations in this energy domain introduces three complications to the methods we have already discussed. First, the p-wave πN interaction is important at these energies in addition to the s-wave interaction. Therefore, the number of coupled equations required for each πd partial wave is increased over that for s-wave interactions only. (Also the number of significant partial waves grows with energy.) Second, the pion energies are relativistic, and the use of a nonrelativistic potential theory like the Faddeev equations, as developed in Section 3.2, may no longer be valid. We defer a discussion of relativistic extensions of the theory to Section 3.4.

A third complication is the resonance in the πN $(\frac{3}{2}, \frac{3}{2})$ channel, at about $T_\pi(\text{lab}) = 200$ MeV. Although it is possible to treat this as a potential resonance, with a nonlocal πN potential, this is not the usual dynamical picture of the resonance, for example, in Chew-Low theory. For a nonpotential treatment of the resonance, a nonpotential version of Faddeev theory is required. [Some recent attempts have been made in this direction (Miz–77 and Rin–77b).]

In spite of these problems, calculations have been performed for πd elastic scattering in this energy region, with a variety of assumptions made concerning how to incorporate the new features into something like Faddeev equations for scattering (Myh–73b, 75, Wol–75, Man–76, Tho–76a, and Rin–77a). In some cases the calculations were simplified by leaving out nucleon spin and/or isospin dependence of amplitudes, to reduce the number of coupled equations (Myh–73b, 75, Wol–75). Results of these calculations cannot be directly compared with experiment. Thomas (Tho–76a) has made the smallest modification for relativity by including only the relativity of the pion kinetic energy—this for a calculation at ∼50 MeV where other relativistic effects may be small. The methods of the more ambitious relativistic treatments are discussed in Section 3.4. Several groups have kept spin and isospin and a number of two-body partial waves, for calculations in the resonance energy region and below (Tho–76a, Man–76, Riv–77, and Rin–77a). These calculations may be considered to be sufficiently realistic to be appropriate for comparison with experimental differential cross sections.

In Fig. 3.4 we show the results of a Faddeev calculation (Tho–76a) for elastic πd scattering at 48 MeV, compared to recent experiment (Axe–76). Figure 3.5 shows similar results for scattering at 180 MeV (Rin–77a) compared to experiment (Nor–71). The calculations appear to represent reasonably well the experimental angular distributions.

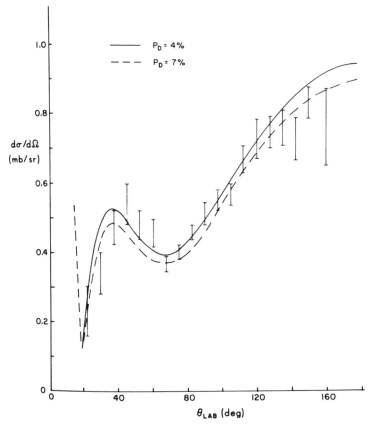

Fig. 3.4 Differential cross section for elastic πd scattering at 48 MeV, calculated by Tho–76a, with experimental data from Axe–76. (P_D refers to the percentage of D-state in the deuteron model.)

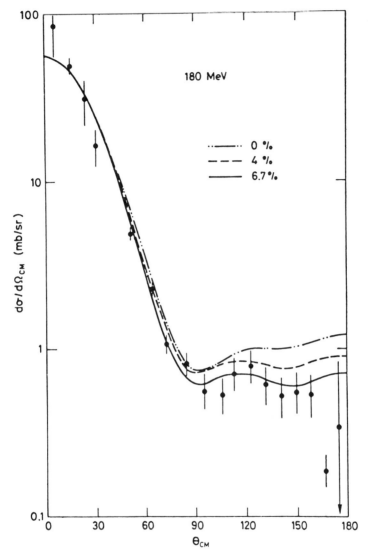

Fig. 3.5 Differential cross section for elastic πd scattering at 180 MeV, calculated by Rin–77a, with deuteron D-state percentage.

97

The results of all Faddeev calculations mentioned agree on some general features of multiple scattering, more or less independently of other details of the theoretical treatment. In particular, they demonstrate that the fixed scatterer approximation is not very accurate for these energies. This appears in two contexts: First, the simple scattering amplitudes in Faddeev theory are not well represented by static kinematics. This is largely because of the strong energy and momentum dependence of the πN scattering amplitude in the 3,3 resonance region, which is treated differently in Faddeev theory than in the fixed scatterer assumption. For example, the three-body kinematics allows for a shift of the πN resonant energy, similar to that discussed above in Eqs. 12–14 in connection with $K^- N$ resonances. Second, the nucleon-nucleon interaction in intermediate states contributes significantly to the scattering. This effect of exciting the deuteron also appears in low energy $K^- d$ scattering, as we have noted. In theories that make the fixed scatterer approximation, this effect is not handled accurately, but it is formally included in the closure approximation.

There have been a number of calculations of πd elastic scattering using the fixed scatterer approximation, for energies $60 \leqslant T_\pi \leqslant 250$ MeV (Gib–71, Car–70, Hoe–74, and Gab–74). With the exception of Gib–71 these calculations include only single and double scattering. In some cases, the small angle approximation that leads to Glauber's result (Eq. 21) is also invoked (Car–70 and Hoe–74). In spite of the apparent inapplicability of these approximations for reasons just discussed, the results of these calculations are in general as good, when compared to present experiment, as the Faddeev results. This may not be a very good test at present; the differences in theoretical angular distributions tend to be largest where the experimental uncertainties tend to be largest, that is, at large angles or at minima in the distributions. In some cases, the errors of the fixed scatterer assumption may be minimized by a particular choice of kinematics (Gab–74). However, one has no reason to trust the accuracy of the approximation for energies below the resonance ($T_\pi < 200$ MeV).

3.4 More Ambitious Approaches to Meson-Deuteron Scattering*

3.4a Relativistic Three-Body Theories

Thus far our treatment of the meson-deuteron problem has been carried out almost exclusively in the framework of nonrelativistic theories. In

---

*This section is a minor digression from the main line of development and is not needed for the understanding of subsequent material.

reality, at many of the energies of interest to us—the $3,3$ resonance region in pion scattering, for instance—the meson is quite relativistic ($v_\pi^{\text{lab}} \sim 0.9$ at the $3,3$ resonance). We therefore turn now to the question of formulating a relativistic version of the Faddeev theory, continuing to ignore, for the moment, the question of particle production, that is, the coupling to channels other than πNN (we treat this topic in Section 3.5 and Chapter 6). The features that we encounter in trying to treat πd scattering relativistically are indicative of what we must expect for πA scattering, $A > 2$ as well.

Since we are prepared to content ourselves with a theory that has potentiallike interactions, that is, that ignores the creation and annihilation of particles, a diagrammatic analysis similar to that based on Fig. 3.3 and leading to Eqs. 55 may again be carried out. This time the diagrams in question are Feynman graphs taken in a ladder approximation so that only the analogs of the processes shown in Fig. 3.3 occur and there are no crossings, particle production, or annihilation. That is to say, we consider the sum of all graphs that are iterations of a particular interaction between two-particle lines, so the graphs can be cut at three-particle intermediate states and the chosen interaction found between these states. The interaction then serves as the iterated "rungs" of the ladder and the particle lines are its uprights, the iterative feature allowing for the generation of an integral equation. While the interaction "rungs" may be quite general, it is clear that once they are fixed as a specific set of exchanges one is no longer dealing with the entirety of the Feynman diagrams. As a consequence, for example, one no longer has a guarantee of unitarity for the resulting ladder amplitude, unlike the case for the amplitude of Eq. 66, say, which is a rewriting of the complete and consistent Schroedinger equation and therefore automatically embodies unitarity. The result of the ladder approximation is a set of equations with the structure of Eq. 55 but now to be taken between states labeled by relativistic four momenta:

$$\langle P'p'q' | T_i(s) | Ppq \rangle = \langle P'p'q' | t_i(s_i) | Ppq \rangle$$

$$+ \sum_{j \neq i} \int \langle P'p'q' | t_i(s_i) | P''p''q'' \rangle$$

$$\times G_0(s) \langle P''p''q'' | T_j(s) | Ppq \rangle$$

$$\times \frac{d^4P''}{(2\pi)^4} \frac{d^4p''}{(2\pi)^4} \frac{d^4q''}{(2\pi)^4}, \qquad \textbf{[89]}$$

which is a generalization of the relativistic Bethe-Salpeter equation (see, for example, Bjo–65 and Bro–69) to the three-body case. In Eq. 89, P, p, and q are four-momenta variables, the definitions of which we do not specify but which are the relativistic analogs of the nonrelativistic variables of Eqs.

56–58 [and indeed can be taken as four-vectors defined just as the three vectors there (Fre–66)]. The role previously played by the energy is now taken by the invariant total energy squared:

$$s = (k_\pi + k_d)^2, \qquad [90]$$

where k_π and k_d are, respectively, the pion and deuteron four-momenta. As before, t_i is the two-body scattering amplitude, taken in a covariant manner for the relativistic theory. The two-particle subsystem energy s_i receives further consideration below. As in Eq. 55, the inhomogeneous driving term in Eq. 89 is suppressed for the $i = 2$ bound deuteron channel. The propagator G_0 in Eq. 89 is further specified in a moment, but at this stage it is determined by the Feynman diagram construction as the relativistic product propagator for the three particles involved. As usual, we can eliminate the overall (four-dimensional) delta function of total energy-momentum conservation by introducing

$$\langle P'p'q' | T_i(s) | Ppq \rangle = (2\pi)^4 \delta^{(4)}(P' - P)\langle p'q' | T_i(s) | pq \rangle, \quad [91]$$

and similarly for t_i, to get

$$\begin{aligned}
\langle p'q' | T_i(s) | pq \rangle = {} & \langle p'q' | t_i(s_i) | pq \rangle \\
& + \sum_{j \neq i} \int \langle p'q' | t_i(s_i) | p''q'' \rangle \\
& \times G_0(p''q''; s)\langle p''q'' | T_j(s) | pq \rangle \\
& \times \frac{d^4 p''}{(2\pi)^4} \frac{d^4 q''}{(2\pi)^4}, \qquad [92]
\end{aligned}$$

where we anticipate working in the three-particle center-of-mass system for which $\mathbf{P} = 0$ and $P_0 = \sqrt{s}$.

Even after partial wave analysis for $\mathbf{p}''$ and $\mathbf{q}''$, which eliminate a polar and azimuthal angular variable for each, or a total reduction of four variables, Eq. 92 still involves four remaining variables. If separable interactions are used, one additional variable is removed and three remain, which is two too many for practical work. Moreover, Eq. 92, which was derived in a ladder approximation, has no guarantee of fulfilling multiparticle unitarity, as we have seen, because we have made a selection from among the totality of Feynman graphs. For a strongly interacting system, this means that we risk major inconsistencies. Blankenbecler and Sugar (Bla–66; see also Ale–65 and Fre–66) showed how to avoid both these difficulties through a judicious model assumption for the propagator G_0. First a dispersion relation is written for this quantity,

$$G_0(s \pm i\eta) = \frac{1}{\pi} \int \frac{\Delta(s')\,ds'}{s' - s \mp i\eta} \qquad [93]$$

where we assume that a small imaginary part is added to s as required. Then the spectral function or discontinuity in G_0, $\Delta(s')$, is determined by taking for it the simplest form consistent with the unitarity requirement for the scattering amplitudes, $T - T^\dagger = 2iT^\dagger T$, or

$$\langle p'q' | T(s + i\eta) | pq \rangle - \langle p'q' | T(s - i\eta) | pq \rangle$$

$$= 2i \int \langle p'q' | T(s - i\eta) | p''q'' \rangle \delta_+ \left(k_1^2 - m_1^2 \right) \delta_+ \left(k_2^2 - m_2^2 \right)$$

$$\times \delta_+ \left(k_3^2 - m_3^2 \right) \langle p''q'' | T(s + i\eta) | pq \rangle \frac{d^4 p''}{(2\pi)^4} \frac{d^4 q''}{(2\pi)^4}$$

$$[94]$$

where $\delta_+ (k^2 - m^2) = \delta(k^2 - m^2)\theta(k_0)$ guarantees the positive energy branch for the on-shell condition. Using Eqs. 92 and 93 in Eq. 94 determines the discontinuity in the propagator, or

$$\Delta(s) = \delta_+ \left(k_1^2 - m_1^2 \right) \delta_+ \left(k_2^2 - m_2^2 \right) \delta_+ \left(k_3^2 - m_3^2 \right), \qquad [95]$$

so that the propagator in this form has an extra integration, but *three* delta functions, thus reducing the number of variables by two. This is a key reason for proceeding along this route since the resulting equations are vastly simpler than the original Bethe-Salpeter formalism. We pay the price, however, of having a propagator that satisfies a necessary requirement for unitarity, but in other respects need not include the correct physics. Nonetheless, when Eqs. 93 and 95 for the propagator are inserted in Eq. 92, there results a set of equations identical in general structure to those for the nonrelativistic problem, apart from the appearance of relativistic normalization and phase space factors, and their solution proceeds in a manner similar to that case.*

*Calculations based on such an approach are given in Man–76 and Riv–77. A closely related formalism due to Aaron, Amado, and Young (Aar–68) has been applied to relativistic πd scattering in Wol–76 and Rin–77a, and an alternative theory has been developed and applied to this problem in Bra–75. See also the review by Tho–76b.

The form of the propagator in Eqs. 93 and 95 is not, of course, the most general, but merely one that satisfies the unitarity condition. Its use in Eq. 94 leads to a special problem in that only when the intermediate states are on the energy shell $\{s' = s$ in Eq. 93, arising from the delta function in $(s' - s \mp i\eta)^{-1} = [P(s' - s)^{-1} \pm i\pi \delta(s' - s)]\}$ does the spectral function of Eq. 95 guarantee that on-mass-shell kinematics prevail, so that in general the prescription for fixing the intermediate state kinematics is ambiguous, as in the formula for selecting the two-particle subsystem scattering energy. The virtue of the procedure based on Eqs. 93–95 is that it leads to linear, one-variable integral equations that are essentially as usable in practical calculations as are the nonrelativistic Faddeev equations, and that satisfy conditions of covariance and unitarity. On the way, however, we have lost crossing symmetry (cf. Section 2.2, for example), since we have restricted ourselves to diagrams in which an intermediate state has the same particles as the initial and final states. (Thus, if we consider the πNN system, the crossed Born diagram has a $\pi\pi NN$ intermediate state, and so forth, not *all* such crossings ever being included within a ladder scheme with an interaction of a given form.) We have also had to reconcile ourselves to using off-energy-shell information concerning the scattering amplitudes involved. A special difficulty that enters for this relativistic, three-body approach is the so-called "clustering problem" (Ale–65, Fre–66; see also Nam–68 and Wei–72). In the nonrelativistic Faddeev equations, Eqs. 55 or 66, the propagator has the structure shown in Eq. 35 and involves $G_0 = (E - k_1''^2/2m_1 - k_2''^2/2m_2 - k_3''^2/2m_3 + i\eta)^{-1}$. If one particle, say 3, does not interact with the other two its energy is subtracted from $E = E(1,2) + k_3^2/2m_3$ and particle 3 factors out of the problem, leaving a two-particle Lippmann-Schwinger equation for the remaining cluster of particles 1 and 2 at the energy available to them $E(1,2)$. In the relativistic case, the inevitable appearance of quadratic energy forms, as in Eqs. 90 and 93, prohibits this clean subtraction, so the presence of a third, noninteracting particle influences the scattering of the other two through more than a trivial apportioning of the total energy. This problem can be circumvented by a particular choice of variables other than a direct generalization of Eqs. 56–59 for four-vectors (Nam–68, Wei–72, and Rin–77a), through the imposition of a particular scheme for going off the energy shell.

3.4b Dispersion Relations

Through the use of dispersion relations we can avoid the need to introduce off-shell information into the dynamical equations and we can retain crossing symmetry and unitarity, but at the expense of having to deal with nonlinear equations in general and losing the possibility of a full multiple scattering treatment. Moreover, the dispersion relation approach is not readily generalizable to more complex targets than the deuteron, since it becomes prohibitive to deal with more than two-particle intermediate states. Hence the technique is in practice restricted to meson-deuteron applications, and at that requires rather severe limitations in the mechanisms treated. Work has been done on forward πd dispersion relations (Iof–56 and Fal–69), but we here briefly consider those theories of elastic meson-deuteron scattering that have attempted to address the broader and more difficult problem of nonforward directions as well (Bre–63 and And–66, who discuss Kd as well as πd scattering, and Sch–67; for dispersion relations of pionic disintegration of the deuteron see Vas–65, Cha–66, and Sch–68a). These dispersion relations are therefore a generalization of the Chew-Low equations of Section 2.4c to systems beyond πN. They base themselves on general features expected to be true for the analyticity structure of the amplitudes in question, which in the case of the theory of Chew and Low for πN scattering could be proved directly from the specific dynamics they assumed.

Schiff and Tran Thanh Van (Sch–67) write a dispersion relation for the πd elastic amplitude h in the channel of total angular momentum J for incident and final orbital angular momenta l and l' ($= J - 1, J, J + 1$ since the deuteron has spin one). This has the form, with the neglect of channel coupling ($l' = l$),

$$h_l(s) = \frac{1}{2\pi i} \int_L \frac{\operatorname{disc} h_l(s')}{s' - s - i\eta}\, ds' + \frac{1}{\pi} \int_{s_P}^{\infty} \frac{\operatorname{Im} h_l(s')}{s' - s - i\eta}\, ds', \qquad [96]$$

where L represents the left-hand cut,* over which the discontinuity in h ($\operatorname{disc} h$) must be taken from a model, and s_P is the physical threshold. ($\operatorname{Im} h_l$ can be related to $|h_l|^2$ through unitarity, whence Eq. 96 could be treated as a nonlinear integral equation.) At threshold $h_l(s)$ must behave like k^{2l} and asymptotically it must vanish. In principle this behavior should emerge from delicate cancellations between the two integrals of Eq. 96, but for an approximate calculation this is achieved by modifying the dispersion relation to read

$$h_l(s) = \frac{k^{2l}}{2\pi i} \int_L \frac{\operatorname{disc} h_l(s')}{k'^{2l}} \frac{ds'}{s' - s - i\eta}$$

$$+ \left(\frac{k^2}{s + s_0}\right)^l \frac{1}{\pi} \int_{s_P}^{\infty} \left(\frac{s' + s_0}{k'^2}\right)^l \frac{\operatorname{Im} h_l(s')}{s' - s - i\eta}\, ds', \qquad [97]$$

*The "left-hand cut" for Chew-Low theory is partially simulated by the pole in the driving term, for example, of Eq. 2.101.

where s_0 is a phenomenological pole in the unphysical region. Equation 97 now represents our main dynamical equation in the dispersion relation approach. Its use requires input concerning the left-hand cut (first integral in Eq. 97) or potentiallike feature presumably arising from the exchange of particles in the t-channel—reliable information on which is usually hard to come by. In addition, one needs information concerning the right-hand cut, or physical region, which involves s-channel transfers or the choice of sufficiently simple intermediate states. From these the full amplitude for each partial wave is calculated through Eq. 97.

In practice, the left-hand cut contribution is then approximated by relatively simple one-particle exchange terms in the crossed channel, so the first integral is replaced by $h_l^B(s)$, in this case as calculated from the lowest-order exchange of a (quite likely fictitious) σ-meson with zero spin and isospin (see Fig. 3.6a). The right-hand cut, or physical region, is

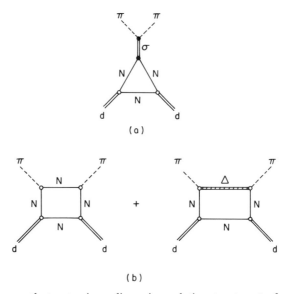

Fig. 3.6 Diagrams that enter in a dispersion relation treatment of πd scattering (Sch–67). (a) The diagram for the left-hand cut is a t-channel diagram; that is, it involves a lateral exchange of a "particle"—in this case a σ-meson that simulates a two-pion resonance—between the pion and deuteron. The diagrams in (b) are for the physical cut in the dispersion relation and are selected so as to limit the intermediate states to those containing two particles, in this case two nucleons (whence true absorption effects are considered in the theory!), or a nucleon and the Δ resonance in the πN system.

approximated with two-particle intermediate states. Here the πd state is omitted to avoid channel coupling. The NN state is included, so the dispersion relation method can incorporate true absorption effects. The NNπ three-particle state is represented by a two-particle NΔ state in which the pion resonates with one of the nucleons as a Δ or 3,3 resonance. These contributions are shown in Fig. 3.6b. The calculational results (Sch–67) tend to show little difference from impulse approximation in the forward direction and deviations of $\lesssim 20\%$ at backward angles in the 3,3 region. Since the formalism is in practice limited to two-particle intermediate states, it is unable to deal with multiple-scattering effects—except as induced by nucleon antisymmetrization—which would also be important at backward directions. This restriction is generally a severe drawback to the dispersion relation approach, preventing, also, its use for $A > 2$ systems. That approach then has as its unique advantage that it works purely with on-shell constructs, namely, the channel amplitudes $h_l(s)$, unlike the methods based on Faddeev equations and their relativistic generalizations. The latter strive for a much more complete dynamical description, especially of multiple-scattering effects, but in so doing must introduce off-energy-shell amplitudes and thus base themselves to some degree on generally poorly known off-energy-shell information.

Before leaving nonmultiple-scattering approaches to meson-deuteron, and especially πd, scattering, we note that theories have also been considered for πd scattering that derive from field theory or elementary particle approaches. Cutkosky (Cut–58) has attempted to develop a consistent and unified theory of πNN systems by representing the interacting two-nucleon state with its meson clouds in terms of dressed single-nucleon states, to which the 2N state reduces when the nucleons are far apart. This is similar in spirit to the London-Heitler method for treating the hydrogen molecule with reference to atomic hydrogen states. Such an approach, if carried out completely, would avoid double-counting of pions in the multiple-scattering and in the nuclear force effects. In practice (Pen–63), however, the technical difficulties of the method force it back to calculational procedures of a more standard multiple-scattering nature. One could imagine related πd methods based on extensions of the theory of Chew and Low for πN scattering discussed in Chapter 2. However, the large physical extent of the nucleus immediately implies that many partial waves must enter in general, and certainly for energies near the 3,3 region for which a Chew-Low type treatment might be appropriate. (We return to such considerations briefly in Chapter 6.) Thus no simple, one-channel analysis is possible except with respect to individual nucleons, and this again underscores the need for methods that treat the meson-nucleus interaction in terms of successive scatterings on the target constituents.

3.5 Meson Absorption

We mentioned in Section 3.1 that in addition to the various scattering reactions of $\pi + d$ and $K + d$, there are also absorption reactions of the types given in Eqs. 1d and 3e:

$$\pi + d \rightarrow N + N \qquad\qquad [98a]$$

$$K^- + d \rightarrow Y + N. \qquad\qquad [98b]$$

These reaction channels differ substantially from the scattering channels—elastic, inelastic, charge- or strangeness-exchange reactions—all of which involve three particles throughout, namely, meson plus two nucleons or meson plus nucleon plus hyperon. Applicable theoretical approaches are given by multiple scattering or Faddeev theory, or their relativistic extensions. For the absorption reactions (Eq. 98), there are three particles initially and two finally. This aspect is not so easily incorporated within ordinary multiple-scattering or Faddeev theory and calls for some new considerations. We postpone a general theoretical discussion of absorption until Section 6.3 and Chapter 7. Here we outline some of the results as they apply to the present problem.

First, let us consider the experimental situation. In Fig. 3.7 we illustrate the total cross section for $\pi^+ + d \rightarrow p + p$, which is given as a function of T_π, and also of the pion momentum $\eta = q/m_\pi$ (in units of m_π). The striking features are the η^{-1} dependence at low energy, characteristic of s-wave absorption at low velocity, and the peak at $T_\pi \simeq 150$ MeV, which is suggestive of the πN resonance $\Delta(1232)$. The magnitude is also interesting (5–12 mb): at low energies this is $\sim 10\%$ of the πd total cross section, and drops to a few percent at $T_\pi \cong 180$ MeV.

In early analyses of the *production* reaction $p + p \rightarrow \pi^+ + d$ just above threshold for production, the total cross section was expressed

$$\sigma(p + p \rightarrow \pi^+ d) = \alpha\eta + \beta\eta^3. \qquad\qquad [99a]$$

In a zero-range theory, the coefficients α and β would characterize the production of pions in s-wave and p-wave, respectively. Using detailed balance (see, e.g., Bla–52) the absorption reaction at low energies takes the form

$$\sigma(\pi^+ d \rightarrow pp) = \frac{2}{3}\left(\frac{p}{m_\pi}\right)^2 \left(\frac{\alpha}{\eta} + \beta\eta\right) \qquad\qquad [99b]$$

where

$$p^2 = M(\varepsilon_\pi(q) + E_d) \qquad\qquad [99c]$$

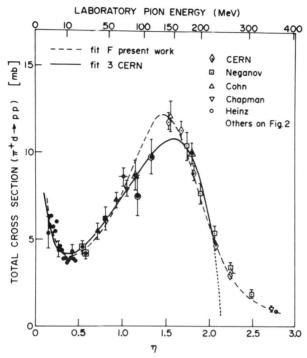

Fig. 3.7 Total cross section for pion absorption by the deuteron, as a function of pion energy T_π or momentum η (in units of $m_\pi c$), from Spu–75. Dashed curve is a fit of Eq. 100 to the data.

is the squared proton c.m. momentum. (The factor $\frac{2}{3}$ is the ratio of number of spin states for pp and πd.) The α-term gives η^{-1} (s-wave) absorption. The energy dependence given by Eq. 99b is not adequate to represent the cross section over any substantial energy range, even below the peak at 150 MeV, unless the coefficients α, β, are allowed to vary with η. A good characterization has been given by Spuller and Measday (Spu–75) who let α be a polynomial in η, and β have a Breit-Wigner resonant shape:

$$\alpha(\eta) = \alpha_0 + \alpha_1\eta + \alpha_2\eta^2 + \alpha_3\eta^3$$

$$\beta(\eta) = \frac{5\pi}{3q^2} \frac{\Gamma_{el}(\eta)\Gamma_r(\eta)}{(E - E_R)^2 + \Gamma_T^2(\eta)/4}, \qquad [100]$$

where E_R is the resonant energy and $\Gamma_{el}, \Gamma_r, \Gamma_T$ are the elastic, reaction, and total widths, all of which are given the dependence on η prescribed by penetration factors (Bla–52). A fit of this function is shown in Fig. 3.7; for this case $\alpha_0 = 0.27 \pm 0.04$ mb and the energy dependence of $\alpha(\eta)$ dominated by $\alpha_1 = -0.4 \pm 0.2$ mb. The resonant energy $E_R = 2183 \pm 8$ MeV corresponds to a πN energy of about 1245 MeV, close to that of $\Delta(1232)$.

Differential cross sections for $\pi^+ d \rightarrow pp$ (and the inverse reaction) for lower pion energies ($T_\pi < 50$ MeV) show an angular distribution of the form ($A + \cos^2 \theta$), characteristic of s- and p-wave pions (see, e.g., Pre–76). For higher energies, for example, $142 < T_\pi < 262$ MeV (Ric–70), $\cos^4 \theta$ dependence has been found, establishing that higher partial waves are involved.

The $K^- d$ absorption reaction (Eq. 98b) has been little studied experimentally. It is known that, for capture at rest ($T_K = 0$), absorption accounts for $\sim 1\%$ of the total reaction rate; the branching ratios into the three modes of Eq. 3e have been measured (Vei–70).

The differential cross section for the absorption reactions (Eq. 98) can be expressed in terms of a transition amplitude

$$\frac{d\sigma}{d\Omega}(\omega_q, \theta) = \frac{Mp}{6\eta(2\pi)^2} \sum_{\alpha\beta} |\langle \mathbf{p}, \alpha | T | \mathbf{q}, \beta \rangle|^2 \qquad [101]$$

where $\mathbf{p}, \alpha$ give the c.m. momentum and spin state of the final NN or YN state, and $\mathbf{q}, \beta$ the initial meson momentum (in meson-deuteron c.m.) and deuteron spin state. Energy conservation relates p^2 to q^2 or $\omega(q)$ by Eq. 99. The sum over spin projections applies to experiments without polarization.

In general it is possible to express the transition amplitude as a matrix element of an absorption operator $\mathcal{Q}$;

$$\langle \mathbf{p}, \alpha | T | \mathbf{q}, \beta \rangle = \langle \psi^{(-)}(\mathbf{p}, \alpha) | \mathcal{Q} | \Psi^{(+)}(\mathbf{q}, \beta) \rangle. \qquad [102]$$

The symbols have the following meaning: $\psi^{(-)}$ is the wave function for two-baryon scattering in the final state (NN or YN) with no meson—only it is given with ingoing (i.e., collapsing spherical) boundary conditions $(-)$; $(\mathbf{p}, \alpha)$ are the final asymptotic (plane wave) c.m. momentum and spin. The wave function $\Psi^{(+)}$ corresponds to the initial three-body scattering state of meson plus two nucleons, with initial asymptotic meson momentum, deuteron spin $(\mathbf{q}, \beta)$. This wave function may be obtained from the solution to the Faddeev equations for meson deuteron scattering, as discussed in Sections 3.2 and 3.3. In terms of the Faddeev t-matrix T_F, given in Eq. 49, the wave function is given by

$$|\Psi^{(+)}(\mathbf{p}, \alpha)\rangle = (1 + G_0 T_F)|\mathbf{p}, \alpha\rangle, \qquad [103]$$

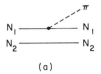

(a)

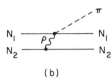

(b)

Fig. 3.8 Representation of the operator $\mathcal{C}$, Eq. 102, responsible for absorption of pions: (a) one-nucleon interaction; (b) two-nucleon interaction mediated by a ρ-meson.

where G_0 is the free three-body propagator (Eq. 44) and $|\mathbf{p},\alpha\rangle$ is the meson-deuteron asymptotic (plane wave) state.

The operator $\mathcal{C}$ is responsible for absorbing the meson. We have already introduced a one-nucleon interaction $H(\mathbf{x})=j(\mathbf{x})\phi(\mathbf{x})$ (Eq. 2.60) that couples $\pi+N\leftrightarrow N$ (see Fig. 3.8a; see also Fig. 2.8). This interaction was the generator of the Chew-Low theory of πN scattering. It is also possible to have two-nucleon operators that contribute to $\mathcal{C}$, for example, if some particle is exchanged between the nucleons during the absorption, as in Fig. 3.8b. We discuss the absorption operator $\mathcal{C}$ in more detail in Chapter 7.

If the interaction of mesons with nucleons were weak, the absorption transition given in Eq. 102 could be treated similarly to the absorption of a photon by the deuteron, that is, photodisintegration. Then the initial state $\Psi^{(+)}$ would be approximated by a product of a meson plane wave and deuteron ground state $\psi_d(\beta)$:

$$|\Psi^{(+)}(\mathbf{q},\beta)\rangle\simeq|\mathbf{q},\psi_d(\beta)\rangle \quad \text{(weak interaction).} \quad [104]$$

For the one-nucleon interaction $H(\mathbf{x})=j(\mathbf{x})\phi(\mathbf{x})$, we may use Eq. 2.63 in order to rewrite Eq. 102 in the form of a transition between nuclear states:

$$\langle\mathbf{p},\alpha|T|\mathbf{q},\beta\rangle\simeq N_K\int\langle\psi^{(-)}(\mathbf{p},\alpha)|j(\mathbf{x})|\psi_d(\beta)\rangle e^{i\mathbf{q}\cdot\mathbf{x}}. \quad [105]$$

This form is illustrative only; the scattering interaction for mesons on nucleons is not, in general, weak, and Eq. 104 is usually a poor approximation.

The full Faddeev wave function in Eq. 103 includes the effects of meson scattering on the two nucleons before absorption. A commonly used approximation to Eq. 103 is given by taking $T_F=t_1+t_3$, that is, the single-scattering approximation.* This is a considerable improvement over Eq. 104. For example, for pions of energy $100 < T_\pi < 200$ MeV, the plane wave approximation (Eq. 104) gives a smooth energy dependence (ω_q) to the transition, while inclusion of πN scattering, even to first order, brings in the ω_q-dependence of the $\Delta(1232)$ resonance. This is necessary to reproduce the energy shape of the cross section for absorption (Fig. 3.7). (Recent calculations to this order are given by Gop–74 and Bra–77; the latter also includes ρ-exchange.)

The absorption transition also has an effect on the scattering of mesons by the deuteron. This is the consequence of processes in which first a meson is absorbed, then the NN (or YN) system propagates (with mutual interaction), and finally a meson is *emitted*, which may scatter as it leaves the remaining NN (or deuteron). The amplitude for this process may be written, in the case of elastic scattering, as

$$\langle \mathbf{q}', \alpha | \Delta T | \mathbf{q}, \beta \rangle = \langle \Psi^{(-)}(\mathbf{q}', \alpha) | \mathcal{Q}^\dagger G^{(2)} \mathcal{Q} | \Psi^{(+)}(\mathbf{q}, \beta) \rangle \qquad [106]$$

where $G^{(2)}$ is the Green function for two interacting nucleons (or $Y + N$),

$$G^{(2)} = (E - H_{NN} + i\eta)^{-1}. \qquad [107]$$

Clearly the hermitian conjugate operator $\mathcal{Q}^\dagger$ is responsible for the emission of a meson. Since the process given by Eq. 106 is not included in the Faddeev scattering amplitude† T_F, the elastic scattering of a meson by a deuteron is given by the sum

$$\langle \mathbf{q}', \alpha | T | \mathbf{q}, \beta \rangle = \langle \mathbf{q}', \alpha | T_F | \mathbf{q}, \beta \rangle + \langle \mathbf{q}', \alpha | \Delta T | \mathbf{q}, \beta \rangle, \qquad [108]$$

where the first term is the momentum and spin matrix elements of Eq. 21, for example.

We may see directly that Eq. 106 preserves the unitarity of the scattering amplitude, in the presence of meson absorption. We consider the forward scattering amplitude—$\mathbf{q}' = \mathbf{q}$ and $\beta = \alpha$—and take the imaginary

---

*There is a technical problem in the calculation of T_F for use in Eqs. 102 and 103. The difficulty is connected with the fact that absorption of the pion (followed by reemission) also occurs in the πN scattering process—for example, the Born terms of Section 2.4. This leads to counting some processes more than once, in the amplitude in Eq. 102, unless compensation is made in T_F, for example, excluding the Born terms from the scattering (see Miz–77). This problem is further discussed in Chapter 6 (see Section 6.1).

†Again, a correction must be made to T_F to eliminate the overcounting of absorption effects (see Chapter 7, Eq. 7.30).

part of Eq. 106:

$$\text{Im}\langle \mathbf{q}, \alpha | \Delta T | \mathbf{q}, \alpha \rangle = \sum_{\gamma} \int \frac{d\mathbf{p}}{(2\pi)^3} |\langle \psi^{(-)}(\mathbf{p}, \gamma) | \mathcal{C} | \Psi^{(+)}(\mathbf{q}, \alpha) \rangle|^2 \delta \left[\frac{p^2}{m} - \omega(q) \right],$$

$$[109]$$

where we have used a complete set of scattering states $\psi^{(-)}(\mathbf{p}, \gamma)$ to evaluate the Green function (Eq. 107). The right-hand side of Eq. 109 is clearly proportional to the total (integrated) cross section for absorption of a meson of momentum $\mathbf{q}$ by a deuteron, as can be seen, for example, by integrating Eq. 101 over solid angles. Thus the pion flux lost by absorption is appropriately accounted for in the imaginary part of the forward elastic scattering amplitude as expected.

An ingenious model that combines absorption and scattering of $\pi + d$ has been developed by Afnan and Thomas (Afn-74). The main idea is to treat the nucleon that has just absorbed a meson (N') as a bound state of a $\pi + N$ ($\pi + N \rightarrow N'$) using an ordinary (separable) potential to bind the pair. In this case, the scattering problem remains a three-body problem throughout, since the two-nucleon state after absorption, $\pi + d \rightarrow N + N'$, is a three-body state in terms of the original $\pi + N + N$. Thus the model may be treated by ordinary Faddeev equation methods. There are some difficulties, however, with the treatment of the identity of particles (N $\neq$ N'); see Miz-77 and Rin-77b.

References

A good modern reference on the three-body problem is the collection edited by Thomas (Tho-77).

4 Mesons and Nuclei — Multiple Scattering Theory

The central issue we now address is how to construct a theoretical framework in which we can use our information concerning the strongly interacting two-particle system of projectile and target constituent (meson and nucleon, in our case) to solve the problem of the interaction between the projectile and the entire complex target, that is, meson-nucleus scattering, reactions, mesic atoms, and so forth. The nature of the problem, and indeed the partial solution we put forth for it, is therefore very similar to that discussed in the previous chapter, except that now we deal with targets having more than two constituents. One might suppose that this extension beyond the deuteron could be dealt with by means of straightforward generalization of the techniques used for π-d scattering, but, as we have seen, the way is not yet clear to carry out such an extension. Therefore we must instead develop an approach that is closer in spirit to the treatment of a many-particle system than to one of two or three particles. There are a number of formal languages in which one could attempt this (field theory, phenomenological approach based on Feynman diagrams, relativistic multiple-scattering formalisms, and so on), among which we choose nonrelativistic multiple scattering theory as our point of departure. We make this selection for several reasons: Such an approach has been applied relatively extensively and has proved itself as a reliable lowest-order approximation to the physics of medium-energy situations. It offers a systematic way of including nonrelativistic higher-order corrections, and, indeed, within the nonrelativistic domain is—at least in principle—a complete theoretical structure by reason of that very limitation. To some degree, it allows one to append various relativistic features while being vastly simpler than relativistic formulations that must from the start cope with the production and annihilation of a whole host of particles, including possibly the projectile itself. Thus, while cognizant of its deficiencies, many of which we try to overcome subsequently in a somewhat patchwork way, we believe that this theory presently represents the most complete systematic and consistent portrayal of the scattering of mesons on nuclei at medium energies. Moreover, it meshes naturally with nonrelativistic descriptions of nuclear structure based on Schroedinger mechanics, which have evolved much further than relativistic ones.

4.1 Optics Approach to Meson-Nucleus Scattering

Since the method we develop is intended to relate to a many-body outlook, it is to be expected that it bears some resemblance to the conventional theory of optics, which also involves the modification of a wave falling on a many-particle system. Indeed it emerges that much of the essential physics of our problem resembles optics, and that one can learn a great deal about medium-energy scattering from the optics analogy. This is especially so because, at the energies under consideration here, semiclassical approximations begin to apply; hence many usual optics results can be taken over en bloc. Thus we begin our discussion of scattering on many-particle systems by considering the relationship of this to an optics problem.

4.1a Index of Refraction, Mean Free Path, and the Optical Potential

Let us consider, as illustrated in Fig. 4.1, a plane wave of propagation vector **k** in vacuum impinging on a semi-infinite medium of scatterers. We suppose, in a manner that is made more precise as we proceed, that the medium of scatterers is reasonably homogeneous as to scatterer density $\rho(\mathbf{r})$ and amplitude f from point to point in the medium, and that the wave propagation in the medium is not radically different from what it was outside. Taking the direction of propagation to be the z-axis, the wave in vacuum is described by $\exp(ikz)$, and, to see what becomes of this wave when it encounters the scatterers, we imagine the medium to be sliced into narrow slabs of thickness d. After traversing the first slab, the wave at position z will be made up of the original front plus the contributions of the scattered wave from each scattering center encountered within the slab:

$$\psi(z) = e^{ikz} + \int_0^\infty f \frac{e^{ikr}}{r} \rho \, d \, 2\pi b \, db, \qquad [1]$$

where we have introduced an impact coordinate **b** perpendicular to the

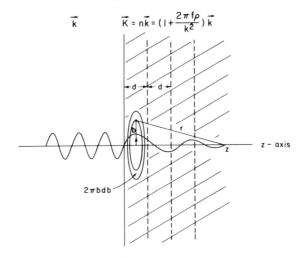

density ρ $(\vec{r})$, scattering amplitudes f

Fig. 4.1 The propagation of a wave with external momentum **k** parallel to the z-axis in a semi-infinite medium of density $\rho(\mathbf{r})$ and scattering amplitudes f. In the medium, the propagation vector **k** is quenched and replaced by **K**, as in Eqs. 4 and 5.

z-axis, and $\mathbf{r}^2 = \mathbf{b}^2 + z^2$, as in Fig. 4.1. We also consider the scattering to be well approximated when taken in the forward direction only and the forward amplitude f to be independent of energy. We can carry out this integration provided the scattering amplitudes f and density ρ do not vary over the slab thickness. To do so we must introduce some form of screening factor, here taken as a decaying exponential, to damp the contribution of distant scatterers and bring about convergence of the integral. Using $b\,db = r\,dr$ and taking the limit of vanishing screening once it is no longer needed, we have

$$\psi(z) = e^{ikz} + 2\pi f\rho d \int_z^\infty e^{ikr - \lambda r}\,dr \underset{\lambda \to 0}{\to} e^{ikz}\left(1 + i\frac{2\pi f\rho d}{k}\right), \qquad [2]$$

where we have assumed that the predominant scattering events are forward, and we have taken out the forward amplitude f as an approximate constant. This wave then impinges on the next slab, and so on, until we reach a point z inside the medium such that N slabs of thickness $z = Nd$ have been more or less successfully negotiated. The value of the wave is then

$$\psi(z) = \lim_{N \to \infty} e^{ikz}\left(1 + i\frac{2\pi f\rho z}{kN}\right)^N = e^{ikz}e^{i(2\pi f\rho z/k)}, \qquad [3]$$

where we have assumed implicitly that the medium is such that scattering from each slab is independent. This result expresses the fact that inside the medium the original external wave of propagation vector **k** has been extinguished in favor of a new wave with propagation vector

$$\mathbf{K} = \left(1 + \frac{2\pi f \rho}{k^2}\right)\mathbf{k} = n\mathbf{k}, \qquad [4]$$

where the proportionality factor n is known as the index of refraction:

$$n = 1 + \frac{2\pi f \rho}{k^2}. \qquad [5]$$

As we have seen, n can be usefully defined provided the "medium" [essentially f and $\rho(\mathbf{r})$] does not change rapidly over many slab widths, effectively measured by the distance K^{-1}, that is, the wave number must not change appreciably over the interval of its own wavelength. That Eq. 3 continues to give propagation in the z-direction is to be expected, of course, because of the symmetry.

In discussing the modified propagation of the wave in the medium it is essential to realize that Eqs. 3–5 do not merely represent a change in the numerical value of the wave propagation vector, but in fact embody a new physical feature in the propagation among the scatterers as opposed to what takes place in the vacuum, namely, absorption. The propagation vector and index of refraction possess imaginary parts, whose presence represents a loss in the foward propagated beam due to encounters with the scatterers. In fact,

$$\text{Im}\,K = k\,\text{Im}\,n = \frac{2\pi\rho}{k}\,\text{Im}\,f = \tfrac{1}{2}\rho\,\sigma_{\text{total}}, \qquad [6]$$

where we have used the optical theorem relating the imaginary part of the forward scattering amplitude f to the total cross section σ_{total} through

$$\text{Im}\,f = \frac{k\,\sigma_{\text{total}}}{4\pi}. \qquad [7]$$

We refer to the absorption incorporated in Eqs. 3–6 by virtue of $\text{Im}\,K \neq 0$ as "optical absorption" as opposed to "true absorption" in which a meson projectile is annihilated. The wave of Eq. 3 is then

$$\psi(z) = e^{ikz \cdot \text{Re}\,n} e^{ikz \cdot \text{Im}\,n} = e^{ik \cdot z} e^{-\rho \sigma_{\text{total}} \cdot z/2}, \qquad [8]$$

and its amplitude squared is

$$|\psi(z)|^2 = e^{-z\rho\sigma_{\text{total}}} = e^{-z/l}, \qquad \qquad [9]$$

where l is defined as the mean free path of the wave in the medium

$$l = \frac{1}{\rho\sigma_{\text{total}}}. \qquad \qquad [10]$$

Its reciprocal represents the probability of a scattering encounter for scatterers distributed with density ρ and having cross sections σ_{total}; σ_{total}, and consequently l, are generally functions of projectile energy. For example, for a pion incident on a semi-infinite nucleus made up of protons and neutrons, the mean free path is

$$l^{-1} = \rho_n \sigma_{\text{total}}(\pi n) + \rho_p \sigma_{\text{total}}(\pi p)$$

$$\approx \rho_0 \left(\frac{N}{A} \sigma_{\text{total}}(\pi n) + \frac{Z}{A} \sigma_{\text{total}}(\pi p) \right), \qquad [11]$$

where we have averaged over nucleon species. This quantity is shown in Fig. 4.2 for equal numbers of protons Z and neutrons N, $A = N + Z$, and a constant nuclear density $\rho_0 = 0.17$ fm^{-3}, the energy dependence of l being induced by the energy dependence of the πN total cross sections. This is especially dramatic due to the vastly increased chance of a scattering at the 3,3 resonance (see Fig. 2.3), which is reflected in the very small mean free path at the relevant energy—a decay length that is very short even on a nuclear distance scale.

If the wave motion we have been discussing so far refers to the motion of a particle as described by the Schroedinger equation, rather than to an optical wave, then by inserting Eq. 3 into the Schroedinger equation we can deduce the potential $\mathcal{V}$ that describes the motion, in contrast to the usual situation where the potential is known and one solves for the wave function. Then

$$(\nabla^2 + k^2)\psi(z) = 2m\mathcal{V}\psi(z) \qquad \qquad [12]$$

for a particle of incident momentum k, and

$$\mathcal{V} = -\frac{1}{2m}(K^2 - k^2) = -\frac{k^2}{2m}(n^2 - 1). \qquad [13]$$

If the effect of the medium is small, that is, ρf is small and n is near unity

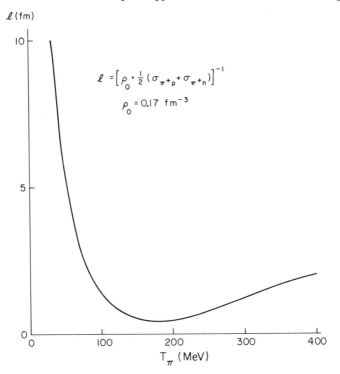

Fig. 4.2 The mean free path l of Eq. 11 for a charged pion in a nuclear medium with equal numbers of protons and neutrons, $N = Z = \frac{1}{2}A$. Here data on the πN total cross sections as a function of pion kinetic energy T_π were taken from Hoh-64. At the πN resonance ($T_\pi \sim 195$ MeV), the mean free path reaches its minimum $l \lesssim 0.5$ fm, as the pion has more and more difficulty plowing through the medium without undergoing a scattering with a nucleon that ejects the pion-nucleus system from the elastic channel.

in Eq. 5, then we can approximate this as

$$\mathcal{V} \cong -\frac{k^2}{m}(n-1) = -\frac{2\pi}{m}f\rho. \qquad [14]$$

This potential will also have an imaginary part reflecting the absorption we have discussed, namely,

$$\mathrm{Im}\,\mathcal{V} = -\frac{2\pi}{m}\mathrm{Im}f\cdot\rho = -\frac{1}{2}v\rho\sigma_{\mathrm{total}} = -\frac{v}{2l}, \qquad [15]$$

where v is the particle velocity. Again, for a nucleus we must average over protons and neutrons, and the potential becomes

$$\mathcal{V} = -\frac{2\pi}{m}\left[\rho_n f_{\pi n} + \rho_p f_{\pi p}\right]. \qquad [16]$$

In the rest of this book we come back to this result in many forms. We have obtained it here in a somewhat heuristic manner for the case of homogeneous nuclear matter with constant densities and for forward scattering amplitudes. This result for the optical potential tells us that if we know the scatterer distributions and their amplitudes for scattering with the projectile—the minimal information we must have to make progress with our program—then we can calculate the propagation of the projectile in the medium. For nuclear matter that result can only involve a changed propagation vector, whose striking feature is an imaginary part depicting absorption of the wave. It is important to note that if there *is* any scattering whatsoever ($\sigma_{\text{total}} \neq 0$), then absorption will be present, as seen from Eq. 9, representing attenuation of the forward beam due to the scattering, or alternatively for a gaseous, uncorrelated medium, the "excitation" of the medium through the recoil of the scatterer.

4.1b   Cross Sections for Absorption and for Elastic Diffractive Scattering

In order to broaden our physical feeling for medium-energy scattering from a many-particle system, before turning to a much more formal treatment, let us attempt further to exploit our rough solution of the problem of the semi-infinite system by joining it in a crude way to the spherical geometry of a more realistic nucleus. We are particularly interested in seeing how the absorption feature manifests itself in this case. We first note that the imaginary part of the optical potential of Eq. 14 will modify our usual notions of the conservation of probability. If we consider a particle flux

$$\mathbf{J} = \frac{1}{2im}\left(\psi^* \nabla \psi - (\nabla \psi^*)\psi\right) \qquad [17]$$

and a particle probability density

$$P = \psi^* \psi, \qquad [18]$$

then in the ordinary situation of a real potential it is easy to use the time-dependent Schroedinger equation to show that these must satisfy a continuity equation. Now this result is slightly changed. The time-dependent Schroedinger equation and its complex conjugate are

$$\left(-\frac{1}{2m}\nabla^2 + \mathcal{V}\right)\psi = i\frac{\partial \psi}{\partial t} \quad \text{and} \quad \left(-\frac{1}{2m}\nabla^2 + \mathcal{V}^*\right)\psi^* = -i\frac{\partial \psi^*}{\partial t},$$

$$[19]$$

whence the construct that usually leads to the continuity equation now obeys

$$\nabla \cdot \mathbf{J} + \frac{\partial P}{\partial t} = \frac{1}{2im}(\psi^* \nabla^2 \psi - (\nabla^2 \psi^*)\psi) + \psi^* \frac{\partial \psi}{\partial t} + \left(\frac{\partial \psi^*}{\partial t}\right)\psi$$

$$= -i\psi^*(\mathscr{V} - \mathscr{V}^*)\psi = 2\psi^*(\operatorname{Im} \mathscr{V})\psi = -\frac{v}{l}|\psi|^2, \quad [20]$$

where we have used Eq. 15. The entire difference between this and the more customary result arises from the presence of a complex potential and the corresponding absorption, as is very explicit in Eq. 20. This means that there is now a sink for the probability of finding the projectile, whose strength is measured by the right-hand side of that equation.

We may ask how this description affects the situation of a projectile impinging on a finite, spherical nucleus of radius R. Of course, our results thus far are not really pertinent for this case, but we have seen that the mean free path for pion propagation near the 3,3 resonance, for instance, is rather less than a nuclear radius, and so, from the perspective of the pion, the nucleus may indeed have a semi-infinite look to it insofar as absorption is concerned. (We have yet to address the wave diffraction aspect of the problem.) The transition probability arising from absorption is then

$$w = -2\operatorname{Im} \mathscr{V} \int_{\substack{\text{nuclear} \\ \text{volume}}} P\, d\mathbf{r} = \frac{v}{l} \int_{\substack{\text{nuclear} \\ \text{volume}}} e^{-D/l}\, d\mathbf{r}, \quad [21]$$

where we have taken explicit account of the fact that the probability rate decreases by introducing a minus sign and we have used Eq. 9, labeling as D the distance of penetration of the projectile into the nucleus, which is approximated by a uniform sphere. The corresponding absorption cross section is obtained by dividing this transition rate by the incident particle flux, which is v if we normalize our continuum wave functions in a unit volume. Then

$$\sigma_{\text{abs}} = \frac{w}{v} = \frac{1}{l} \int_{\substack{\text{nuclear} \\ \text{volume}}} e^{-D/l}\, d\mathbf{r}, \quad [22]$$

where this absorption cross section includes the contributions arising from all possible inelastic nuclear excitations, and is sometimes called the reaction cross section.

To evaluate the absorption cross section we make use of the cylindrical geometry shown in Fig. 4.3, from which $D = z + (R^2 - b^2)^{1/2}$ for a given

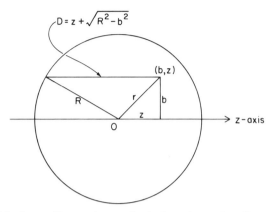

Fig. 4.3 Cylindrical coordinates in a spherical nucleus of radius R and constant density. The impact parameter is **b** and z is the axis coordinate along the direction of propagation.

impact parameter b and axial distance z. Then

$$\sigma_{abs} = \frac{2\pi}{l} \int_0^R b\,db \int_{-\sqrt{R^2-b^2}}^{\sqrt{R^2-b^2}} dz\, e^{-(z+\sqrt{R^2-b^2})/l}$$

$$= \pi R^2 \left\{ 1 - \frac{1}{2}\left(\frac{l}{R}\right)^2 \left[1 - \left(1 + \frac{2R}{l}\right)e^{-2R/l}\right]\right\}. \qquad [23]$$

This result for the absorption cross section pertains in a semiclassical, high energy limit, since it assumes a nucleus of infinite size relative to wave effects measured by the wavelength $\lambdabar = 1/k$ and hence makes no reference to the wave nature of the problem whatsoever. In fact, at low energies the wavelength is very long and determines the absorption cross section to be (see, for example, Sch–68b, Bla–52)

$$\sigma_{abs} = \pi\lambdabar^2 = \frac{\pi}{k^2}. \qquad [24]$$

We can provide an extrapolation to this opposite limiting case by replacing $R \to R + \lambdabar = R + 1/k$ in Eq. 23, which will then automatically embody Eq. 24 at low energies. On the other hand, at high energies where $R \gg \lambdabar$, σ_{abs} is determined by a pure geometric factor πR^2 and the ratio of the radius to the mean free path. If R/l is large, that is, we have a truly geometrical situation, then the absorption cross section is just

$$\sigma_{abs} \xrightarrow[R \gg \lambdabar, l]{} \pi R^2. \qquad [25]$$

In fact, this situation is somewhat achieved for pion-nucleus scattering near the 3,3 resonance (see Fig. 4.2), and even for $l = \frac{1}{2}R$, which from Eq. 10 implies $\sigma_{\text{total}}^{\pi N} \lesssim 40$ mb, we have $\sigma_{\text{abs}} \approx 0.9 \pi R^2$. This means that at high energies in an even moderately absorptive situation (for example, for nucleon-nucleon scattering at a few hundred MeV $\sigma_{\text{total}}^{NN} \sim 40$ mb and this condition will apply), the geometric aspect of the absorption cross section tends to dominate and the details of the πN interaction serve only to determine a slightly modified "effective" absorption radius for our uniform sphere model, through the action of the factor in braces in Eq. 23.

Absorption is one aspect of pion-nucleus scattering at medium energies that has an analog in optics. The other major feature of this process for which that is the case is diffractive behavior of the scattering in this energy region, which we now explore in the heuristic, semiclassical language we have been using. The scattering of a wave $\psi(\mathbf{r})$ from the optical potential $\mathcal{V}(\mathbf{r})$ introduced in Eqs. 12–14 is described by the Schroedinger equation in the integral form,

$$\psi(\mathbf{r}) = e^{i\mathbf{k}\cdot\mathbf{r}} - \frac{2m}{4\pi} \int G(\mathbf{r},\mathbf{r}') \mathcal{V}(\mathbf{r}')\psi(\mathbf{r}') \, d\mathbf{r}', \qquad [26]$$

where we have introduced the Green function satisfying

$$(\nabla^2 + k^2) G(\mathbf{r},\mathbf{r}') = -4\pi\delta(\mathbf{r} - \mathbf{r}'), \qquad [27]$$

the solution of which for outgoing waves is

$$G(\mathbf{r},\mathbf{r}') = \frac{e^{ik|\mathbf{r}-\mathbf{r}'|}}{|\mathbf{r}-\mathbf{r}'|}. \qquad [28]$$

As always in scattering problems, we are interested in the asymptotic region where the wave function is given by the sum of the incident plane wave and an outgoing spherical wave. The latter has an angle-dependent coefficient $F(\theta)$ that is the scattering amplitude, now for scattering on the full nucleus,

$$\psi(\mathbf{r}) \xrightarrow[r\to\infty]{} e^{i\mathbf{k}\cdot\mathbf{r}} + F(\theta)\frac{e^{ikr}}{r}. \qquad [29]$$

Inserting the asymptotic form for Eq. 28 into Eq. 26, we get

$$\psi(\mathbf{r}) \xrightarrow[r\to\infty]{} e^{i\mathbf{k}\cdot\mathbf{r}} - \frac{2m}{4\pi} \frac{e^{ikr}}{r} \int e^{-i\mathbf{k}'\cdot\mathbf{r}'} \mathcal{V}(\mathbf{r}')\psi(\mathbf{r}') \, d\mathbf{r}', \qquad [30]$$

where $\mathbf{k}' = k\hat{\mathbf{r}}$ is the direction in which we are sampling the scattered wave.

This yields the standard result for the scattering amplitude

$$F(\theta) = -\frac{2m}{4\pi} \int e^{-i\mathbf{k}'\cdot\mathbf{r}} \mathcal{V}(\mathbf{r}')\psi(\mathbf{r}')\,d\mathbf{r}', \qquad [31]$$

to which we now apply our results for semi-infinite nuclear matter on the assumption that we are at sufficiently high energy for the wavelength λ to be much smaller than the relevant nuclear dimensions. In the same spirit as earlier, we also assume that the optical potential is constant over the nuclear volume. Then, using Eq. 8 and the geometry of Fig. 4.3, we have

$$F(\theta) = -\frac{2m}{4\pi} \mathcal{V} \int_{\substack{\text{nuclear}\\\text{volume}}} e^{-i\mathbf{k}'\cdot\mathbf{r}} e^{i[\mathbf{k}\cdot\mathbf{r}' + (n-1)kD(\mathbf{r}')]}\,d\mathbf{r}', \qquad [32]$$

where $\psi(\mathbf{r}')$ has been written so as to correspond to a plane wave $\exp(i\mathbf{k}\cdot\mathbf{r}')$ until it begins to penetrate into the nuclear medium. The penetration distance is again $D(\mathbf{r}') = z' + (R^2 - b'^2)^{1/2}$. Evaluating this result in our cylindrical geometry first requires the introduction of the cylindrical Bessel function, here arising from the angular integration that is just an integral representation for that function (Mor–53):

$$J_0(\xi) = \frac{1}{2\pi} \int_0^{2\pi} e^{i\xi\sin\phi}\,d\phi. \qquad [33]$$

Then we assume that most of the scattering is forward, as is appropriate for a large system, since the scattering mainly involves $|\mathbf{k} - \mathbf{k}'| \lesssim 1/R$, and we take the z-axis along the "common" direction of $\mathbf{k}$ and $\mathbf{k}'$. The angular integration is then defined with respect to a direction perpendicular to this axis: thus the transverse momentum transfer enters. Taking

$$\boldsymbol{\Delta} \equiv \mathbf{k} - \mathbf{k}', \quad \Delta_\perp \approx k\sin\theta \approx 2k\sin\tfrac{1}{2}\theta \qquad [34]$$

for scattering angle θ, we have

$$F(\theta) = -m\mathcal{V} \int_0^R b'\,db'\,J_0(\Delta_\perp b')e^{i(n-1)k\sqrt{R^2-b'^2}} \times \int_{-\sqrt{R^2-b'^2}}^{\sqrt{R^2-b'^2}} dz'\,e^{i(n-1)kz'},$$

$$= \frac{im\mathcal{V}}{(n-1)k} \int_0^R b'\,db'\,J_0(\Delta_\perp b')\left[e^{2i(n-1)k\sqrt{R^2-b'^2}} - 1 \right]$$

$$= -ik \int_0^R b'\,db'\,J_0(\Delta_\perp b')\left[e^{2i\chi(b')} - 1 \right], \qquad [35]$$

where the last step follows from Eq. 14 and the identification of the phase parameter

$$\chi(b') = (n-1)k\sqrt{R^2 - b'^2} \ . \qquad [36]$$

As we see in Section 4.4, this final result for the scattering amplitude is very general in the high energy limit, and the phase parameter $\chi(b')$ is a continuous analog of the phase shift $\delta_L(E)$, with the relevant impact parameter fixed from $b' \sim L/k$, and the assumption that the energy dependence of the phase shift in the high energy limit enters only in the way implied by that relationship.

For a large nucleus and in a highly absorptive region, one may neglect the exponential in Eq. 35 and carry out the remaining integral to obtain (Gra–65, p. 634)

$$F(\theta) \approx i\frac{k}{\Delta_\perp} RJ_1(\Delta_\perp R), \qquad [37]$$

and the corresponding differential cross section

$$\frac{d\sigma_{\text{elastic}}}{d\Omega} = |F(\theta)|^2 = \pi R^2 \cdot \frac{1}{\pi}\left[\frac{kJ_1(\Delta_\perp R)}{\Delta_\perp}\right]^2, \qquad [38]$$

a result exhibiting a characteristic diffraction pattern, with a fall-off in momentum transfer governed by $1/R$. We can also estimate the angle-integrated elastic cross section in the high energy limit $k \to \infty$ using Eq. 34 and a definite integral over the Bessel function (Gra–65, p. 692):

$$\sigma_{\text{elastic}} = \int \frac{d\sigma_{\text{elastic}}}{d\Omega}\, d\phi\, d\cos\theta = \pi R^2 \cdot 2\int_0^\infty J_1^2(\Delta_\perp R)\frac{d\Delta_\perp}{\Delta_\perp}$$

$$= \pi R^2. \qquad [39]$$

Thus, at high energies, both the absorption and the elastic cross sections approach the geometric limit of πR^2 so that the total cross section is equally shared between one channel—the elastic one—and the sum over all the others. With regard to the elastic cross section we must note, much as we did for the absorption cross section, that in a highly absorptive regime a great deal is preordained by geometry, and, indeed, it becomes something of a challenge to find ways to ferret out more detailed information in the face of the strong determining influence of strong absorption and geometry.

The optical analogy has served us well in providing a qualitative assessment for scattering on nuclei at fairly high energy. One could push it further by exploring more detailed optical phenomena and searching for their nuclear parallels (a well-known interesting example of such a situation being the Lorentz-Lorenz effect to which we return later). One can also exploit the nuclear version of this analogy, as presented here, for its reliability as a semiquantitative predictor for, say, pion-nucleus scattering. However, it is worth noting that there are a host of clear deficiencies in our approach up to now, which make it far more desirable to proceed to more rigorous and more complete formulations of the problem. The most glaring omission to this point has been in the treatment of quantal effects, which so far have been dealt with very cavalierly and clearly require a better effort. Furthermore, we have thus far contented ourselves with treating a uniform, homogeneous distribution of scatterers, while it is clear that one of the main features we wish to examine in the nuclear case is the matter of inhomogeneities in the density and multiparticle clusters and correlations in the nucleus. We must also expect to contend with a phenomenon in pion-nucleus scattering that has no simple analog in classical optics (or, for that matter in nucleon-nucleon scattering). This is the possibility of the *true* absorption of the pion, as distinguished from "absorption" in the sense we have meant in the context of the imaginary part of the index of refraction or optical potential or in discussing the mean free path. The true absorption refers to the fact that the pion, as a boson, can be absorbed on a multinucleon system, giving up its energy in the form of nucleon kinetic energy or radiation. This mechanism as yet finds no place in our scheme and indeed only begins to receive systematic treatment in Chapter 6.

A last point with regard to which our present problem differs sharply from conventional optics is lodged in Eqs. 13 and 14, which we combine as

$$K^2 = k^2 + 4\pi\rho f(k; K), \qquad [40]$$

where we have made explicit that the scattering amplitude that enters depends on the projectile external energy (or k) and, at each scattering event, may also depend on the internal momentum K. Hence Eq. 40 may represent a subtle problem in self-consistency in determining the internal momentum. This problem is aggravated substantially by the fact that we do not usually have a very general prescription for assigning the various kinematical dependences of an on-shell scattering amplitude as to energy variables and momentum variables discussed in Chapter 2. Thus we have to exploit what may prove to be overly restrictive models for this purpose.

The simple picture given in this subsection is dominated by considerations of geometry, optical absorption, and diffractive features of scattering. We focus on the scattering mechanism of the meson rather than on ways in which the physical properties of the scattering medium—the nucleus—arise in the theory. For nuclear applications we require far more refined treatments that are the topics of the remainder of this chapter; these treatments include the effects of a nonuniform medium and inelastic processes, and incorporate energy- and angle-dependent scattering amplitudes, their off-energy-shell features, and the effects of the medium on them.

Before leaving heuristic discussions we prepare for this more complete treatment by characterizing in a rough and preliminary way how nuclear structure enters the multiple-scattering problem. We do this in the context of a fixed-scatterer or frozen-nucleus approximation (see also Eqs. 175–185 below and the surrounding discussion). That is, we suppose that the A target nucleons are fixed at positions $r_1, \ldots, r_A$ and the projectile scatters from this fixed assemblage; subsequently we average these nucleon positions over the A-particle nuclear wave function. We may anticipate that the wave function for the full system of target nucleons plus meson with coordinate r will have the approximate form

$$\Psi^{(+)}(r_1, \ldots, r_A ; r) = \Phi_0(r_1, \ldots, r_A)\phi_k(r)$$

$$- \frac{2m}{4\pi} \int dr' \frac{e^{ik|r-r'|}}{|r-r'|} \sum_{i=1}^{A} v_i(r', r_i)\Psi^{(+)}(r_1, \ldots, r_A ; r'),$$

$$[41]$$

where we have indicated in the superscript on the system wave function $\Psi^{(+)}$ that we are taking an outgoing spherical wave for the projectile in the asymptotic region. The projectile enters with momentum k, and in the initial asymptotic region the projectile and target have separated wave functions ϕ_k and Φ_0, the latter referring to the nucleus in the ground state. Propagation from scattering center to scattering center is described by the Green function $e^{ik|r-r'|}/|r-r'|$, where we approximate the internal meson wave number by k. The interaction potential of the meson with each nucleon is v_i, $i = 1, 2, \ldots, A$, taken here in a local form.

As we see in greater detail at the end of Section 4.3, the separation of the nuclear and the projectile coordinates in the driving term of Eq. 41 leads to a similar separation in the full solution

$$\Psi^{(+)}(r_1, \ldots, r_A ; r) = \Phi_0(r_1, \ldots, r_A)\psi_k^{(+)}(r), \qquad [42]$$

where $\psi_k^{(+)}$ refers to the projectile wave as distorted by the A fixed scatters. The usual result for the scattering amplitude as involving the interaction taken between distorted and asymptotic waves then gives for the meson-nucleus scattering

$$\langle \mathbf{k}'|f|\mathbf{k}\rangle = -\frac{2m}{4\pi}\int d\mathbf{r}_1\cdots d\mathbf{r}_A\,\Phi_0^\dagger(\mathbf{r}_1,\ldots,\mathbf{r}_A)$$

$$\times\int d\mathbf{r}\,\phi_{\mathbf{k}'}^\dagger(\mathbf{r})\sum_{i=1}^{A}v_i(\mathbf{r},\mathbf{r}_i)\psi_{\mathbf{k}}^{(+)}(\mathbf{r})\Phi_0(\mathbf{r}_1,\ldots,\mathbf{r}_A). \quad [43]$$

This form makes the physics of the fixed-scatterer approximation very explicit: To the extent that that approximation is valid (a point to which we return at the end of Section 4.3), all of the nuclear physics input is by means of the ground state nuclear wave function. The full interaction of the projectile with each nucleon is eventually averaged over nucleon positions with respect to that wave function.

Of course, Eq. 43 does not yet constitute a completely satisfactory solution to the multiple-scattering problem, since its evaluation requires knowledge of $\psi_{\mathbf{k}}^{(+)}(\mathbf{r})$, which in this formulation is now the main unknown in the problem. Moreover, the interactions $v(\mathbf{r},\mathbf{r}_i)$ may, for example, involve strong, short-range parts that make its use in Eq. 43 problematical. We see in great detail in the next section that to lowest order the combination $v_i\psi_{\mathbf{k}}^{(+)}$ can be replaced by $t_i\phi_{\mathbf{k}}$, where t_i is the t-matrix for projectile scattering on the ith nucleon. The t_i, $i=1,2,\ldots,A$, may be better behaved and better known than the v_i, and the multiple-scattering problem has now been reduced—at least in a crude approximation—to a nuclear ground state average over the sum of them. We refer to terms involving ground state averages of this sort as *coherent*, since they embody coherent wavelets for the projectile on the various scattering centers. If the scattering transition matrices t_i do not contain spin or isospin variables, the coherent averaging is simply an integration over the nuclear ground state density. As we see in the next two sections, in the fixed-scatterer situation the higher-order corrections for the πA amplitude, and for the πA optical potential from which the amplitude can be obtained, introduce nuclear information in the form of two- three-,...,A-particle correlation functions. These contain the effects of the nuclear excited states or of the granularity of the medium.

In a more general situation where appeal is not made to the fixed-scatterer approximation, the full nuclear dynamics comes into play during the multiple-scattering process, as in the Faddeev approach to πd scattering in Chapter 3. In the approach of Watson to the multiple-scattering problem, which we discuss in the next two sections, we indeed have a formalism that allows us to go beyond the fixed-scatterer assumption, and hence the nuclear dynamics permeates the discussion, although in a form somewhat different from that of Faddeev theory. In particular, the nuclear interactions do not enter on as symmetric a footing with the meson-nucleon interactions as was the case there, and this is natural since for πd scattering the "nuclear interactions" are again two-body NN scattering while for πA with $A > 2$ they already involve a many-body problem. Last, at the end of this chapter we discuss the approach of Glauber to the multiple-scattering problem. This is presumably applicable only at rather high energies. It then rests in an intrinsic way on the fixed-scatterer assumption, but it may have a range of validity beyond that of projectile-nucleon potential interactions.

4.2 The Watson Series: Theoretical Development

4.2a The Multiple-Scattering Amplitude

For a systematic, quantal approach to the nonrelativistic multiple-scattering problem, we start by defining the governing Hamiltonian* as

$$H = H_0(\mathbf{r}_1, \mathbf{r}_2, \ldots, \mathbf{r}_A; \mathbf{r}) + V(\mathbf{r}_1, \mathbf{r}_2, \ldots, \mathbf{r}_A; \mathbf{r})$$

$$= \left[H_N(\mathbf{r}_1, \mathbf{r}_2, \ldots, \mathbf{r}_A) + K(\mathbf{r}) \right] + \sum_{i=1}^{A} v_i(\mathbf{r}, \mathbf{r}_i), \qquad [44]$$

where the variables $\{\mathbf{r}_1, \mathbf{r}_2, \ldots, \mathbf{r}_A\}$ refer to nucleon coordinates for the nuclear Hamiltonian H_N involving A nucleons, the spin and isospin degrees of freedom being left implicit, and $\mathbf{r}$ is the projectile variable contained in the projectile free Hamiltonian K and in its interaction with the ith nucleon v_i. Already at this stage of the problem we have limited our treatment severely by implying a restriction to nucleon and projectile degrees of freedom, without the possibility, for example of absorption for a boson projectile. At present, we have in mind a description based on scattering potentials only, and this constraint is with us throughout the present chapter.[†] We write the Lippmann-Schwinger equation for the

*We note that this division of the Hamiltonian is satisfactory for processes that leave the final nucleus in a bound state, but would be problematical for knockout reactions where in the final state the interaction between the outgoing nucleon and the core enters on a different footing in modifying that scattering state.
[†]This is somewhat less of a restriction than it appears since we see in Chapter 6 that a larger class of theories can be handled in the forms used here through an effective Schroedinger equation.

system in the form

$$T = V + VG_0T = \sum_{i=1}^{A} v_i + \sum_{i=1}^{A} v_i G_0 T, \qquad [45]$$

where

$$G_0 \equiv \frac{1}{E - H_0 + i\eta} = \frac{1}{E - H_N - K + i\eta}, \qquad \eta \to 0^+. \qquad [46]$$

Note that the amplitude T refers both to nuclear coordinates and the meson variable, as in Eqs. 41–43, and matrix elements of it will have the structure $\langle \mathbf{k}' \nu | T | \mathbf{k} \mu \rangle$, where $\mathbf{k}$ and $\mathbf{k}'$ are labels for the projectile and ν, μ denote nuclear states for the space of the A nucleons.

The division of the Hamiltonian into an unperturbed part $H_0 = H_N + K$ and a perturbation $V = \Sigma v_i$ implies that we have some sort of solution to the nuclear Hamiltonian H_N and that we are prepared to cope with the extreme complexity of its inclusion in the propagator, which has the effect of burying the nuclear dynamics in the heart of the multiple-scattering treatment. We permit ourselves this luxury now with the knowledge that we later apply the theory in a regime where the nuclear effects can be approximated as small; in particular, at sufficiently high energies we may be able to invoke nuclear closure, use only an average H_N, and introduce nuclear properties through the ground state density and correlation functions. Note that Eq. 46 is based on a rather different attitude from that which guided us in Chapter 3 (particularly in Section 3.2) in dealing with πd scattering. There we took the nucleon-nucleon interaction, that is, the nuclear Hamiltonian, on the same footing as the πN interaction, that is, the scattering potentials, and the propagator involved only free-particle Hamiltonians. Here, of course, such an approach would be hopeless. Thus, while the structure of the equations which we now obtain are very similar to the Faddeev equations of the three-body case, their significance is quite different. The formal development on which we now embark was first set forth by Watson and his collaborators (see the references at the end of this chapter).

Once again, as in Chapter 3, we recast the Lippmann-Schwinger equation (Eq. 45) in a form that permits us first to sum exactly all the projectile rescatterings on a given target nucleon. There are several reasons for doing this in the present context: We know we are dealing with strong interactions: thus it is plausible that we gain by first summing over individual nucleon contributions and verifying the validity of this description before proceeding to look for many-body effects. In fact, if the projectile-nucleon system were to contain a hard-core interaction, this would be obligatory, since matrix elements of the scattering potentials would then be infinite and we would have first to solve for the t-matrix elements in the two-particle subsystem to get a well-behaved result for use in the next phase of the problem. (This is competely analogous to what is done for the nuclear many-body problem, where just such a difficulty arises due to the hard core of the nucleon-nucleon interaction. Indeed, at this stage, the main difference between these two formalisms has to do with the effects of the Pauli principle, which damps the consequences of the interaction between two bound nucleons at large distances, thus producing the healing phenomenon in the two-nucleon wave function, while, for the many-body scattering problem, the wave function is truly modified at large distances where it is phase shifted.) At a technical level, the prior summation over the scattering on each single nucleon replaces a possibly poorly known, supposedly nonrelativistic, interaction potential v by the transition matrix t, which we may know better or be able to parametrize from experimental information. This may be a somewhat risky procedure, since by means of it we cut loose from the base of our theory in dynamical equations— however inadequate, as, for example, in the use of a nonrelativistic Schroedinger equation for a relativistic problem (cf. Section 3.5)—and risk inconsistencies. Nonetheless, such an approach may be justifiable when there is no convincing alternative; we see for the pion case that the need to include field theory aspects of the problem even at relatively low energies tends to weaken our confidence in the existence of a conventional πN potential even in that regime. It is therefore gratifying that we can make reasonable progress in describing pion-nucleus scattering before we have to commit ourselves in enormous detail to the nature of the πN interaction. In short, it is possible that the Watson multiple-scattering theory is better than its origins in the Schroedinger equation; we see a more concrete example of this in Chapter 6 where true absorptive effects are treated through an effective Schroedinger equation.

It is quite easy to see how to carry out a systematic, preliminary summation of scatterings on the individual nucleons by using the diagrams of Fig. 4.4. From there it is clear that we can order the diagrams in terms of the last (leftmost) nucleon to undergo an interaction with the pion. By summing over all possible diagrams in which the pion interacts only with nucleon "1," we obtain t_1, the amplitude for pion scattering on nucleon 1 —though *always* with particle propagation involving the *full nuclear* Hamiltonian! A similar result follows for each of the other nucleons, and then we can go on to look at diagrams that at the end again involve all possible interactions between the pion and nucleon 1 only, but just before that have an interaction with nucleon 2 or 3 or 4, and so on (but, of course, not with nucleon 1 since we have already summed that in the previous term).

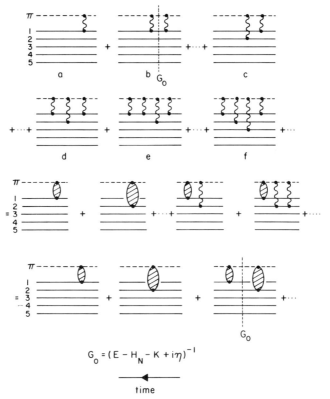

Fig. 4.4 Diagrams that enter in a multiple-scattering theory (to be read from right to left). When all possible scatterings (wavy lines) are summed, the various two-particle amplitudes t (blobs) enter, with intermediate propagation $G_0 = (E - H_N - K + i\eta)^{-1}$ involving the projectile and the interacting nuclear system. Interaction must then take place with a *different* nucleon before the projectile again scatters on a given nucleon, since all uninterrupted successive scatterings on the same nucleon have gone into producing the full amplitude t.

Continuing in this way we produce diagrams with arbitrary numbers of t-matrices in arbitrary orderings with respect to the nucleon on which the interactions take place, but we *never* have two adjacent t-matrices for the *same* nucleon. That is to say, to avoid double counting of terms summed to produce the t-matrix for a given nucleon there must always take place an intervening scattering on a different nucleon before we again allow scattering from the given nucleon.

Formally, this result is easily obtained from Eq. 45 by noting from there that*

$$T = \sum_{i=1}^{A} T_i, \qquad [47]$$

with the definition

$$T_i \equiv v_i + v_i G_0 T. \qquad [48]$$

This last construct simply represents the sum of all diagrams for which the last interaction is on the ith nucleon. By using Eq. 47 in Eq. 48, we have

$$T_i = v_i + v_i G_0 \sum_{j=1}^{A} T_j, \qquad [49]$$

and we select for special consideration the term in the sum with $j = i$, subtracting that contribution from both sides of the equation to get

$$(1 - v_i G_0) T_i = v_i + v_i G_0 \sum_{j \neq i} T_j. \qquad [50]$$

We now divide through formally by $(1 - v_i G_0)$, which is equivalent to summing the class of diagrams with all orders of final interactions with nucleon i. Defining

$$t_i \equiv (1 - v_i G_0)^{-1} v_i, \qquad [51]$$

we have

$$T_i = t_i + t_i G_0 \sum_{j \neq i} T_j, \qquad [52]$$

which represents an integral equation for that part of the full amplitude in which the last scattering takes place on the ith nucleon. (If $i = 1$, this involves terms a, b, d, e, f in Fig. 4.4.) Equation 52 can essentially be read

---

*Despite the formal similarity of the multiple-scattering theory here to the theory that lead to Faddeev equations in Section 3.2, we note that the compactness problem does not arise in the present context—as it did in Chapter 3—because we are not trying to solve an integral equation directly for T (and we retain everywhere the effects of the nuclear Hamiltonian in binding the nucleons and assuring connectedness).

off from Fig. 4.4, since, if we extract terms whose last interactions are on i, we are left with the same amplitudes for last interactions on everything but i, as shown in Eq. 52. An integral equation for the construct of Eq. 51 can be determined by reversing in it the steps we have just performed, that is, multiply by $(1 - v_i G_0)$ and add $v_i G_0 t_i$ to both sides of the equation to obtain

$$t_i = v_i + v_i G_0 t_i. \qquad [53]$$

This result is the Lippmann-Schwinger equation for scattering of the projectile on a nucleon in interaction with the other $(A - 1)$ nucleons of the target; it differs from the corresponding equation for scattering on the free nucleon by reason of the nuclear Hamiltonian H_N which lurks in the propagator G_0. Last, we sum over the contributions from all nucleons, as stated in Eq. 47, to get

$$T = \sum_{i=1}^{A} t_i + \sum_{i=1}^{A} t_i G_0 \sum_{j \neq i} T_j, \qquad [54]$$

and by iterating this we obtain

$$T = \sum_{i=1}^{A} t_i + \sum_{i=1}^{A} t_i G_0 \sum_{j \neq i} t_j + \sum_{i=1}^{A} t_i G_0 \sum_{j \neq i} t_j G_0 \sum_{k \neq j} t_k + \cdots. \qquad [55]$$

This is an infinite series of scattering terms, for single, double, triple,..., and so on scatterings, in which there are no two successive scatterings on the same nucleon; that is, $j \neq i$, $k \neq j$,..., and so on but we *can* have $k = i$ in the third term, and so forth. We note that the central results of Eqs. 52–55 — usually known as the Watson series for the projectile-nucleus *scattering amplitude* (see the references at the end of this chapter)—are quite general in that they can in principle refer to inelastic as well as to elastic scattering. Furthermore, they apply for *any* form of the propagator G_0 in Eq. 45, provided only that the *same* propagator appears throughout.

The somewhat artificial transition operator t_i of Eq. 53, for the interaction of a pion with nucleon i that is itself interacting with the rest of the nucleons in the nucleus, can be related—still within the confines of a potential interaction theory, of course—to the amplitude t_i^{free} for truly free scattering* from nucleon i at an equivalent energy E_π such that the

---

*For free πN scattering, the operator K in the propagator of Eq. 5 includes both pion and nucleon kinetic energies, as in Eqs. 2.7 and 2.41: $K = K_\pi + K_N$. For some purposes we may later omit the nucleon recoil term by taking $K = K_\pi$, thus defining a fixed-nucleon t-matrix.

contribution of the nuclear Hamiltonian H_N is balanced by $E_N = E - E_\pi$ in Eq. 53,:

$$t_i^{\text{free}} = v_i + v_i g_0 t_i^{\text{free}}, \quad g_0 = \frac{1}{E_\pi - K + i\eta}. \qquad [56]$$

To do so, we use the formal solution to the inverse problem for Eq. 56, that is, for the potential in terms of the amplitude,

$$v_i = t_i^{\text{free}} \left(1 + g_0 t_i^{\text{free}}\right)^{-1}, \qquad [57]$$

to write, from Eq. 53,

$$
\begin{aligned}
t_i &= v_i(1 + G_0 t_i) = t_i^{\text{free}}\left(1 + g_0 t_i^{\text{free}}\right)^{-1}(1 + G_0 t_i) \\
&= t_i^{\text{free}}\left(1 + g_0 t_i^{\text{free}}\right)\left(1 + g_0 t_i^{\text{free}}\right)^{-1}(1 + G_0 t_i) \\
&\quad - t_i^{\text{free}} g_0 t_i^{\text{free}}\left(1 + g_0 t_i^{\text{free}}\right)^{-1}(1 + G_0 t_i) \\
&= t_i^{\text{free}} + t_i^{\text{free}} G_0 t_i - t_i^{\text{free}} g_0 v_i(1 + G_0 t_i) = t_i^{\text{free}} + t_i^{\text{free}} G_0 t_i - t_i^{\text{free}} g_0 t_i \\
&= t_i^{\text{free}} + t_i^{\text{free}}(G_0 - g_0)t_i. \qquad [58]
\end{aligned}
$$

This is a completely general result for any two operators satisfying a Lippmann-Schwinger equation having the same interaction potential v_i but with any two propagators. Naturally, the final expression in Eq. 58 is especially useful, and therefore the entire procedure we have given for the Watson series is most easily applied if the difference between G_0 and g_0 is not too great so that $t_i \approx t_i^{\text{free}}$ that is, the amplitude is given approximately by the free nucleon amplitude. This is so if nuclear effects do not matter a great deal, as may be the case, for example, if the projectile energy is much greater than the nuclear binding energy (see Section 4.3). It must be emphasized, however, that even if Eq. 58 is not subject to *easy* use in a perturbative sense, it still yields together with Eq. 55 a complete solution to the multiple-scattering problem.

A remark on nuclear antisymmetry.

There are two ways of handling the identity of nucleons, as it affects the propagator G_0 in Eq. 46 in the Watson expansion: one is to use anti-symmetrized target states, and the second is to treat the nucleons in G_0 as if they were distinguishable. It is well-known that for a symmetrical interaction such as V of Eq. 44, the two methods give equivalent transition matrix elements between antisymmetric initial and final states, as long as the entire series is kept. However, the two treatments lead to different arrangements of the multiple scattering expansion, and therefore to somewhat different approximation methods. It is conventional in the Watson expansion to ignore antisymmetry in G_0, which means that intermediate states of mixed symmetry will occur. This is implied, for example, in the approximation following Eq. 58, that $t_i \simeq t_{free}$: the free propagator g_0 does not require that the struck nucleon have a particular symmetry relation to the other target nucleons. In practice, the kind of calculation implied by ignoring the antisymmetry can at best be carried out in a single-particle model of the nucleus. Alternatively, we may explicitly include target antisymmetry; we shall return to this in the next subsection (p. 143).

4.2b   The Optical Potential

Our motivation for introducing the Watson series in place of the original Lippmann-Schwinger equation involving potentials had to do with our expectations that for many problems of interest the potentials would immediately lead to divergent results; in the case of a hard-core interaction this was directly and obviously the case, since the matrix elements of v_i were then infinite. Thus the ultimate justification for using the Watson series for the amplitude T in place of the Lippmann-Schwinger equation with potentials rests, in good measure, on the question of whether, in so doing, we succeed in producing improved convergence in our calculational procedure. The series of Eq. 55 for T also has no guarantee, or even strong assurance, of convergence. In fact, for normal nuclear densities it generally diverges, and its direct use has thus been confined to the deuteron case (discussed in Chapter 3), where its applicability follows from the diffuseness of the deuteron, and to use in light nuclei with Monte Carlo techniques (Gib–76 and other references noted there). We therefore wish to follow the procedure, which leads to the optical potential, whereby we sum over an infinite subseries of Eq. 55, hopefully identifying the largest and most divergent pieces for exact summation. By successive application of this approach, we develop an algorithm for solving the multiple-scattering problem in terms of an expansion in single-nucleon distributions (densities), two-particle correlations, three-particle clusters, and so on. If the higher-order clusters do not greatly matter, then we can expect that this will be a convergent method. We emphasize, however, that the whole somewhat lengthy procedure on which we now embark can really only receive its justification—and at that a partial one—when we actually estimate the magnitude of higher-order terms to establish whether they are truly negligible.

The general concept of the optical potential here is the same as that which we discussed in a heuristic way in Section 4.1, namely, we attempt to construct a potential, in the projectile variables alone, such that when used in a Schroedinger equation the resulting scattering amplitude is an exact solution to the multiple-scattering series for elastic πA scattering. The optical potential thus has the character of an expectation value of some operator in the nuclear space with respect to the target state, leaving available the projectile variable in the relevant space for which the one-particle Schroedinger equation is to be written, for example, $\langle \mathbf{k}'0|v|\mathbf{k}0\rangle$ for the operator v, projectile meson momenta $\mathbf{k}$ and $\mathbf{k}'$ (for work in momentum space), and a nuclear expectation value for the ground state 0 involving A nucleon variables.

Expansion: first order.

In order to discuss as lucidly as possible the physics of the optical potential, we start by simplifying the structure of our dynamic equations in an innocent-looking but significant way: We note in Eqs. 54 and 55 the crucial instruction not to return for further scattering on the same nucleon before first visiting another. This instruction is very central to our entire procedure, as we have seen, since it allows for a consistent summing of all interactions on a given nucleon before going on to the next. Nonetheless, we now relax this restriction on the summation, allowing the inclusion of all terms and among them those involving successive scattering on the same nucleon. This we do on the grounds that it includes one extra term in the summation of the same character as the $A - 1$ terms that are already present in it, and is only an error of order $1/A$, which for our present purposes is supposed to be small, since we focus on "large" nuclei. In general, one would anticipate that dropping the restriction on the summations is legitimate wherever average field concepts apply, that is, if there are no dominant effects arising from correlations and no great irregularities in the strength of the interaction from one scattering center to another. However, in view of the fact that this approximation explicitly violates the consistency of our approach, it is especially urgent to check its validity. This we do in two different ways below—both of them exact. The first involves formal manipulations of the full series and the other exploits the special consideration that we are dealing with equivalent target particles so that we can everywhere antisymmetrize and then exhibit that each term makes a like contribution. Since each of these lines of approach has its own nonnegligible complications, we prefer to start with the crude method of dropping the restrictions on summing over particles.

We therefore replace Eq. 52 by

$$T_i \cong t_i + t_i G_0 \sum_{\text{all } j} T_j, \quad [\text{approximate}, O(1/A)], \qquad [59]$$

and sum over the index using Eq. 47 to get

$$T = \mathfrak{T} + \mathfrak{T} G_0 T, \quad [\text{approximate}, O(1/A)] \qquad [60]$$

where

$$\mathfrak{T} = \sum_{i=1}^{A} t_i. \qquad [61]$$

At this state we are interested in calculating *elastic* scattering, that is, events in which the nucleus is in the ground state after, as well as before, the fact; therefore we take the ground state expectation value of the scattering amplitude T with respect to the nuclear variables. We introduce Greek letter labels for the nuclear states and a shorthand notation for these states such that

$$|0\rangle = |\Phi_0(\mathbf{r}_1, \mathbf{r}_2, \ldots, \mathbf{r}_A)\rangle\rangle, \quad |\alpha\rangle = |\Phi_\alpha(\mathbf{r}_1, \mathbf{r}_2, \ldots, \mathbf{r}_A)\rangle\rangle, \ldots \qquad [62]$$

The ground state expectation value of Eq. 60 becomes

$$\langle 0|T|0\rangle = \langle 0|\mathfrak{T}|0\rangle + \langle 0|\mathfrak{T} G_0 T|0\rangle$$

$$= \langle 0|\mathfrak{T}|0\rangle + \sum_\nu \langle 0|\mathfrak{T}|\nu\rangle G_0(\nu)\langle \nu|T|0\rangle, \qquad [63]$$

where in the second version we have introduced a complete set of nuclear states and noted that the propagation is diagonal with respect to these:

$$\langle \nu'|G_0|\nu\rangle = \left\langle \nu' \left| \frac{1}{E - H_N - K + i\eta} \right| \nu \right\rangle = \left\langle \nu \left| \frac{1}{E - E_\nu^{\text{nucl}} - K + i\eta} \right| \nu \right\rangle \delta_{\nu'\nu}$$

$$\equiv G_0(\nu)\delta_{\nu'\nu}. \qquad [64]$$

Note that both the propagator and the full scattering amplitude remain operators in the projectile space after taking the nuclear expectation value (see Eq. 44). This is as it should be, since ultimately we wish to arrive at equations that describe the motion of the projectile after we have *eliminated* nuclear degrees of freedom. In Eq. 63 there are two classes of nuclear matrix elements that enter, namely, those that are diagonal, $\nu = 0$, and those that are nondiagonal, $\nu \neq 0$. The former are expected to be larger than the latter because of the anticipated good overlap of the nuclear ground state wave function with itself, at least at small momentum transfers. These diagonal nuclear ground state expectation values thus play a preferred role in our approach and we often label them with a subscript c for "coherent," since they pertain to the elastic case where the projectile wavelets from various scattering centers contribute coherently. In the forward direction this coherence reflects itself in terms in which the contributions of the A nucleons add with the same sign, as opposed to inelastic matrix elements that generally tend to vanish in the forward direction (if no spin or isospin operators are present) due to orthogonality of the nuclear wave functions. We thus divide the right-hand side of that equation and write

$$\langle 0|T|0\rangle = \langle 0|\mathfrak{T}|0\rangle + \langle 0|\mathfrak{T}|0\rangle \frac{1}{E-K+i\eta}\langle 0|T|0\rangle$$

$$+ \left[\sum_{\nu \neq 0} \langle 0|\mathfrak{T}|\nu\rangle G_0(\nu)\langle \nu|T|0\rangle \right], \qquad [65]$$

where the term in square brackets is expected to be small, and we attempt an expansion in order of off-diagonal matrix elements in the nuclear space (which will ultimately lead to an expansion in clusters). We note further that in Eq. 65 we have taken the arbitrary zero of energy in such a way that the nuclear ground state energy vanishes: $E_0^{nucl} = 0$. The "leading" terms of Eq. 65 satisfy an equation of the form

$$\langle 0|T|0\rangle = \mathcal{V}_c + \mathcal{V}_c \frac{1}{E-K+i\eta}\langle 0|T|0\rangle, \qquad [66]$$

with

$$\mathcal{V}_c = \mathcal{V}_c^{(1)} \equiv \langle 0|\mathfrak{T}|0\rangle = \left\langle 0 \left| \sum_{i=1}^{A} t_i \right| 0 \right\rangle. \qquad [67]$$

Equation 66 has precisely the form that we seek for the solution of our problem. It is a Lippmann-Schwinger equation for the projectile in the case of elastic scattering with a known equivalent potential—the optical potential—given by Eq. 67, at least in this approximation. Equation 67 is an important standard result that arises essentially in every multiple-scattering approach as a low-order approximation. It is, in fact, identical to Eqs. 14 or 16, as we see later, and therefore embodies in it the qualitative physical features that we have discussed in Section 4.1.

Higher orders.

Of course, Eq. 67 is only a lowest-order result due to our neglect of the last term on the right-hand side of Eq. 65 involving off-diagonal matrix elements in the nuclear space. We must now turn to a consideration of its effects (still holding in abeyance the entire question of the $1/A$ error committed in neglecting the $i \neq j \neq k \dots$ etc. in our sums over particle indices). Let us therefore use Eq. 60 to write down the description of an off-diagonal case,

$$\langle \nu | T | \mu \rangle = \langle \nu | \mathfrak{I} | \mu \rangle + \sum_\lambda \langle \nu | \mathfrak{I} | \lambda \rangle G_0(\lambda) \langle \lambda | T | \mu \rangle, \qquad [68]$$

which we now need, in Eq. 65, for $\mu = 0$ and $\nu \neq 0$. Then we have

$$\langle 0 | T | 0 \rangle = \langle 0 | \mathfrak{I} | 0 \rangle + \langle 0 | \mathfrak{I} | 0 \rangle G_0(0) \langle 0 | T | 0 \rangle$$

$$+ \sum_{\nu \neq 0} \langle 0 | \mathfrak{I} | \nu \rangle G_0(\nu) \langle \nu | \mathfrak{I} | 0 \rangle$$

$$+ \sum_{\nu \neq 0} \langle 0 | \mathfrak{I} | \nu \rangle G_0(\nu) \sum_\lambda \langle \nu | \mathfrak{I} | \lambda \rangle G_0(\lambda) \langle \lambda | T | 0 \rangle. \qquad [69]$$

Apart from our minor $1/A$ approximation, this result is exact, being merely the diagonal matrix element of the iteration of our master equation (Eq. 60). To continue our chain of approximations in some reasonably systematic way, we now select for separate consideration those terms in which *no* matrix elements $\langle \alpha | \dots | 0 \rangle$ diagonal with the ground state appear. With this separation and a minor reordering of the terms, Eq. 69 becomes

$$\langle 0 | T | 0 \rangle = \left[\langle 0 | \mathfrak{I} | 0 \rangle + \sum_{\nu \neq 0} \langle 0 | \mathfrak{I} | \nu \rangle G_0(\nu) \langle \nu | \mathfrak{I} | 0 \rangle \right]$$

$$+ \left[\langle 0 | \mathfrak{I} | 0 \rangle + \sum_{\nu \neq 0} \langle 0 | \mathfrak{I} | \nu \rangle G_0(\nu) \langle \nu | \mathfrak{I} | 0 \rangle \right] G_0(0) \langle 0 | T | 0 \rangle$$

$$+ \left\{ \sum_{\nu \neq 0} \sum_{\lambda \neq 0} \langle 0 | \mathfrak{I} | \nu \rangle G_0(\nu) \langle \nu | \mathfrak{I} | \lambda \rangle G_0(\lambda) \langle \lambda | T | 0 \rangle \right\}. \qquad [70]$$

In this exact (up to $1/A$) result, we now ignore the last term in braces as having a lower number of diagonal ground state matrix elements, and again note that we have an equation of the form of Eq. 66, this time with the next-order correction in the optical potential

$$\mathcal{V}_c = \mathcal{V}_c^{(2,0)} \equiv \langle 0 | \mathfrak{I} | 0 \rangle + \sum_{\nu \neq 0} \langle 0 | \mathfrak{I} | \nu \rangle \frac{1}{E - E_\nu^{\text{nucl}} - K + i\eta} \langle \nu | \mathfrak{I} | 0 \rangle.$$

$$[71]$$

As we see, this correction embodies two-particle correlation effects in the optical potential, one of the important features we sought in our improved treatment, and is therefore a key result.

Unfortunately, there is at best an ambiguity and at worst an inconsistency in this second-order result. In Eq. 70 we ignored the last term because it involved fewer appearances of the ground state than the previous terms, but, in fact, it involves no fewer appearances of *diagonal* terms, since the part of the summation with $\nu = \lambda$ has such pieces. Moreover, there is not necessarily any reason to believe that diagonal matrix elements with respect to states other than the ground state $\langle \nu | \ldots | \nu \rangle$, $\nu \neq 0$, will be any smaller than the corresponding ground state expectation value $\langle 0 | \ldots | 0 \rangle$; on the contrary, one would tend to expect them to be comparable to it. To examine the consequences of incorporating this effect, we must imagine iterating Eq. 60 successively, extracting those terms that finally lead back to the ground state, since we seek an equation for $\langle 0 | T | 0 \rangle$, and then, for the *second*-order optical potential, retaining terms that involve no more than *two* off-diagonal matrix elements. This natural extension of Eq. 71 then gives

$$\mathcal{V}_c^{(2,1)} \equiv \langle 0 | \mathcal{T} | 0 \rangle + \sum_{\nu \neq 0} \langle 0 | \mathcal{T} | \nu \rangle G_0(\nu) \big[1 + \langle \nu | \mathcal{T} | \nu \rangle G_0(\nu)$$
$$+ \langle \nu | \mathcal{T} | \nu \rangle G_0(\nu) \langle \nu | \mathcal{T} | \nu \rangle G_0(\nu) + \cdots \big] \langle \nu | \mathcal{T} | 0 \rangle. \quad [72]$$

(If this result is not already obvious, it will be in a moment when we look at still higher orders.) The geometric series in the brackets is easily summed to give an equation for a modified propagator

$$G_0^{(1)}(\nu) = G_0(\nu) \big[1 + \langle \nu | \mathcal{T} | \nu \rangle G_0(\nu) + \langle \nu | \mathcal{T} | \nu \rangle G_0(\nu) \langle \nu | \mathcal{T} | \nu \rangle G_0(\nu) + \cdots \big]$$
$$= G_0(\nu) + G_0(\nu) \langle \nu | \mathcal{T} | \nu \rangle G_0^{(1)}(\nu), \quad [73]$$

as can be trivially verified by iterating this equation. This result again is of a very commonly encountered type and has a general formal solution

$$G_0^{(1)}(\nu) = \big[1 - G_0(\nu) \langle \nu | \mathcal{T} | \nu \rangle \big]^{-1} G_0(\nu)$$
$$= \left[1 - \frac{1}{E - E_\nu^{\text{nucl}} - K + i\eta} \langle \nu | \mathcal{T} | \nu \rangle \right]^{-1} \frac{1}{E - E_\nu^{\text{nucl}} - K + i\eta}$$
$$= \frac{1}{E - E_\nu^{\text{nucl}} - K - \langle \nu | \mathcal{T} | \nu \rangle}, \quad [74]$$

where we have dropped the $i\eta$ mnemonic since $\langle \nu | \mathcal{T} | \nu \rangle$ generally has a (negative) nonvanishing imaginary part (cf. Eq. 15). Thus we conclude

$$\mathcal{V}_c^{(2,1)} = \langle 0|\mathfrak{T}|0\rangle + \sum_{\nu \neq 0} \langle 0|\mathfrak{T}|\nu\rangle \frac{1}{E - E_\nu^{\text{nucl}} - K - \langle \nu|\mathfrak{T}|\nu\rangle} \langle \nu|\mathfrak{T}|0\rangle,$$

$$[75]$$

which is to be contrasted with Eq. 71 and is a key result like it. The physical difference between the two lies in the fact that in the optical potential of Eq. 71 the projectile propagates as if free, while in Eq. 75 it propagates in the *presence* of the (lowest-order) optical potential, so that various elastic, non-ground-state encounters between the inelastic collisions are included. Of course, the question as to which of these formulations is "better" can only be determined, in principle, by examining the higher-order terms, which will naturally be different in the two cases, and by seeing which series for $\mathcal{V}_c$ converges faster; formalism 1, for $\mathcal{V}_c^{(2,1)}$, has a prejudice in its favor over formalism 0, for $\mathcal{V}_c^{(2,0)}$, since it sums once and for all over an infinite subset of diagonal terms that otherwise must crop up later, though it does so at the cost of complicating considerably the description of the propagation in the optical potential, and therefore its evaluation.

[For nearly all practical purposes, the optical potentials $\mathcal{V}_c^{(1)}$, $\mathcal{V}_c^{(2,0)}$, and $\mathcal{V}_c^{(2,1)}$ are all that can be readily considered for realistic meson-nucleus calculations, and it is not our purpose here to explore multiple-scattering theories *per se* in great detail. Nonetheless, we would like to pursue our iteration scheme somewhat further in order to complete the systematization of the possible optical potential series. We therefore iterate Eq. 70 yet again, and, to save repetition, we even do so twice, going immediately to fourth order, where a new feature presents itself. Thus

$$\langle 0|T|0\rangle = \left[\langle 0|\mathfrak{T}|0\rangle + \sum_{\nu \neq 0} \langle 0|\mathfrak{T}|\nu\rangle G_0(\nu)\langle \nu|\mathfrak{T}|0\rangle \right.$$

$$+ \sum_{\nu \neq 0} \sum_{\mu \neq 0} \langle 0|\mathfrak{T}|\nu\rangle G_0(\nu)\langle \nu|\mathfrak{T}|\mu\rangle G_0(\mu)\langle \mu|\mathfrak{T}|0\rangle$$

$$\left. + \sum_{\nu \neq 0} \sum_{\mu \neq 0} \sum_{\lambda \neq 0} \langle 0|\mathfrak{T}|\nu\rangle G_0(\nu)\langle \nu|\mathfrak{T}|\mu\rangle G_0(\mu)\langle \mu|\mathfrak{T}|\lambda\rangle G_0(\lambda)\langle \lambda|\mathfrak{T}|0\rangle \right]$$

$$\times \left(1 + \frac{1}{E - K + i\eta} \langle 0|T|0\rangle \right) + \text{fifth-order terms}, \qquad [76]$$

where the fourth-order optical potential is trivially read off, by comparison with Eq. 66, as the term in brackets. We have by now seen that the third term there with $\mu = \nu$ and the fourth term with $\lambda = \mu = \nu$ can be viewed as contributing to the second-order optical potential in formalism 1, $\mathcal{V}_c^{(2,1)}$, or to the third- and fourth-order optical potentials in formalism 0. The new element enters when we consider the fourth-order term. There we can take the result as given for $\mathcal{V}_c^{(4,0)}$, prohibit all diagonal terms for $\mathcal{V}_c^{(4,1)}$, in which case the diagonal terms are in the second- and third-order pieces, as we just noted, or we can prohibit diagonal terms *and* terms in which we return to the same state after *every other* step, that is, prohibit *second-order diagonal* terms of the form

$$\sum_{\mu \neq \nu} \langle \nu | \mathcal{T} | \mu \rangle G_0(\mu) \langle \mu | \mathcal{T} | \nu \rangle. \qquad [77]$$

Then we would distinguish terms in the fourth-order optical potential and arrive at

$$\mathcal{V}_c^{(4,2)} \Rightarrow \langle 0 | \mathcal{T} | 0 \rangle + \sum_{\nu \neq 0} \langle 0 | \mathcal{T} | \nu \rangle G_0^{(2)}(\nu) \langle \nu | \mathcal{T} | 0 \rangle$$

$$+ \sum_{\nu \neq 0} \sum_{\mu \neq 0, \nu} \langle 0 | \mathcal{T} | \nu \rangle G_0^{(2)}(\nu) \langle \nu | \mathcal{T} | \mu \rangle G_0^{(2)}(\mu) \langle \mu | \mathcal{T} | 0 \rangle$$

$$+ \sum_{\nu \neq 0} \sum_{\mu \neq 0, \nu} \sum_{\lambda \neq 0, \mu, \nu} \langle 0 | \mathcal{T} | \nu \rangle G_0^{(2)}(\nu) \langle \nu | \mathcal{T} | \mu \rangle G_0^{(2)}(\mu)$$

$$\times \langle \mu | \mathcal{T} | \lambda \rangle G_0^{(2)}(\lambda) \langle \lambda | \mathcal{T} | 0 \rangle, \qquad [78]$$

where the propagator

$$G_0^{(2)}(\nu) \equiv \frac{1}{E - E_\nu^{\text{nucl}} - K - \mathcal{V}_c^{(2,0)}(\nu)}$$

$$= \left\{ E - E_\nu^{\text{nucl}} - K - \langle \nu | \mathcal{T} | \nu \rangle - \sum_{\mu \neq \nu} \langle \nu | \mathcal{T} | \mu \rangle G_0(\mu) \langle \mu | \mathcal{T} | \nu \rangle \right\}^{-1}$$

$$[79]$$

involves the *second*-order optical potential for the excited state ν (which might be approximated by the second-order, formalism 0, ground state optical potential of Eq. 71). The arrow in Eq. 78 is meant to imply that we go through the same procedure as for Eqs. 72–74 to sum the iterated result of including terms of the type given in Eq. 77, the only one that we can show explicitly to fourth order is that appearing as the last term in the second-order of Eq. 78. The use of this formalism 2 would be appropriate for physical situations where we felt it inadequate to include only first-order coherent terms $\langle \nu | \ldots | \nu \rangle$—coherent because they leave the target in the same state, whence the subscript on $\mathcal{V}_c$— and wished also to sum second-order coherent terms. This would be the case if we encountered important two-particle correlation terms that appreciably modified the lowest-order optical potential. In this same fashion, we can continue to generate this hierarchy of approximations to the optical potential, incorporating even higher orders of the potential itself in the propagators used to generate it at each level. This culminates ultimately in a series that forbids *any* return at any stage to a nuclear level once encountered and involves that full potential itself in the propagators. This is, of course, a formidable problem in self-consistency for any practical calculation, a consideration that forces us pretty much to restrict our attention to formalisms 0 and 1, involving no optical potential or only the lowest-order one (mercifully containing *no* propagator in its evaluation); this has the effect of projecting out the usable theories from among the embarrassing wealth of formal results.]

A^{-1} corrections.

We now return to examine the question of our $1/A$ omissions in going from Eq. 52 to Eq. 59 or Eq. 54 to Eq. 60, and in the process provide a concise derivation of formalism 0. To do this, we first define a projection operator that projects on to the nuclear ground state,

$$P_0|0\rangle = |0\rangle, \qquad (1-P_0)|0\rangle = 0, \qquad P_0^2 = P_0, \qquad [\mathbf{80}]$$

so that $(1 - P_0)$ projects on to all nuclear levels other than the ground state, as required in all our prescriptions for the optical potentials. In terms of this, we define an operator that is to satisfy a Lippmann-Schwinger equation in the form

$$\mathcal{V} = V + VG_0(1 - P_0)\mathcal{V}, \qquad V = \sum_{i=1}^{A} v_i \qquad [\mathbf{81}]$$

(note that G_0 and P_0 are both diagonal in the nuclear space and therefore commute). Equation 81 is to be compared, first, with Eq. 45 for the scattering amplitude T. They have the same inhomogeneous term V, and differ only in their propagators (and, indeed, not much in those!). Thus, the conditions under which Eq. 58 follows from Eqs. 53 and 56 apply, and we can immediately write that

$$T = \mathcal{V} + \mathcal{V}[G_0 - G_0(1 - P_0)]T = \mathcal{V} + \mathcal{V}G_0 P_0 T. \qquad [\mathbf{82}]$$

If we now take the expectation value of this with respect to the nuclear ground state, we have

$$T_c \equiv \langle 0| T |0\rangle = \langle 0| \mathcal{V} |0\rangle + \langle 0| \mathcal{V} G_0 P_0 T |0\rangle$$
$$= \langle 0| \mathcal{V} |0\rangle + \langle 0| \mathcal{V} |0\rangle G_0 \langle 0| T |0\rangle, \qquad [83]$$

where we have used the property of the projection operator P_0 to allow only the ground state among the intermediate nuclear states. Comparing Eq. 83 with Eq. 66 immediately gives us the optical potential as $\mathcal{V}_c = \langle 0| \mathcal{V} |0\rangle$, the nuclear ground state expectation value of the operator solution of Eq. 81.

Next we compare Eq. 81 with Eq. 45 and note that the treatment of Eqs. 47–55 for the latter pertain also for the former, again with only a difference in the propagator, so that we can immediately write down a Watson series for the optical potential:

$$\mathcal{V} = \sum_{i=1}^{A} t_i' + \sum_{i=1}^{A} \sum_{j \neq i} t_i' G_0 (1 - P_0) t_j'$$

$$+ \sum_{i=1}^{A} \sum_{j \neq i} \sum_{k \neq j} t_i' G_0 (1 - P_0) t_j' G_0 (1 - P_0) t_k' + \ldots, \qquad [84]$$

with

$$t_i' = v_i + v_i G_0 (1 - P_0) t_i' \qquad [85]$$

replacing Eq. 53. Equation 84 achieves our original goal of an exact result for the optical potential. We anticipate that the differences between t_i and t_i', as generated, say from Eq. 58, may not be vast. In particular, at high energies we expect the dominant physical mechanism to be quasi-free knock-out, and thus the nuclear ground state should play a minor role. (Note that for the full nucleus the nuclear ground state plays an important preferred role, but for scattering on a single nucleon it is merely one state out of a vast number.) With $t_i' \approx t_i$ *and* our restriction $i \neq j \neq k \ldots$ replaced, Eq. 84 is just the equivalent generalization to all orders of Eq. 71 for the formalism 0 optical potential, thus giving us a more satisfactory result for that construct. We emphasize again, however, that even if t_i' and t_i differ appreciably, our theory is still complete in the sense that we can in principle use Eq. 58 to calculate the former from the latter for any given case.

Antisymmetry of target particles.

To complete our refinements with regard to the $1/A$ error of omission, we now exploit an incidental feature of our particular problem: the equivalence of the nucleons in the target and the consequent antisymmetry of the nuclear wave functions. The important role that antisymmetrization can play in the multiple-scattering formalism was first noted by Kerman, McManus, and Thaler (see the references at the end of this chapter). We incorporate this feature by introducing an antisymmetrization projection operator $\mathcal{C}$ whose action on a complete set of many-particle states is to project onto the space of the fully antisymmetric states and to annihilate all others.* Of course, we anticipate that in our physical applications the initial and final nuclear states of relevance are antisymmetrized with respect to the target nucleons. We now define a modified t-operator for scattering on an individual nucleon from among the target constituents, modifying Eq. 53 to read

$$\tau_i = v_i + v_i \frac{1}{E - H_0 + i\eta} \, \mathcal{C}\tau_i = v_i + v_i \, G_0 \, \mathcal{C}\tau_i. \qquad [\mathbf{86}]$$

As in the case of the other projection operators we have considered in the nuclear space, $\mathcal{C}$ commutes with G_0, since the latter is of necessity symmetric in the nucleon variables. Furthermore, we have now seen repeatedly that we can always generate a multiple-scattering series, such as Eq. 55 and its accompanying results, provided only that we use the *identical* propagator in it as in the original Lippmann-Schwinger equation pertaining to the single nucleon, in this case Eq. 86. We emphasize the importance of this absolute requirement for consistency in treating both equations for t_i and for T with the *same* propagator! Furthermore, when taken between antisymmetric nuclear states the Lippmann-Schwinger equation (Eq. 45) $T = \Sigma v_i + \Sigma v_i \, G_0 \, T$ is equivalent to $T = \Sigma v_i + \Sigma v_i \, G_0 \mathcal{C} T$ due to the symmetry of $V = \Sigma v_i$. Then Eq. 55 becomes

$$T = \sum_{i=1}^{A} \tau_i + \sum_{i=1}^{A} \tau_i \, G_0 \, \mathcal{C} \sum_{j \neq i} \tau_j + \sum_{i=1}^{A} \tau_i \, G_0 \, \mathcal{C} \sum_{j \neq i} \tau_j \, G_0 \, \mathcal{C} \sum_{k \neq j} \tau_k + \dots,$$

$$[\mathbf{87}]$$

and we can now argue that, in inserting a complete set of intermediate states and taking T between antisymmetric states, the matrix elements we encounter involve a bilinear structure of wave functions, each of which is antisymmetric so that their product is symmetric. Then the matrix elements of each τ_l, $l = 1, 2, \dots, A$, are identical, and the condition of no successive scattering on the same nucleon before visiting another $i \neq j \neq k$ can be

*See remarks at end of Section 4.2a.

honored with a mere counting factor:

$$T = \sum_{i=1}^{A} \tau_i + \left(\frac{A-1}{A}\right) \sum_{i=1}^{A} \tau_i G_0 \mathcal{C} \sum_{j=1}^{A} \tau_j$$
$$+ \left(\frac{A-1}{A}\right)^2 \sum_{i=1}^{A} \tau_i G_0 \mathcal{C} \sum_{j=1}^{A} \tau_j G_0 \mathcal{C} \sum_{k=1}^{A} \tau_k + \dots \qquad [88]$$

This can be cast into the form of Eq. 54 as

$$T = \sum_{i=1}^{A} \tau_i + \left(\frac{A-1}{A}\right) \sum_{i=1}^{A} \tau_i G_0 \mathcal{C} T, \qquad [89]$$

which we note to be exact, but almost identical in structure to the approximate Eq. 60, with t_i replaced, of course, by τ_i, and with the additional factor $(A-1)/A$. Naturally, we anticipate that at high energies τ_i may differ little from t_i; that is, we can exploit Eq. 58 in that same way in which we argued $t_i' \approx t_i$ below Eq. 85, and $(A-1)/A \approx 1$ to order $1/A$, thus justifying our approximation in using Eq. 60.

To cast Eq. 89 into a form in which it may be easily solved using our earlier techniques, we introduce

$$\mathcal{T}' \equiv \frac{A-1}{A} \sum_{i=1}^{A} \tau_i, \qquad T' = \frac{A-1}{A} T, \qquad [90]$$

for which

$$T' = \mathcal{T}' + \mathcal{T}' G_0 \mathcal{C} T', \qquad [91]$$

and then note that $\mathcal{T}'$, which is a completely symmetric operator in the particle indices, must commute with the antisymmetrization operator $\mathcal{C}$, and $\mathcal{C}^2 = \mathcal{C}$, so that the expectation value of T' for the antisymmetric nuclear states we have in mind will not require the $\mathcal{C}$-instruction (but the feature of antisymmetrization will be preserved by the structure of the equations). Then

$$T' = \mathcal{T}' + \mathcal{T}' G_0 T' \quad \text{(for use in } \langle \beta | T' | \alpha \rangle, \text{ with } antisymmetric \ |\alpha\rangle, |\beta\rangle),$$
$$[92]$$

which is identical to Eq. 60, but exact. Thus the results of Eqs. 63–79 now can be taken over fully to yield an optical potential formalism for T', which is *not* the scattering amplitude but is trivially related to it through

Eq. 90, and which becomes equal to it for large A. In formalism 0, for example, the optical potential for T' is

$$\mathcal{V}_c' = \left(\frac{A-1}{A}\right)\langle 0| \sum_{i=1}^{A} \tau_i + \left(\frac{A-1}{A}\right)\sum_{i=1}^{A}\sum_{j=1}^{A} \tau_i G_0 \mathcal{C}(1-P_0)\tau_j$$

$$+ \left(\frac{A-1}{A}\right)^2 \sum_{i=1}^{A}\sum_{j=1}^{A}\sum_{k=1}^{A} \tau_i G_0 \mathcal{C}(1-P_0)\tau_j G_0 \mathcal{C}(1-P_0)\tau_k + \cdots |0\rangle,$$

$$[93]$$

where we note that the $\mathcal{C}$-instruction could again just as well be omitted. We note that at the level of Eqs. 54 and 55 there was no allowable exact way to incorporate this enormous simplification in the structure of our dynamic equations, except through an approximation as in Eqs. 59 and 60, nor in fact would it even be correct to do so if, for example, there were some special distinguishing feature involving only some of the nucleons. (Since we have implied antisymmetrization over all the nucleons, we have also supposed that we use the isospin formalism in this context so that protons and neutrons are equivalent, but have different isospin projection.)

We could also generate an expression for the optical potential in formalism 0 embodying the antisymmetrization feature directly from Eq. 81, and in parallel to our treatment there. Again noting that we ultimately take the $\mathcal{V}$-operator between antisymmetric states and that $V = \sum_i v_i$ is completely symmetric, we can rewrite Eq. 81 after verifying as in Eq. 83 that $\mathcal{V}_c = \langle 0|\mathcal{V}|0\rangle$ is indeed the required optical potential, as

$$\mathcal{V} = V + V G_0(1-P_0)\mathcal{V} = V + V G_0 \mathcal{C}(1-P_0)\mathcal{V}$$
$$\text{(for use between antisymmetric } |0\rangle), \qquad [94]$$

changing nothing. Then, just as we argued for Eq. 81, we compare with Eq. 45 and the manipulations of Eqs. 47–55 to obtain immediately the exact result

$$\mathcal{V} = \sum_{i=1}^{A} \tau_i' + \sum_{i=1}^{A}\sum_{j\neq i} \tau_i' G_0 \mathcal{C}(1-P_0)\tau_j'$$

$$+ \sum_{i=1}^{A}\sum_{j\neq i}\sum_{k\neq j} \tau_i' G_0 \mathcal{C}(1-P_0)\tau_j' \mathcal{C}(1-P_0)\tau_k' + \dots. \qquad [95]$$

We now exploit the antisymmetrization operator and the antisymmetry of $|0\rangle$ to "symmetrize" the summation instruction, after which, for our matrix elements between antisymmetric nuclear states, $\mathcal{C}$ is no longer needed, and we can write

$$\mathcal{V} = \sum_{i=1}^{A} \tau_i' + \left(\frac{A-1}{A}\right) \sum_{i=1}^{A} \sum_{j=1}^{A} \tau_i' G_0 (1 - P_0) \tau_j'$$

$$+ \left(\frac{A-1}{A}\right)^2 \sum_{i=1}^{A} \sum_{j=1}^{A} \sum_{k=1}^{A} \tau_i' G_0 (1 - P_0) \tau_j' G_0 (1 - P_0) \tau_k' + \ldots,$$

$$[96]$$

again our exact result for which we have now paid the small price that τ_i' satisfies the modified equation

$$\tau_i' = v_i + v_i G_0 \mathcal{Q} (1 - P_0) \tau_i'. \qquad [97]$$

This price is expected to be small because, again, at high energies we anticipate that $\tau_i' \approx t_i$, that is, that major differences will not result from the limitations imposed on the intermediate propagation for the "single-nucleon" amplitude. This is the case if all the nucleons contribute more or less equivalently so that the "averaged" τ_i' that results from the $\mathcal{Q}$-instruction is approximately equal to any given π_i'. This is manifestly a reasonable expectation for our case of equivalent nucleons in *elastic* scattering, but may be much less valid in inelastic processes where particular nucleons (for example, the outermost ones) may play a special role. Note that in the approach of Eq. 94 the indirect relationship of the optical potential to the scattering amplitude through factors of $(A-1)/A$ of Eqs. 90–93 is avoided.

Before turning to the formal treatment of *in*elastic processes, which completes our construction of the basic theoretical apparatus in the multiple-scattering approach, we note once again that our welter of forms for the optical potential merely represents a rearrangement of the terms in our original Lippmann-Schwinger equation for the projectile-target system. It is by no means clear at this stage that we have gained much—in any given application—by these manipulations. This will become the case only if we can establish that the series for the optical potentials lead to accelerated convergence. We examine this point and the physical content of the series below in Section 4.3.

4.2c The Multiple-Scattering Series for Inelastic Processes

We now develop a formal treatment of the multiple-scattering series for application to inelastic meson-nucleus scattering. Unlike the elastic case, this involves dividing the reaction mechanism into two parts: the distortion of the wave describing the incoming and outgoing meson, and the "hard" interactions that produce the desired nuclear transition and have no analog in the elastic case. Although our formal manipulations in large measure closely resemble those for coherent (=elastic) processes, the underlying justifications of the approximations we invoke may be less well founded due to the complicated interaction between the dynamic (nuclear transitions) and static (meson wave distortion) parts of the problem. We start from the formal results of Eqs. 86–92, which, as operator relations, are valid also for inelastic processes, whereas our various arguments concerning antisymmetry required only initial and final antisymmetric nuclear states—not diagonal matrix elements. (We note in advance, however, that the treatment of τ_i, Eq. 86, as opposed to t_i, Eq. 53, requires care, since the special role played by some of the nucleons in an inelastic process makes the conclusion $\tau_i \approx t_i$, based on the "averaging" effect of the $\mathcal{Q}$-instruction, nonobvious.)

The dynamic equation for the projectile in the presence of a transition between initial nuclear state $|0\rangle$ and final state $|\alpha\rangle$ is then

$$\langle\alpha|T'|0\rangle = \langle\alpha|\mathcal{T}'|0\rangle + \langle\alpha|\mathcal{T}'G_0 T'|0\rangle, \qquad [98]$$

and we now proceed in parallel to the scheme used in Eqs. 63–79, developing an approximation in successively higher orders of diagonal nuclear matrix elements. Thus we examine

$$\langle\alpha|T'|0\rangle = \langle\alpha|\mathcal{T}'|0\rangle + \langle\alpha|\mathcal{T}'|0\rangle G_0(0)\langle0|T'|0\rangle$$
$$+ \langle\alpha|\mathcal{T}'|\alpha\rangle G_0(\alpha)\langle\alpha|T'|0\rangle$$
$$+ \sum_{\nu\neq0,\alpha} \langle\alpha|\mathcal{T}'|\nu\rangle G_0(\nu)\langle\nu|T'|0\rangle, \qquad [99]$$

singling out for special consideration *both* the initial and the final states. For our lowest-order approximation, we ignore the last term here on the grounds that it involves *two* nondiagonal nuclear matrix elements, and recast the resulting equation in the form

$$[1-\langle\alpha|\mathcal{T}'|\alpha\rangle G_0(\alpha)]\langle\alpha|T'|0\rangle \cong \langle\alpha|\mathcal{T}'|0\rangle[1+G_0(0)\langle0|T'|0\rangle], \qquad [100]$$

or

$$\langle\alpha|T'|0\rangle \cong [1-\langle\alpha|\mathcal{T}'|\alpha\rangle G_0(\alpha)]^{-1}\langle\alpha|\mathcal{T}'|0\rangle[1+G_0(0)\langle0|T'|0\rangle]. \qquad [101]$$

The physical meaning of Eq. 101 is easily extracted. If we consider the amplitude taken between unperturbed (plane wave) meson states $\langle \phi_{\mathbf{k'}}; \alpha | T' | \phi_{\mathbf{k}}; 0 \rangle$, then on the right-hand side of Eq. 101 we have in the initial state

$$\phi_{\mathbf{k}} + G_0(0)\langle 0 | T' | 0 \rangle \phi_{\mathbf{k}} \equiv \psi_{\mathbf{k}}, \qquad [102]$$

which is the projectile wave function as distorted by the *complete* elastic optical potential since, from Eqs. 66 and 90–93,

$$\langle 0 | T' | 0 \rangle = \mathcal{V}_c' + \mathcal{V}_c' G_0(0)\langle 0 | T' | 0 \rangle, \qquad [103]$$

whence

$$
\begin{aligned}
G_0(0)\langle 0 | T' | 0 \rangle &= \frac{1}{E - K + i\eta}\left(\mathcal{V}_c' + \mathcal{V}_c' \frac{1}{E - K + i\eta} \langle 0 | T' | 0 \rangle \right) \\
&= \frac{1}{E - K + i\eta}\left(\mathcal{V}_c' + \mathcal{V}_c' \frac{1}{E - K - \mathcal{V}_c' + i\eta} \mathcal{V}_c' \right) \\
&= \frac{1}{E - K - \mathcal{V}_c' + i\eta} \mathcal{V}_c'. \qquad [104]
\end{aligned}
$$

Similarly,

$$\psi_{\mathbf{k}} = \phi_{\mathbf{k}} + \frac{1}{E - K - \mathcal{V}_c' + i\eta} \mathcal{V}_c' \phi_{\mathbf{k}} = \phi_{\mathbf{k}} + \frac{1}{E - K + i\eta} \mathcal{V}_c' \psi_{\mathbf{k}}. \qquad [105]$$

This last is just the Schroedinger equation for the meson wave function as distorted by the complete optical potential. On the other hand, the left side of the inelastic matrix element $\langle \phi_{\mathbf{k'}}; \alpha | T' | \phi_{\mathbf{k}}; 0 \rangle$ involves, from Eq. 101,

$$\langle \phi_{\mathbf{k'}} | [1 - \langle \alpha | \mathcal{T}' | \alpha \rangle G_0(\alpha)]^{-1} \equiv \langle \chi_{\mathbf{k'}}^{(-)} |. \qquad [106]$$

This is the hermitian conjugate of the final state wave function with boundary conditions involving a collapsing spherical wave (as appropriate for the final state), but it is distorted only by the *lowest-order* optical potential of Eqs. 90–93, as can be seen from noting

$$\langle \phi_{\mathbf{k'}} | = \langle \chi_{\mathbf{k'}}^{(-)} | - \langle \chi_{\mathbf{k'}}^{(-)} | \langle \alpha | \mathcal{T}' | \alpha \rangle G_0(\alpha), \qquad [107]$$

or

$$\langle \chi_{\mathbf{k'}}^{(-)} | = \langle \phi_{\mathbf{k'}} | + \langle \chi_{\mathbf{k'}}^{(-)} | \langle \alpha | \mathcal{T}' | \alpha \rangle G_0(\alpha), \qquad [108]$$

as expected.* The asymmetry between the projectile wave distortions on the left and right sides of the inelastic matrix element is a necessary, essential feature of the result if we are to avoid double counting of intermediate states, as our derivation in Eqs. 98–101 shows; of course, the roles of the distortions on the left and on the right can be switched by an alternate route for the manipulations, but the difference between them is not eliminated. This result is usually referred to as the *distorted wave impulse approximation* (DWIA), since it leads to

$$\langle \phi_{\mathbf{k}'}; \alpha | T' | \phi_{\mathbf{k}}; 0 \rangle = \langle \chi_{\mathbf{k}'}^{(-)}; \alpha | \mathfrak{I}' | \psi_{\mathbf{k}}; 0 \rangle, \qquad [109]$$

with $\mathfrak{I}' \equiv (A-1)A^{-1}\Sigma_i \tau_i$, as in Eq. 90. This is just the matrix element of the sum of single-nucleon amplitudes taken between asymmetrically distorted meson waves. (In practice, the final projectile wave is often taken as distorted by the nuclear ground state potential, that is, α is replaced by 0 in Eqs. 106–108.) Operators of the form $\Sigma_i \tau_i$ are referred to as involving an "impulse approximation," since we envision the possibility of approximating the τ_i by free-nucleon amplitudes based on the notion that the impulse given to a nucleon in the high-energy event is much larger than momentum changes induced by the action of the interaction between that nucleon and the others in the target. We do not mean by "impulse approximation" a restriction to one scattering event, since we retain initial and final multiple-scattering distortions in Eq. 109.

Equation 109 represents a central, useful result for calculation of inelastic reactions, which is usually obtained in one form or another in any theory of such processes. Its main assumption is that inelastic reactions proceed through many scattering steps that distort the projectile wave but basically leave the nucleus in the same state, and through only one step, which involves the excitation or "hard" inelastic scattering, in question. Such an assumption need not be correct (and in a moment we explore corrections to it—see Eq. 112), but is likely to have some validity for high energy scattering in forward directions, that is, for large systems.

We can now proceed to exhibit corrections to the lowest-order DWIA result of Eq. 109. To do so, we again iterate the system of Eqs. 98–101 by retaining the last term in Eq. 99, into which we insert

$$\langle \nu | T' | 0 \rangle = \langle \nu | \mathfrak{I}' | 0 \rangle + \langle \nu | \mathfrak{I}' | 0 \rangle \, G_0(0) \langle 0 | T' | 0 \rangle$$
$$+ \langle \nu | \mathfrak{I}' | \nu \rangle \, G_0(\nu) \langle \nu | T' | 0 \rangle$$
$$+ \sum_{\mu \neq 0, \nu} \langle \nu | \mathfrak{I}' | \mu \rangle \, G_0(\mu) \langle \mu | T' | 0 \rangle, \qquad [110]$$

*Note that the final state $|\chi_{\mathbf{k}'}^{(-)}\rangle$ is distorted by the *conjugate* of the complex potential $\langle \alpha | \mathfrak{I}' | \alpha \rangle$; compare Sch–68b. Moreover, due to the distortion by complex energy-dependent potentials neither the set $|\psi_{\mathbf{k}}\rangle$ nor the set $|\chi_{\mathbf{k}'}^{(-)}\rangle$ is necessarily orthonormal.

with the last term of Eq. 110—which is threefold nondiagonal in Eq. 99—again neglected; its inclusion would lead to second-order corrections in the optical potential distorting the meson waves and to third-order corrections in the transition operator. Then, from Eq. 110,

$$\langle \nu | T' | 0 \rangle \cong [1 - \langle \nu | \mathcal{T}' | \nu \rangle \, G_0(\nu)]^{-1} \langle \nu | \mathcal{T}' | 0 \rangle [1 + G_0(0) \langle 0 | T' | 0 \rangle],$$

$$[111]$$

which inserted into Eq. 99 yields

$$\langle \alpha | T' | 0 \rangle \cong [1 - \langle \alpha | \mathcal{T}' | \alpha \rangle \, G_0(\alpha)]^{-1}$$

$$\times \left\{ \langle \alpha | \mathcal{T}' | 0 \rangle + \sum_{\nu \neq 0, \alpha} \langle \alpha | \mathcal{T}' | \nu \rangle \frac{1}{E - E_\nu^{\text{nucl}} - K - \langle \nu | \mathcal{T}' | \nu \rangle} \right.$$

$$\left. \times \langle \nu | \mathcal{T}' | 0 \rangle \right\} [1 + G_0(0) \langle 0 | T' | 0 \rangle].$$

$$[112]$$

In this generalization of our previous key result of Eq. 109, the initial and final factors have the same interpretation in terms of asymmetrically distorted meson waves as before. But the transition operator now has a second-order term involving two-step, doubly nondiagonal encounters between which the meson propagates in a medium that distorts its wave through the lowest-order optical potential with approximately the propagation $G_0^{(1)}(\nu)$ of Eq. 74. This scheme can then be iterated in an obvious way, though in practice it is exceedingly difficult to take into account corrections beyond those already exhibited here. It also has a natural generalization to processes—such as the (π^+, π^-) reaction—which must of necessity involve two-step mechanisms already in their lowest order. We note further that the corrections to elastic scattering, as in $\mathcal{V}_c^{(2,0)}$ of Eq. 71, involve comparing a product of two nondiagonal matrix elements to a diagonal one. For inelastic processes, the corrections immediately compare two nondiagonal terms to a *non*diagonal one so that the approximation scheme is intrinsically less promising.

Last, we mention that the antisymmetrization prescriptions can be removed from the transition operator of Eq. 112; in the diagonal, wave-distorting factors, we have already examined the consequences of these prescriptions in the context of elastic scattering. To see this, we use Eqs.

58, 53, 86, and 90 to write the brace in Eq. 112 as

$$\{\cdots\} = \frac{A-1}{A}\langle\alpha|\sum_{i=1}^{A}[t_i + t_i(G_0\mathcal{Q} - G_0)\tau_i]|0\rangle$$

$$+\left(\frac{A-1}{A}\right)^2 \sum_{\nu\neq 0,\alpha}\langle\alpha|\sum_{i=1}^{A}\tau_i|\nu\rangle\frac{1}{E-E_\nu^{\mathrm{nucl}}-K-\langle\nu|\mathcal{J}'|\nu\rangle}\langle\nu|\sum_{j=1}^{A}\tau_j|0\rangle$$

$$\cong \frac{A-1}{A}\langle\alpha|\sum_{i=1}^{A}t_i|0\rangle + \frac{A-1}{A^2}\langle\alpha|\sum_{i=1}^{A}t_iG_0\sum_{j=1}^{A}t_j|0\rangle$$

$$-\frac{A-1}{A}\langle\alpha|\sum_{i=1}^{A}t_iG_0t_i|0\rangle$$

$$+\left(\frac{A-1}{A}\right)^2 \sum_{\nu\neq 0,\alpha}\langle\alpha|\sum_{i=1}^{A}t_i|\nu\rangle\frac{1}{E-E_\nu^{\mathrm{nucl}}-K-\langle\nu|\mathcal{J}'|\nu\rangle}\langle\nu|\sum_{j=1}^{A}t_j|0\rangle,$$

$$[113]$$

where we have approximated τ_i by the bound-nucleon amplitude t_i of Eq. 53 in the second-order term, leaving further differences to third order, and have exploited the antisymmetry of $|\alpha\rangle$ or $|0\rangle$ and

$$\mathcal{Q}\tau_i\mathcal{Q} = \frac{1}{A}\sum_{j=1}^{A}\tau_j\mathcal{Q}. \qquad [114]$$

We further ignore the distortion of the meson wave in the intermediate propagation in Eq. 113, explicitly subtract out the forbidden states $\nu=0,\alpha$, and use closure, to get

$$\{\cdots\} \cong \frac{A-1}{A}\langle\alpha|\sum_{i=1}^{A}t_i|0\rangle$$

$$+\left[\frac{A-1}{A}\langle\alpha|\sum_{i=1}^{A}\sum_{j\neq i}t_iG_0t_j|0\rangle\right.$$

$$\left.-\left(\frac{A-1}{A}\right)^2\sum_{\nu=0,\alpha}\langle\alpha|\sum_{i=1}^{A}t_i|\nu\rangle G_0(\nu)\langle\nu|\sum_{j=1}^{A}t_j|0\rangle\right].$$

$$[115]$$

The antisymmetry of $|0\rangle$ and $|\alpha\rangle$ and the use of Eq. 90 allows us to write finally

$$\langle \phi_{k'}; \alpha | T | \phi_k; 0 \rangle \cong \langle \chi_{k'}^{(-)}; \alpha | \left[\sum_{i=1}^{A} t_i + \sum_{i=1}^{A} \sum_{j \neq i} t_i G_0 (1 - P_0 - P_\alpha) t_j \right] | \psi_k; 0 \rangle,$$

[116]

where P_0 and P_α are projection operators for the state indexed. The result of Eq. 116 has eliminated the intermediate antisymmetry instruction with a corresponding compensation in the second-order term. The possibility of removing the antisymmetrizing $\mathcal{C}$-instruction is present also in the elastic case, as can easily be seen in the same manner as here, but beginning with Eq. 93 to second order. (This was first discussed in an appendix to Fes–71, where it was also shown that the distorting term $\langle \nu | \mathcal{T}' | \nu \rangle$ in the intermediate propagation of Eq. 113 can be taken into account exactly.) We again caution the reader that all of these manipulations involving the instructions $\mathcal{C}$ and $i \neq j \neq k \dots$, though they appear to involve innocent differences of order $1/A$, actually go to fairly deep aspects of the structure of the theory. They must be treated carefully for consistency, and, especially where inelastic processes are present that can select out "preferred" nucleons, they can make sizeable numerical differences in calculations. In a language of schematic diagrams, for example, as in Fig. 4.5, it is easy to see that the effect of the antisymmetrization in the single-scattering term must be to introduce double-scattering features there through the exchange, so that only a combined, consistent treatment of both these terms can be correct. The fact that equivalent results can be obtained from consistent application of the theory *with* or *without* antisymmetrization of intermediate states provided the initial and final states are properly antisymmetrized is merely the manifestation here of a well-known general principle for Feynman diagrams, many-body theories, and the like. It is well to bear in mind that if the formalism *without* the $\mathcal{C}$-instruction is chosen, then consideration may *not* be restricted to antisymmetric intermediate states on the grounds that only these are "physical," and the practitioner must steel himself to include nuclear configurations from

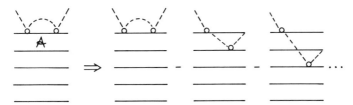

Fig. 4.5 The effect of antisymmetrizing among the target nucleons in an intermediate state subtracts exchange diagrams that include double-scattering processes.

which any right-thinking person will instinctively—but incorrectly—recoil with horror. Conversely, if the $\mathcal{C}$-formalism is used, then its consequences in τ_i (Eq. 86) *must* also be explicitly considered.

4.3 Applications of the Watson Theory: Explicit Forms and Estimates of Convergence

4.3a More Explicit Expressions for the Lowest-Order Optical Potential

In order to make use of the formal results of the previous section, we wish first to dress them in more explicit and detailed forms and then to make estimates of the rate of convergence of the expansion of the optical potential in orders of clusters and of the validity of approximations for the πN amplitude in the nuclear medium. We have seen in Section 4.2 that the first-order result for the optical potential is, in its structure, essentially independent of the particular formalism selected. From Eqs. 67, 84, or 95, it is given by

$$\mathcal{V}_c^{(1)} = \langle 0| \sum_{i=1}^{A} t_i(E)|0\rangle. \qquad [117]$$

This must be evaluated for an appropriate choice of nuclear ground state wave functions, and then used in a one-particle Schroedinger equation —or, completely equivalently, a Lippmann-Schwinger equation—in the projectile variables for the meson-nucleus elastic scattering amplitude. Let us first consider doing this in the momentum space of the projectile variable; thus we seek

$$\langle \mathbf{k}', \mathbf{P}'|\mathcal{V}_c^{(1)}|\mathbf{k}, \mathbf{P}\rangle = \sum_{i=1}^{A} \int \frac{d\mathbf{p}_1}{(2\pi)^3} \frac{d\mathbf{p}_2}{(2\pi)^3} \cdots \frac{d\mathbf{p}_i}{(2\pi)^3} \frac{d\mathbf{p}_i'}{(2\pi)^3} \cdots \frac{d\mathbf{p}_A}{(2\pi)^3}$$
$$\times \Phi_0^\dagger(\mathbf{p}_1, \mathbf{p}_2, \ldots, \mathbf{p}_i', \ldots, \mathbf{p}_A)\langle \mathbf{k}', \mathbf{p}_i'|t_i(E)|\mathbf{k}, \mathbf{p}_i\rangle$$
$$\times \Phi_0(\mathbf{p}_1, \mathbf{p}_2, \ldots, \mathbf{p}_i, \ldots, \mathbf{p}_A) \qquad [118]$$

for pion and nucleus of initial and final momentum $\mathbf{k}, \mathbf{P}, \mathbf{k}', \mathbf{P}'$, where we have made explicit the fact that the relevant (free) scattering on each nucleon takes place with momentum values appropriate to that nucleon. The nucleon momenta are ultimately averaged in Eq. 118 with the weighting of nuclear momentum space wave functions. Momentum is, of course, conserved in the encounter with the ith nucleon when it is considered as free, so

$$\langle \mathbf{k}', \mathbf{p}_i'|t_i|\mathbf{k}, \mathbf{p}_i\rangle = (2\pi)^3 \delta(\mathbf{k}' + \mathbf{p}_i' - \mathbf{k} - \mathbf{p}_i)t_i(\mathbf{k}', \mathbf{k}; \mathbf{p}_i; E) \qquad [119]$$

(cf. Eq. 2.22), which we envision using to eliminate $\mathbf{p}'_i$. The resulting expression,

$$\langle \mathbf{k}', \mathbf{P}' | \mathcal{V}_c^{(1)} | \mathbf{k}, \mathbf{P} \rangle = \sum_{i=1}^{A} \int \frac{d\mathbf{p}_1}{(2\pi)^3} \cdots \frac{d\mathbf{p}_i}{(2\pi)^3} \cdots \frac{d\mathbf{p}_A}{(2\pi)^3}$$

$$\Phi_0^\dagger(\mathbf{p}_1, \ldots, \mathbf{p}_i + \mathbf{k} - \mathbf{k}', \ldots, \mathbf{p}_A) t_i(\mathbf{k}', \mathbf{k}; \mathbf{p}_i; E) \Phi_0(\mathbf{p}_1, \ldots, \mathbf{p}_i, \ldots, \mathbf{p}_A),$$

$$[120]$$

then gives us a complete kinematical averaging over the nucleon motion, which can in principle be carried out in applying our formalism to meson-nucleus scattering.

If for the moment we ignore nucleon spin and isospin effects, the results of Eqs. 118–120 can be cast in the form

$$\langle \mathbf{k}', \mathbf{P}' | \mathcal{V}_c^{(1)} | \mathbf{k}, \mathbf{P} \rangle = \int \rho(\mathbf{p}', \mathbf{p}) \langle \mathbf{k}', \mathbf{p}' | t(E) | \mathbf{k}, \mathbf{p} \rangle \frac{d\mathbf{p}'}{(2\pi)^3} \frac{d\mathbf{p}}{(2\pi)^3}, \qquad [121]$$

where the nuclear momentum-space single-nucleon density is

$$\rho(\mathbf{p}', \mathbf{p}) \equiv \sum_{i=1}^{A} \int \Phi_0^\dagger(\mathbf{p}_1, \ldots, \mathbf{p}_{i-1}, \mathbf{p}', \mathbf{p}_{i+1}, \ldots, \mathbf{p}_A)$$

$$\times \Phi_0(\mathbf{p}_1, \ldots, \mathbf{p}_{i-1}, \mathbf{p}, \mathbf{p}_{i+1}, \ldots, \mathbf{p}_A)$$

$$\times \frac{d\mathbf{p}_1}{(2\pi)^3} \cdots \frac{d\mathbf{p}_{i-1}}{(2\pi)^3} \frac{d\mathbf{p}_{i+1}}{(2\pi)^3} \cdots \frac{d\mathbf{p}_A}{(2\pi)^3}. \qquad [122]$$

This result makes the nature of the averaging over the nucleon momenta particularly explicit. It also makes it clear that there are several possible sources of nonlocality* in the first-order optical potential when we carry out a Fourier transform to write it in configuration space. Such nonlocalities must occur whenever there is dependence in Eq. 122 on $\mathbf{k}$ and $\mathbf{k}'$ other than in the combination $\mathbf{k} - \mathbf{k}'$. The sources of nonlocality include many forms of explicit dependence on the meson momenta $\mathbf{k}$ and $\mathbf{k}'$ in the πN center-of-mass system (leading, for example, to the Kisslinger potential discussed in the next chapter—see Eqs. 5.5–5.8); dependence on the nucleon momenta $\mathbf{p}$ and $\mathbf{p}'$ is then induced by the transformation from the πN c.m. system to the particular meson-nucleon frame for each nucleon

---

*For the consequences of such nonlocalities on the consistency of the fixed-scatterer approximation, see Ben-76.

(cf. Eqs. 5.22–5.24), whereas dependence on meson momentum is induced through the energy dependence of t in Eq. 121 (also discussed in Chapter 5; see Eqs. 5.29–5.38).

For convenience, it is sometimes well to introduce a further approximation: At medium energies the amplitude t_i may have relatively little dependence on $\mathbf{p}_i$ simply because of the high momentum $\mathbf{k}$ of the incident projectile. For pions on nuclei in the $3,3$ resonance region this is *not* the case, since $k_{\text{res}} \sim 304$ MeV/c while $p_i \lesssim p_F \lesssim 270$ MeV/c, where p_F is the nuclear Fermi momentum. But the operative combination must be $\mathbf{k} - (\omega/M)\mathbf{p}_i \propto \mathbf{v}_\pi - \mathbf{v}_N$, the pion-nucleon relative velocity, in order to achieve Galilean invariance. Thus the nucleon momentum effects are suppressed by a factor $\omega/M \sim \frac{1}{3}$ for pion total energy ω appropriate to the $3,3$ region. We may exploit this fact to drop* the dependence of the amplitude on nucleon momentum, using for this the amplitude for a free nucleon at rest, that is, in the laboratory frame. The dependence on spin and isospin degrees of freedom may remain,

$$\langle \mathbf{k}', \mathbf{p}'|t_i|\mathbf{k}, \mathbf{p}\rangle = \langle \mathbf{k}'|t_i|\mathbf{k}\rangle = \sum_\alpha \langle \mathbf{k}'|t^\alpha|\mathbf{k}\rangle O_i^\alpha$$

$$= \langle \mathbf{k}'|t^{ss}|\mathbf{k}\rangle + \langle \mathbf{k}'|t^{vs}|\mathbf{k}\rangle \, \boldsymbol{\sigma}_i \cdot \hat{\mathbf{n}}$$

$$+ \langle \mathbf{k}'|t^{sv}|\mathbf{k}\rangle \vec{I} \cdot \vec{\tau}_i + \langle \mathbf{k}'|t^{vv}|\mathbf{k}\rangle \, \boldsymbol{\sigma}_i \cdot \hat{\mathbf{n}} \vec{I} \cdot \vec{\tau}_i \quad [123]$$

(cf. Eqs. 2.10 and 2.16b), where $\hat{\mathbf{n}} \equiv \overline{\mathbf{k}' \times \mathbf{k}}$, $\vec{I}$ and $\vec{\tau}$ are meson and nucleon isospin operators, and $\boldsymbol{\sigma}_i$ is the nucleon spin operator (for nucleon projectiles, projectile spin also enters, of course). We have decomposed in terms of nucleon spin and isospin scalar and vector components. Then

$$\langle \mathbf{k}', \mathbf{P}'|\mathcal{V}_c^{(1)}|\mathbf{k}, \mathbf{P}\rangle = A \sum_\alpha \langle \mathbf{k}'|t^\alpha|\mathbf{k}\rangle \rho^\alpha(\mathbf{k} - \mathbf{k}'), \quad [124]$$

where

$$\rho^\alpha(\mathbf{k} - \mathbf{k}') \equiv \frac{1}{A} \sum_{i=1}^A \int \frac{d\mathbf{p}_1}{(2\pi)^3} \cdots \frac{d\mathbf{p}_A}{(2\pi)^3} \Phi_0^\dagger(\mathbf{p}_1, \ldots, \mathbf{p}_i + \mathbf{k} - \mathbf{k}', \ldots, \mathbf{p}_A)$$

$$\times O_i^\alpha \Phi_0(\mathbf{p}_1, \ldots, \mathbf{p}_i, \ldots, \mathbf{p}_A)$$

$$= \int e^{i(\mathbf{k} - \mathbf{k}') \cdot \mathbf{r}} \rho^\alpha(\mathbf{r}) \, d\mathbf{r}. \quad [125]$$

---

*Alternatively, this dependence on $\mathbf{p}_i$ may be treated by expanding in a power series (see for example Mac–72, 73 and the discussion in Section 5.1). If p_i is measured by the nuclear Fermi momentum, then the resulting imprecision in k is on the order of 50 MeV/c to 70 MeV/c, which is less than or comparable to uncertainties arising from the width of the $3,3$ resonance. Terms of the form $(\omega/M)\,\mathbf{p}_i$ also have an impact on inelastic πA scattering as discussed at the end of Section 5.5.

Here,

$$\rho^\alpha(\mathbf{r}) \equiv \frac{1}{A} \sum_{i=1}^{A} \int d\mathbf{r}_1 \cdots d\mathbf{r}_A \, \Phi_0^\dagger(\mathbf{r}_1, \ldots, \mathbf{r}_A) \delta(\mathbf{r} - \mathbf{r}_i) O_i^\alpha \Phi_0(\mathbf{r}_1, \ldots, \mathbf{r}_A)$$

[126]

are the nuclear spin/isospin scalar and vector densities, normalized such that

$$\int \rho^\alpha(\mathbf{r}) \, d\mathbf{r} = \frac{1}{A} \langle \sum_{i=1}^{A} O_i^\alpha \rangle.$$

[127]

In particular, for the scalar density,

$$\int \rho^{ss}(\mathbf{r}) \, d\mathbf{r} = 1.$$

[128]

For spin and isospin saturated targets, which are almost always a good approximation (order $1/A$) for heavier nuclei,

$$\int \rho^\alpha(\mathbf{r}) \, d\mathbf{r} \approx O(1/A), \qquad \alpha = vs, sv, vv.$$

[129]

Note also that from these results, for example, Eq. 125, the single-scattering can only support momentum transfers $|\mathbf{k} - \mathbf{k}'| \lesssim 1/R$, with R the nuclear radius. This is expressed in $\rho^\alpha(\mathbf{k} - \mathbf{k}')$ as the fall-off in momentum space at large arguments of the Fourier transform of the density, or, in terms of the physics in configuration space, as the interference between the unperturbed wavelets emerging from the various points in the medium. This interference is stronger for the less coherent waves of larger momentum-transfer scattering. Note that, as in Eqs. 124–126, we can always anticipate a factor $e^{i(\mathbf{k}-\mathbf{k}')\cdot\mathbf{r}_i}$ accompanying each $\langle \mathbf{k}'|t_i|\mathbf{k} \rangle$; it is useful to bear this in mind for later purposes, as in Eq. 138. Since the spin vector parts of the amplitude cannot contribute for small angles, the preference for forward scattering means that they are unimportant for much of the pertinent range. If we restrict our attention to the scalar density, we have from Eqs. 124 and 125

$$\langle \mathbf{k}', \mathbf{P}' | \mathcal{V}_c^{(1)} | \mathbf{k}, \mathbf{P} \rangle = A \langle \mathbf{k}'|t|\mathbf{k} \rangle \int e^{i(\mathbf{k}-\mathbf{k}')\cdot\mathbf{r}} \rho^{ss}(\mathbf{r}) \, d\mathbf{r}.$$

[130]

Again, in the expectation that most scattering is forward, we can easily read off the configuration space result for the lowest-order optical potential

$$\mathcal{V}_c^{(1)}(\mathbf{r}) = t_0 A \rho(\mathbf{r}) = -\frac{4\pi}{2m} f_0 A \rho(\mathbf{r}), \qquad \int \rho(\mathbf{r}) \, d\mathbf{r} = 1, \qquad [131]$$

exactly as expected from Eq. 14, with t_0 or f_0 the forward transition operator or amplitude; attention to the isospin variables for elastic, non-charge-exchange scattering then easily gives Eq. 16.

Before leaving these first-order results it is worthwhile to re-examine the absorptive feature of this optical potential. We note that

$$\operatorname{Im} \mathcal{V}_c^{(1)}(\mathbf{r}) = A \rho(\mathbf{r}) \cdot \operatorname{Im} t_0. \qquad [132]$$

The fact that the absorptive part of the optical potential arises from the imaginary part of the πN amplitude immediately suggests that optical absorption comes about because of incoherent processes, presumably chiefly nucleon knockout. The imaginary part of the amplitude, easily obtained from the Lippmann-Schwinger equation in the form*

$$t = v + v \frac{1}{E_\pi - K - v + i\eta} v, \qquad [133]$$

is, for a real interaction,

$$
\begin{aligned}
\operatorname{Im} t_0 = \operatorname{Im} \langle \mathbf{k} | t(E_\pi) | \mathbf{k} \rangle &= -\pi \langle \mathbf{k} | v \, \delta(E_\pi - K - v) v | \mathbf{k} \rangle \\
&= -\pi \sum_{\mathbf{k}'} |\langle \mathbf{k} | v | \psi_{\mathbf{k}'}^{(+)} \rangle|^2 \delta(E_\pi - E_{\mathbf{k}'}) \\
&= -\pi \sum_{\mathbf{k}'} |\langle \mathbf{k} | t | \mathbf{k}' \rangle|^2 \delta(E_\pi - E_{\mathbf{k}'}), \qquad [134]
\end{aligned}
$$

where $\psi_{\mathbf{k}'}^{(+)}, E_{\mathbf{k}'}$ refer to the perturbed projectile states, whose scattering spectrum $E_{\mathbf{k}'}$ starts naturally from $E_{\mathbf{k}'} = 0$; hence Eq. 134 has a (negative) contribution for all positive E_π. This contribution is well known to be given by unitarity or the optical theorem (cf. Eq. 15) in terms of the total cross section, whose threshold is at zero energy so that the threshold for absorption is also at $E_\pi = 0$. In reality, of course, there can be no absorption until the projectile has sufficient energy to raise the target out of its ground state. This is an infinitesimal demand for nuclear matter, but not for real nuclei. The trouble arises from the use of the impulse approximation, without which Eqs. 133 and 134 would be replaced according to Eq.

*Here we ignore nucleon motion: $K = K_\pi$; see Eq. 56 and note on p. 132.

85 or Eq. 97:

$$t' = v + v \frac{1}{E - K - H_N - v + i\eta} (1 - P_0) v \qquad [135]$$

and

$$\operatorname{Im} t'_0 = \operatorname{Im} \langle \mathbf{k}0 | t'(E) | \mathbf{k}0 \rangle = -\pi \langle \mathbf{k}0 | v \delta(E - K - H_N - v)(1 - P_0) v | \mathbf{k}0 \rangle$$

$$= -\pi \sum_{\mathbf{k}', \alpha \neq 0} |\langle \mathbf{k}0 | v | \psi_{\mathbf{k}'}^{(+)} \alpha \rangle|^2 \delta(E - E_{\mathbf{k}'} - E_\alpha^{\text{nucl}}). \qquad [136]$$

Here the $\operatorname{Im} t'_0$ differs as expected from zero for $E > E_1^{\text{nucl}} - E_0^{\text{nucl}}$, the excitation of the first excited nuclear state. Note that for a strictly correct result here the $(1 - P_0)$ prescription is essential,* though, of course, for all practical purposes the medium-energy theory is applied in a domain where there is always plenty of energy available to excite the nucleus; thus the formal concern may not enter severely. Furthermore, for meson interactions there is an imaginary part to the optical potential even at zero energies due to the possibility of true absorption, which we have not thus far included (see Chapters 6 and 7).

4.3b Higher Orders in the Optical Potential

Now we turn to exhibiting the higher-order terms in the optical potential, examining their structure, properties and, in particular, convergence features, and considering some of the underlying, deeper limitations of our present approach. We start with the second-order term, which has the dual virtue of possessing an essentially equivalent form for the various possible choices of formalism in Section 4.2 and of allowing an approximate, closed-form estimate. From Eq. 84 in formalism 0 the second-order optical potential in momentum space with use of the impulse approximation[†] is

$$\langle \mathbf{k}' | \mathcal{V}_c^{(2,0)} | \mathbf{k} \rangle = \langle \mathbf{k}', 0 | \mathcal{V}^{(2,0)} | \mathbf{k}, 0 \rangle$$

$$= \sum_{i=1}^{A} \sum_{j \neq i} \sum_{\alpha \neq 0} \int \langle \mathbf{k}', 0 | t_i | \mathbf{k}'', \alpha \rangle \frac{1}{E - E_\pi(k'') - E_\alpha^{\text{nucl}} + i\eta}$$

$$\times \langle \mathbf{k}'', \alpha | t_j | \mathbf{k}, 0 \rangle \frac{d\mathbf{k}''}{(2\pi)^3}, \qquad [137]$$

---

*A complete investigation also requires consideration of the second-order term in the optical potential, which guarantees the appearance of only physical, antisymmetrized nuclear states in Eq. 136.

[†]We use this language exclusively to refer to the approximation $t \approx t^{\text{free}}$ discussed at the beginning of Section 4.3 and *not* to mean—as was done in some of the older literature—the further assumption of only single scattering, under which this sentence would contain a contradiction in terms.

where we have suppressed the uninteresting labels of total nuclear momentum $\mathbf{P}$ and $\mathbf{P}'$. We first use a result analogous to those in Eqs. 118–126 to put in evidence the plane wave factors attached to each amplitude t_i in nuclear $\mathbf{r}$-space, and then define a mean nuclear excitation energy $\overline{E^{\text{nucl}}}$ such that, with this replacing E_α^{nucl} in Eq. 137, closure there will be valid in the sum over the nuclear states α. Then in configuration space

$$
\langle \mathbf{k}' | \mathcal{V}_c^{(2,0)} | \mathbf{k} \rangle = \int \Phi_0^\dagger(\mathbf{r}_1,\ldots,\mathbf{r}_A) \sum_{i=1}^A \sum_{j \neq i} \langle \mathbf{k}' | t_i | \mathbf{k}'' \rangle
$$

$$
\times \frac{e^{i(\mathbf{k}''-\mathbf{k}')\cdot\mathbf{r}_i} e^{i(\mathbf{k}-\mathbf{k}'')\cdot\mathbf{r}_j}}{\left(E - \overline{E^{\text{nucl}}}\right) - E_\pi(k'') + i\eta}(1 - P_0)
$$

$$
\times \langle \mathbf{k}'' | t_j | \mathbf{k} \rangle \Phi_0(\mathbf{r}_1,\ldots,\mathbf{r}_A)\,d\mathbf{r}_1\ldots d\mathbf{r}_A \frac{d\mathbf{k}''}{(2\pi)^3}. \qquad [138]
$$

The meaning and validity of the closure argument that produced $\overline{E^{\text{nucl}}}$ is discussed at the end of this subsection; for the moment it suffices to say that the argument is only useful if $\overline{E^{\text{nucl}}}$ is known through some simple, physical argument, and it then presumably involves some fairly low nuclear excitation energy. We attempt to make our discussion more transparent at this stage by retaining only spin/isospin scalar parts of the amplitudes t, bearing in mind that we can always generalize our results to include spin and isospin, as in Eqs. 123–126. With this simplification, Eq. 138 becomes

$$
\langle \mathbf{k}' | \mathcal{V}_c^{(2,0)} | \mathbf{k} \rangle = A(A-1)\int \langle \mathbf{k}' | t | \mathbf{k}'' \rangle \frac{e^{i(\mathbf{k}''-\mathbf{k}')\cdot\mathbf{r}'} e^{i(\mathbf{k}-\mathbf{k}'')\cdot\mathbf{r}}}{\left(E - \overline{E^{\text{nucl}}}\right) - E_\pi(k'') + i\eta} \frac{d\mathbf{k}''}{(2\pi)^3}
$$

$$
\times \langle \mathbf{k}'' | t | \mathbf{k} \rangle \left[P^{(2)}(\mathbf{r}',\mathbf{r}) - \rho(\mathbf{r}')\rho(\mathbf{r}) \right] d\mathbf{r}'\,d\mathbf{r}, \qquad [139]
$$

where the second term in the brackets represents the subtraction of the nuclear ground state contribution. The first term in brackets is the nuclear two-particle correlation function

$$
P^{(2)}(\mathbf{r}',\mathbf{r}) \equiv \frac{1}{A(A-1)} \int \Phi_0^\dagger(\mathbf{r}_1,\ldots,\mathbf{r}_A) \sum_{i=1}^A \sum_{j \neq i} \delta(\mathbf{r}'-\mathbf{r}_i)\delta(\mathbf{r}-\mathbf{r}_j)
$$

$$
\times \Phi_0(\mathbf{r}_1,\ldots,\mathbf{r}_A)\,d\mathbf{r}_1\ldots d\mathbf{r}_A, \qquad [140]
$$

representing the probability of finding one nucleon at $\mathbf{r}$ and a second at $\mathbf{r}'$.

The normalization of the correlation function is such that

$$\int P^{(2)}(\mathbf{r}',\mathbf{r})\,d\mathbf{r} = \frac{1}{A}\int \Phi_0^\dagger(\mathbf{r}_1,\ldots,\mathbf{r}_A)\sum_{i=1}^{A}\delta(\mathbf{r}'-\mathbf{r}_i)\Phi_0(\mathbf{r}_1,\ldots,\mathbf{r}_A)\,d\mathbf{r}_1\ldots d\mathbf{r}_A$$

$$\equiv \rho(\mathbf{r}') \qquad\qquad [141]$$

and

$$\int P^{(2)}(\mathbf{r}',\mathbf{r})\,d\mathbf{r}\,d\mathbf{r}' = \int \rho(\mathbf{r}')\,d\mathbf{r}' = 1. \qquad [142]$$

Equation 139 now shows explicitly that we are dealing with an expansion in orders of correlation. We can anticipate that the third-, fourth-,... order terms in the optical potential involve triple-, quadruple-,... correlations. Naturally, our hope is that at some point higher-order correlations become unimportant so that our series for the optical potential can be truncated to a good approximation. Although we exhibit the third-order term, for example, only somewhat later on, it requires little imagination on the part of the reader to anticipate that it is exceedingly difficult to evaluate, both technically—from the expression for $\mathcal{V}_c^{(3)}$ itself—and because we know very little about three-nucleon correlations in nuclei (or, indeed, about anything beyond one-nucleon distributions, i.e., densities). Thus we wish to make an estimate of the second-order effect not only for its own sake but also to give us some inkling as to the conditions under which our optical potential series may be reasonably truncated. To accomplish this, we first restrict ourselves to forward scattering $\langle\mathbf{k}|t|\mathbf{k}\rangle\equiv t_0$ on the usual grounds that as we move to larger momentum transfers the cross section will drop. We also introduce for convenience a function $C(\mathbf{r}',\mathbf{r})$ to describe the correlations, such that

$$\rho(\mathbf{r}')\rho(\mathbf{r})C(\mathbf{r}',\mathbf{r}) = P^{(2)}(\mathbf{r}',\mathbf{r}) - \rho(\mathbf{r}')\rho(\mathbf{r}); \qquad [143]$$

since strong, repulsive correlations, for example, would require $P^{(2)}(\mathbf{r},\mathbf{r})=0$ by forbidding two nucleons to be at the same point, we would have under those conditions $C(\mathbf{r},\mathbf{r})=-1$. Thus we have

$$\langle\mathbf{k}'|\mathcal{V}_c^{(2,0)}|\mathbf{k}\rangle \cong A(A-1)t_0^2 \int e^{-i\mathbf{k}'\cdot\mathbf{r}'}\left[\int \frac{e^{i\mathbf{k}''\cdot(\mathbf{r}'-\mathbf{r})}}{\dfrac{p^2}{2m}-\dfrac{k''^2}{2m}+i\eta}\frac{d\mathbf{k}''}{(2\pi)^3}\right]$$

$$\times e^{i\mathbf{k}\cdot\mathbf{r}}\rho(\mathbf{r}')\rho(\mathbf{r})C(\mathbf{r}',\mathbf{r})\,d\mathbf{r}'\,d\mathbf{r}, \qquad [144]$$

where we have also introduced explicitly the nonrelativistic kinematics for the meson energy, and

$$\frac{p^2}{2m} \equiv E - \overline{E^{\text{nucl}}} \qquad [145]$$

is the propagation energy available for the intermediate meson.

The propagator in the square bracket $G(\mathbf{r}',\mathbf{r})$ is perfectly susceptible to evaluation for its configuration space form, leading to the well-known

$$G(\mathbf{r}',\mathbf{r}) = -\frac{2m}{4\pi} \frac{e^{ip|\mathbf{r}'-\mathbf{r}|}}{|\mathbf{r}'-\mathbf{r}|}, \quad p = \sqrt{2m\left(E - \overline{E^{\text{nucl}}}\right)}, \qquad [146]$$

but this structure proves slightly inconvenient for purposes of our estimate of $\mathcal{V}_c^{(2,0)}$. Instead, we base our rough evaluation on the assumption that we are primarily interested in fairly high energies, for which our form factors guarantee predominantly forward scattering, and for which our propagator is reasonably approximated by

$$G(\mathbf{r}',\mathbf{r}) = \int \frac{e^{i\mathbf{k}''\cdot(\mathbf{r}'-\mathbf{r})}}{\dfrac{p^2}{2m} - \dfrac{k''^2}{2m} + i\eta} \frac{d\mathbf{k}''}{(2\pi)^3} = \int \frac{e^{i\mathbf{k}''\cdot(\mathbf{r}'-\mathbf{r})}}{\dfrac{1}{2m}(\mathbf{p}+\mathbf{k}'')\cdot(\mathbf{p}-\mathbf{k}'') + i\eta} \frac{d\mathbf{k}''}{(2\pi)^3}$$

$$\cong \int \frac{e^{i\mathbf{k}''\cdot(\mathbf{r}'-\mathbf{r})}}{\mathbf{v}\cdot(\mathbf{p}-\mathbf{k}'') + i\eta} \frac{d\mathbf{k}''}{(2\pi)^3}, \qquad [147]$$

where, in the last step, we have assumed that, due to the high-energy nature of the scattering, the main contribution in the integral comes from forward directions and on-shell scattering $\mathbf{k}'' \approx \mathbf{p} = m\mathbf{v}$. The direction of $\mathbf{p}$ or $\mathbf{v}$ is assumed to be given by that of $\frac{1}{2}(\mathbf{k}+\mathbf{k}')$, the average initial and final meson momenta (which are anyway supposed to be nearly parallel here). We take this direction to be the z-axis for purposes of integrating the approximate, linearized propagator

$$G(\mathbf{r}',\mathbf{r}) \cong v^{-1} \int \frac{e^{i\mathbf{k}_\perp''\cdot(\mathbf{b}'-\mathbf{b})}}{(p - k_z'') + i\eta} e^{ik_z''(z'-z)} \frac{d^2\mathbf{k}_\perp'' \, dk_z''}{(2\pi)^3}$$

$$= -iv^{-1}\delta^{(2)}(\mathbf{b}'-\mathbf{b})e^{ip(z'-z)}\theta(z'-z), \qquad [148]$$

where $\mathbf{k}_\perp''$ refers to the component of $\mathbf{k}''$ perpendicular to the z-axis or $\mathbf{v}$, k_z'' is parallel to it, and, similarly for $\mathbf{b}'$, $\mathbf{b}$ and z',z, the corresponding components of $\mathbf{r}'$ and $\mathbf{r}$. The result of Eq. 148 is the propagator in the eikonal, high-energy approximation (which we find to be useful in the context of Glauber theory in Section 4.4). Before inserting it into our expression for the second-order optical potential, we note

$$\mathbf{k}\cdot\mathbf{r}-\mathbf{k}'\cdot\mathbf{r}' = \tfrac{1}{2}(\mathbf{k}+\mathbf{k}')\cdot(\mathbf{r}-\mathbf{r}')+(\mathbf{k}-\mathbf{k}')\cdot\tfrac{1}{2}(\mathbf{r}+\mathbf{r}')$$
$$= p(z-z')+(\mathbf{k}-\mathbf{k}')\cdot\tfrac{1}{2}(\mathbf{b}+\mathbf{b}'), \qquad [149]$$

where we have assumed $\overline{E^{\text{nucl}}}\ll E$ so that $p\approx k=k'$. Then Eq. 144, with Eqs. 148 and 149, gives

$$\langle \mathbf{k}'|\mathcal{V}_c^{(2,0)}|\mathbf{k}\rangle \approx -iv^{-1}A(A-1)t_0^2\int d^2\mathbf{b}\int_{-\infty}^{\infty}dz\, e^{i(\mathbf{k}-\mathbf{k}')\cdot\mathbf{b}}\rho(\mathbf{b},z)$$
$$\times\int_z^{\infty}dz'\,\rho(\mathbf{b},z')C(\mathbf{b},z';\mathbf{b},z). \qquad [150]$$

We now assume the nucleus to be sufficiently large that the correlation function only depends, in the main, on the relative separation between the nucleon pair

$$C(\mathbf{r}',\mathbf{r}) = C(|\mathbf{r}'-\mathbf{r}|), \qquad [151]$$

and not on their position or orientation in the nucleus. Then, assuming the densities ρ do not change rapidly over the small interval in which C has support $[\rho(\mathbf{b},z')\approx\rho(\mathbf{b},z)]$, we obtain

$$\langle \mathbf{k}'|\mathcal{V}_c^{(2,0)}|\mathbf{k}\rangle \approx -iv^{-1}A(A-1)t_0^2\int d^2\mathbf{b}\int_{-\infty}^{\infty}dz\, e^{i(\mathbf{k}-\mathbf{k}')\cdot\mathbf{b}}\rho^2(\mathbf{b},z)R_{\text{corr}},$$
$$[152]$$

where

$$R_{\text{corr}} \equiv \int_0^{\infty} C(\rho)\,d\rho \qquad [153]$$

is defined as the two-particle correlation length.

The result of Eq. 152 is to be compared with the first-order optical potential in Eqs. 130 or 131, for which purposes the configuration space result—obtained by stripping off the Fourier transform integral—is slightly more concise. Then

$$\mathcal{V}_c^{(1)}(r)+\mathcal{V}_c^{(2,0)}(r)\approx \mathcal{V}_c^{(1)}(r)\left[1+\frac{2\pi i}{k}f_0(A-1)\rho(r)R_{\text{corr}}\right]. \qquad [154]$$

Let us estimate the second (correction) term in the brackets for central nuclear densities $A\rho(0)\sim\frac{1}{6}\mathrm{fm}^{-3}$, a correlation length appropriate to a hard-core interaction

$$C(\rho)=\begin{cases} -1 & \rho<\rho_{corr}\approx0.4\ \mathrm{fm} \\ 0 & \rho>\rho_{corr} \end{cases};\qquad [155]$$

that is, $R_{corr}=-0.4$ fm, and the πN 3,3 resonance region, for which $f_0\sim\frac{4}{3}i/k_{res}$, $k=k_{res}\cong304$ MeV$/c$. The bracket is then $[1+0.23]$; that is, at resonance the second-order term as derived from a hard-core correlation effect increases* the optical potential by some 20% or 25%. This would appear to be a rather smaller effect than the correction to the impulse approximation (discussed in the next subsection), the latter arising from rescatterings on the *same* nucleon in the presence of the medium. Our present correction involves, of course, a different nucleon; hence the nucleon density and correlation function are introduced. In fact, since the correction must also involve the scattering amplitude, the form of Eq. 154 is pretty well dictated by physical considerations in advance, apart from the specific numerical coefficient. The appearance of the projectile momentum in the denominator assures us that with higher energies we are less and less in need of higher-order corrections to $\mathcal{V}_c$.

Our numerical estimate here of a 20% correction to the lowest-order optical potential for the 3,3 region is not terribly reliable. This comes about first because the high-energy approximations we have used to make the estimate are really intended for use in regions where many partial waves contribute "smoothly," and not, as here, where one channel dominates completely.† We have allowed ourselves this crudity only because it yields a very convenient, local form for the second-order optical potential; almost any effort to refine this estimate replaces it instantly by a full-blown calculation of a nonlocal potential. Yet another sign of the deficiencies of the estimate lies in Eq. 153, which suggests that if the correlations are such that in this linear integral averaging $R_{corr}=0$ (for example, because attractive medium-range correlations balance repulsive short-range ones), then there are no correlation effects, which is surely an oversimplification. In any event, this rough evaluation is likely to be a serious overestimate, at least for some circumstances, since it uses the central density $A\rho(0)$, which pertains to a region where the nucleus is

---

*Note that Pauli correlations would also tend in general to be negative or repulsive as for the hard-core situation.

†There are indicators (Gar–78) that it may in fact be more appropriate to estimate the higher-order corrections with a properly suppressed Lorentz-Lorenz form, as in Section 5.1e.

highly absorptive (see Section 4.1) and which is made still more absorptive by the second-order effect; thus few mesons will emerge from that part of the nucleus. Instead we see more pions that have scattered nearer the surface where the density is substantially lower and second-order corrections correspondingly less. However, our present method of estimate, which is inappropriate at the nuclear surface, cannot be used reliably to estimate them. For inelastic processes, however, the particular class of events that involve more deeply bound nucleons requires knowledge of the distorted pion wave in the nuclear interior so that our estimate must be seen as a warning for these cases. Last, we remark that so long as we are prepared to content ourselves with eikonal, forward approximations—which may be less true as the meson encounters more nucleons—the two-particle estimate generalizes in a qualitative way for the higher-order terms of the optical potential series, implying that unless there are unusually large many-nucleon correlation features in the nucleus we can, up to a point, content ourselves with the first- and second-order terms. At larger angles, the higher-order terms enter successively, though generally in these regions they are accompanied by other uncertain features in the theory (see Chapter 5).

To complete a systematic exploration of the optical potential series, we must now examine the third-order term, where two new elements enter. In formalism 0 the expression for this is obtained from Eq. 84, in which we again apply closure and assume no spin or isospin dependence in the t's to obtain, in parallel to Eqs. 137–139,

$$\langle \mathbf{k}' | \mathcal{V}_c^{(3,0)} | \mathbf{k} \rangle = \langle \mathbf{k}' | \mathcal{V}_c^{(3,0)} | \mathbf{k} \rangle_{3\text{-particle}} + \langle \mathbf{k}' | \mathcal{V}_c^{(3,0)} | \mathbf{k} \rangle_{2\text{-particle}},$$

$$[156]$$

with the three-particle term

$$\langle \mathbf{k}' | \mathcal{V}_c^{(3,0)} | \mathbf{k} \rangle_{3\text{-particle}} = A(A-1)(A-2) \int \ldots \int e^{-i\mathbf{k}' \cdot \mathbf{r}'} \langle \mathbf{k}' | t | \mathbf{k}''' \rangle$$

$$\times \mathcal{G}_{\mathbf{k}'''}(\mathbf{r}'' - \mathbf{r}') \frac{d\mathbf{k}'''}{(2\pi)^3} \langle \mathbf{k}''' | t | \mathbf{k}'' \rangle$$

$$\mathcal{G}_{\mathbf{k}''}(\mathbf{r}' - \mathbf{r}) \frac{d\mathbf{k}''}{(2\pi)^3} \langle \mathbf{k}'' | t | \mathbf{k} \rangle e^{i\mathbf{k} \cdot \mathbf{r}}$$

$$\times \Big[P^{(3)}(\mathbf{r}'', \mathbf{r}', \mathbf{r}) - P^{(2)}(\mathbf{r}'', \mathbf{r}')\rho(\mathbf{r}) - \rho(\mathbf{r}'')P^{(2)}(\mathbf{r}', \mathbf{r})$$

$$+ \rho(\mathbf{r}'')\rho(\mathbf{r}')\rho(\mathbf{r}) \Big] d\mathbf{r}'' \, d\mathbf{r}' \, d\mathbf{r} \qquad [157]$$

and the *two*-particle term

$$\langle \mathbf{k}' | \mathfrak{V}_c^{(3,0)} | \mathbf{k} \rangle_{\text{2-particle}} = A(A-1) \int \cdots \int e^{-i\mathbf{k}'\cdot\mathbf{r}} \langle \mathbf{k}' | t | \mathbf{k}''' \rangle \, \mathcal{G}_{\mathbf{k}'''}(\mathbf{r}''-\mathbf{r}') \frac{d\mathbf{k}'''}{(2\pi)^3}$$

$$\times \langle \mathbf{k}''' | t | \mathbf{k}'' \rangle \, \mathcal{G}_{\mathbf{k}''}(\mathbf{r}'-\mathbf{r}) \frac{d\mathbf{k}''}{(2\pi)^3} \langle \mathbf{k}'' | t | \mathbf{k} \rangle e^{i\mathbf{k}\cdot\mathbf{r}}$$

$$\times \left[P^{(2)}(\mathbf{r}'',\mathbf{r}')\delta(\mathbf{r}''-\mathbf{r}) - P^{(2)}(\mathbf{r}'',\mathbf{r}')\rho(\mathbf{r}) \right.$$

$$\left. - \rho(\mathbf{r}'')P^{(2)}(\mathbf{r}',\mathbf{r}) + \rho(\mathbf{r}'')\rho(\mathbf{r}')\rho(\mathbf{r}) \right] d\mathbf{r}'' \, d\mathbf{r}' \, d\mathbf{r}.$$

$$[158]$$

In these equations we have introduced for notational convenience

$$\mathcal{G}_{\mathbf{k}}(\mathbf{r}'-\mathbf{r}) \equiv \frac{e^{i\mathbf{k}\cdot(\mathbf{r}'-\mathbf{r})}}{\left(E - \overline{E^{\text{nucl}}}\right) - E_{\pi}(k) + i\eta}. \qquad [159]$$

There naturally appears in the three-particle term the correlation function for nucleon triplets,

$$P^{(3)}(\mathbf{r}'',\mathbf{r}',\mathbf{r}) \equiv \frac{1}{A(A-1)(A-2)} \int \Phi_0^\dagger(\mathbf{r}_1,\ldots,\mathbf{r}_A)$$

$$\times \sum_{i=1}^{A} \sum_{j\neq i} \sum_{\substack{k\neq j \\ k\neq i}} \delta(\mathbf{r}''-\mathbf{r}_i)\delta(\mathbf{r}'-\mathbf{r}_j)\delta(\mathbf{r}-\mathbf{r}_k)\Phi_0(\mathbf{r}_1,\ldots,\mathbf{r}_A)\,d\mathbf{r}_1\cdots d\mathbf{r}_A,$$

$$[160]$$

as well as the two-particle correlation function $P^{(2)}(\mathbf{r}',\mathbf{r})$ and the one-particle density $\rho(\mathbf{r})$ that enter due to the instruction in Eq. 84 to subtract the ground state in constructing the optical potential. The triple-correlation function is obviously normalized so that

$$\int P^{(3)}(\mathbf{r}'',\mathbf{r}',\mathbf{r})\,d\mathbf{r} = P^{(2)}(\mathbf{r}'',\mathbf{r}'), \qquad [161]$$

and so forth.

The first new ingredient in the third-order term is the appearance of a two-particle part, arising from the possibility that the meson may impinge on nucleon *i*, then visit *j*, and then return to *i*, which is certainly possible in our multiple-scattering formalism. This class of scattering is often termed "reflections" and is, of course, present for all orders of scattering higher than double scattering. The terms it introduces are of the same order as $\mathcal{V}_c^{(2)}$ unless the amplitudes t are themselves small (i.e., the scattering amplitude f must be small relative to the internucleon separation distance that arises from the integration of $\mathcal{G}_{\mathbf{k}}(\mathbf{r}'-\mathbf{r})$ of Eq. 159 over the intermediate momentum $\mathbf{k}$). Thus the reflection terms—while of order $1/A$ relative to the leading n-body term of structure t^n in $\mathcal{V}_c^{(n)}$—are not necessarily negligible, and they destroy the neatness of our understanding of the optical potential series in terms of a correlation expansion. Moreover, it is clear that for a system of A nucleons the $(A+1)$-order contribution from the infinite optical potential series, and all of the infinite number of subsequent terms, involves only such reflections, since they inevitably require the revisiting of previously encountered nucleons. Of course, one can imagine a regrouping of the multiple-scattering series such that the reflections will be included as contributions to irreducible two-body, three-body,...,A-body terms, though this does not solve the problem of their evaluation.

The second new ingredient to appear beyond second order is the possibility of different kinds of correlation functions. In particular, had we chosen to work in formalism 1 then our second-order optical potential would be changed from $\mathcal{V}_c^{(2,0)}$ to $\mathcal{V}_c^{(2,1)}$, and, according to Eq. 78 for instance, we would prohibit intermediate returns not only to the ground state but also to any previously aroused intermediate state. The bracket of Eq. 157 would correspondingly be replaced by the more symmetric combination

$$[\quad]\rightarrow\big[\,P^{(3)}(\mathbf{r}'',\mathbf{r}',\mathbf{r})-\rho(\mathbf{r})P^{(2)}(\mathbf{r}'',\mathbf{r}')$$

$$-\rho(\mathbf{r}')P^{(2)}(\mathbf{r}'',\mathbf{r})-\rho(\mathbf{r}'')P^{(2)}(\mathbf{r}',\mathbf{r})+2\rho(\mathbf{r}'')\rho(\mathbf{r}')\rho(\mathbf{r})\,\big],$$

$$[\,162\,]$$

provided we make the somewhat optimistic assumption that all the diagonal, but not necessarily ground-state, densities

$$\rho_\alpha(\mathbf{r})\equiv\frac{1}{A}\int\Phi_0^\dagger(\mathbf{r}_1,\ldots,\mathbf{r}_A)\sum_{i=1}^{A}\delta(\mathbf{r}-\mathbf{r}_i)\Phi_\alpha(\mathbf{r}_1,\ldots,\mathbf{r}_A)\,d\mathbf{r}_1\ldots d\mathbf{r}_A$$

$$[\,163\,]$$

are equal for all α. The combination in brackets in Eq. 157 and the more appealing, symmetric version of Eq. 162 are often referred to as "true" correlation functions because in them the trivial third-order correlations induced by the second- and first-order distributions have already been subtracted. As a consequence, for them, as well as for the true second-order correlations in the bracket of Eq. 139, the integration over any of the variables $\mathbf{r}, \mathbf{r}', \ldots$ that appear gives zero*. In spite of this, one should note that for any given physical situation or model it may be exceedingly difficult to impose a two-particle correlation without also inducing higher-order correlations as well. Thus the pairwise antisymmetrization of the Pauli principle does not merely bring about two-particle correlations, and the introduction of a two-particle hard-core correlation, acting pairwise between the particles, would cause $P^{(2)}(\mathbf{r}, \mathbf{r}) = P^{(3)}(\mathbf{r}, \mathbf{r}', \mathbf{r}) = \ldots = 0$, since no two particles can be at the same point and thus inevitably leads to a nonvanishing *true* triple-correlation. This can be seen trivially from the form of Eq. 162, say, with $\mathbf{r} = \mathbf{r}' = \mathbf{r}''$ in which case the term with $\rho^3(\mathbf{r})$ survives in the true triple-correlation function.

4.3c Corrections for the Single-Nucleon Amplitudes t_i

We have seen in the formal development of the Watson multiple-scattering theory of Section 4.2 that an integral part of the development is the amplitude for scattering of the projectile on a nucleon while the latter is interacting with other nucleons in the nucleus. Indeed, depending on the way in which the multiple-scattering series were treated, this amplitude entered as $t_i, t_i', \tau_i,$ or τ_i' of Eqs. 53, 85, 86, or 97, all of which differed from the free meson-nucleon amplitude by reason of modifications in their propagators arising from the presence of the many-nucleon system. In this respect, the present theory is very similar to Brueckner theory for nuclear matter. In the latter one considers the nucleon-nucleon interaction as modified by the presence of the nuclear medium; in our present approach the medium has its primary impact on the πN interaction for only one of the participants in the interaction, namely, the nucleon. Like Brueckner theory, the multiple-scattering formalism we are considering is a complete theory in the context of potential scattering and therefore, in principle, we have laid out a consistent, exhaustive algorithm for calculating in multiple-scattering situations. Of course in practice the full treatment of the modified πN Lippmann-Schwinger equations, Eqs. 53, 85, 86, and 97, may be tedious (or even impossible), and it is important to see to what degree they can be approximated by simpler forms, bearing in mind that ultimately the theory may require us to address the solution of the modified Lippmann-Schwinger equation at least in a rough form. The

---

*Note from Eq. 55, for instance, that the series for the amplitude T also involves an expansion in correlations, but not true correlations due to the lack of the ground state projection operator P_0. It therefore holds much less promise of convergence than does the series for $\mathcal{V}$.

simplest approximation for t_i, t_i', τ_i, and τ_i' would naturally be to take the free amplitude. This is known as the impulse approximation, since it rests on the notion that in the meson collision at high energies the impulse transferred to the nucleon is sufficiently great that the small energies of nuclear binding do not appreciably modify the free πN amplitude. Thus we can anticipate that this approximation will at best be valid at high energies, although once again we emphasize that the formalism we have developed can be applied at low energies if one is willing to treat it in its entirety.

We now turn to an estimate of the error made in using the free projectile—nucleon amplitude in place of the t_i in Eq. 53. We use Eq. 58 to relate these two amplitudes, first noting that the propagator with the nuclear Hamiltonian* can be related to that without through the general formula

$$\frac{1}{E-K-H_N+i\eta} = \frac{1}{E_\pi-K+i\eta}$$

$$+ \frac{1}{E_\pi-K+i\eta}\left[H_N-(E-E_\pi)\right]\frac{1}{E-K-H_N+i\eta}$$

$$[164]$$

or, in terms of the notation introduced in Eqs. 46 and 56,

$$G_0 = g_0 + g_0\left[H_N-(E-E_\pi)\right]G_0. \qquad [165]$$

Equations 164 or 165 are easily established by multiplying through from the right with the denominator of G_0. Then Eq. 58 can be rewritten [†] as

$$t_i = t_i^{\text{free}} + t_i^{\text{free}}g_0\left[H_N-(E-E_\pi)\right]G_0 t_i. \qquad [166]$$

For purposes of evaluating the correction term, that is, Δ in

$$t_i = t_i^{\text{free}} + \Delta, \quad \Delta \equiv t_i^{\text{free}}g_0\left[H_N-(E-E_\pi)\right]G_0 t_i, \qquad [167]$$

we assume that the mean nuclear excitation energies

$$R \equiv \langle H_N-(E-E_\pi)\rangle \qquad [168]$$

*Here we use $K=K_\pi$, as in Eq. 133. See Eq. 56 and note, and especially Lan–78.
[†]Equation 166 can also be cast into the (exact) form $t_i = t_i^{\text{free}} + t_i^{\text{free}}g_0 R g_0 t_i$, with $R=[H_N-(E-E_\pi)] + [H_N-(E-E_\pi)][E-K-H_N+i\eta]^{-1}[H_N-(E-E_\pi)]$, but for our approximation scheme in the estimate here this has no special advantage.

that enter are everywhere small. Then we estimate, in a very rough way,

$$\Delta \cong R \cdot t_i^{\text{free}} g_0^2 t_i^{\text{free}}, \qquad [169]$$

which for projectile states of momenta $\mathbf{k}, \mathbf{k}'$ involves

$$\langle \mathbf{k}'|\Delta|\mathbf{k}\rangle \cong R \cdot \int \langle \mathbf{k}'|t_i^{\text{free}}|\mathbf{k}''\rangle \frac{d\mathbf{k}''/(2\pi)^3}{(E_\pi - E(k'') + i\eta)^2} \langle \mathbf{k}''|t_i^{\text{free}}|\mathbf{k}\rangle$$

$$\rightarrow \frac{2\pi R}{(2\pi)^3} \cdot \int_{E(0)}^{\infty} \left[t_i^{\text{free}}(E_\pi, E'') \right]^2 \frac{k''^2 \left[\dfrac{dE(k'')}{dk''} \right]^{-1} dE(k'')}{\left[E_\pi - E(k'') + i\eta \right]^2}$$

$$[170]$$

where we have assumed p-wave angular dependence in the t's and restricted our attention to the on-shell forward case, $\mathbf{k} \approx \mathbf{k}', |\mathbf{k}| = k_0 \equiv (2mE_\pi)^{1/2}$. Then assuming that the main contribution in the integrand comes from the neighborhood of the double pole in order to extend the range of integration, and using the calculus of residues, we obtain for the fractional correction

$$\frac{\langle \Delta \rangle}{\langle t^{\text{free}}(E_\pi) \rangle} \sim \frac{iR}{2\pi t^{\text{free}}(E_\pi)} \frac{d}{dE_\pi} \left\{ \left[t^{\text{free}}(E_\pi) \right]^2 k_\pi^2 \left(\frac{dE_\pi}{dk_\pi} \right)^{-1} \right\}$$

$$\xrightarrow[\substack{\text{relativistic} \\ \text{kinematics}}]{} \frac{1}{\pi} iR \left[t^{\text{free}}(E_\pi) \right] E_\pi k_\pi$$

$$\times \left[\frac{1}{2E_\pi} \left(1 + \frac{E_\pi^2}{k_\pi^2} \right) + \frac{d}{dE_\pi} \log t^{\text{free}}(E_\pi) \right], \qquad [171]$$

where in the last stage we have chosen to use relativistic kinematics, $E_\pi^2 = k_\pi^2 + m^2$, as is generally appropriate for our meson applications. As a case for numerical estimate, we select pion-nucleus scattering in the region of the 3,3 resonance. In evaluating the "mean nuclear excitation energy" R we must bear in mind that in the free scattering we have assumed the target nucleon to be fixed; that is, we have not included its kinetic energy operator in the propagator (or, alternatively, we can think of incorporating that operator into K and omitting kinetic energies from H_N). Thus $\langle H_N \rangle$ really refers to the nuclear average potential energy, which is of the order of 40 MeV. Provided this average potential is not very rapidly varying we can anticipate that in many circumstances we can balance some of its effects by a judicious choice of E_π, the energy at which we evaluate the amplitude for free meson-nucleon scatterings.*

One way to accomplish this is by creating an artificial three-particle problem, involving the meson, the "active" nucleon, and the remaining nuclear core, in order to carry out an effective kinematic averaging, as discussed in Chapter 3. This includes the main binding effects, thus yielding a reliable t_i for use in the multiple-scattering series; it does not of course cope with modifications due to antisymmetrization or the projection of the nuclear ground state as required for t_i', τ_i, or τ_i'. In this way we reduce the effects of the binding so that R is measured by the average binding energy ~ 8 MeV. (In general, one anticipates for many systems that the average potential energy, kinetic energy, or binding energy are all quite comparable, but this is not the case for nuclei where the first two nearly cancel each other and the third is correspondingly smaller. Hence we have the need to allow for the possibility of some "tuning" of the relevant projectile energy to compensate for binding effects. See Tab–76 and Lan–78.) The second piece of Eq. 171 for the resonant situation can be estimated by taking the energy-dependence of the amplitude as

$$t^{\text{free}}(E_\pi) \sim \frac{1}{E_{\text{res}} - E_\pi - i(\Gamma/2)} \; ; \qquad [172]$$

at the energy of the resonance ($E_\pi = E_{\text{res}} \sim 335$ MeV, $k_\pi = k_{\text{res}} \sim 304$ MeVc), the pion is quite relativistic and Eq. 171 can be written as[†]

$$\left| \frac{\langle \Delta \rangle}{\langle t^{\text{free}} \rangle} \right|_{\text{res}} \sim 2|k_{\text{res}} f(E_{\text{res}})| R \left| \frac{1}{E_\pi} + \frac{2i}{\Gamma} \right| . \qquad [173]$$

The last factor represents the collision time delay in the form of the projectile time $1/E_\pi$ and the resonance decay time $2/\Gamma$. The collision is "impulsive" if this time is much shorter than the time modes of the bound system $1/R$. The 3,3 resonance is especially demanding, since the first factor here is of order unity: the forward scattering amplitude on resonance is $f(E_{\text{res}}) = \frac{4}{3} i / k_{\text{res}}$, while the width is not vastly greater than nuclear binding energies; hence

$$\left| \frac{\langle \Delta \rangle}{\langle t^{\text{free}} \rangle} \right| \sim 0.3. \qquad [174]$$

---

*Indeed, if it were a strictly constant potential, we could exactly cancel its effects by selecting a different zero for the energy scale; that is, we could modify the numerical value of E_π.
[†] For relativistic situations the nonrelativistic $t = -(4\pi/2m)f$ is replaced with $t = -(4\pi/2E_\pi)f$.

Thus we must expect nonnegligible effects from nuclear binding in correcting the impulse approximation result, though we reassure the reader that our crude estimate here need not be taken overly seriously, and that the true test of our approximations must ultimately lie in the success of their confrontation with experiment, while the specific validity of the impulse aspect, among others, can best be ascertained by a more refined and realistic evaluation than that given here.* (In this regard, see Rev–73, Ded–76, Mai–76, Lan–78, and references therein.) For general applications of the formalism at medium-energies the situation is much more satisfactory both because the amplitudes are smaller and because they are usually not as rapidly varying in energy.

Before leaving the question of modifications to the single-particle amplitude t_i, we note the existence of a hierarchy of corrections known as "reflections," or "local" or "effective" field corrections. These are corrections to a given πN amplitude in the medium, say, for nucleon i, arising from cases in which the projectile scatters on i and then travels to one or more nucleons ultimately returning to the original i, perhaps repeating such paths many times (Fol–69). This can be thought of as modifying the field incident on scatterer i, whence the last two names. The effects of such reflections have been estimated in models and are found not to be especially small (Koc–74, Kei–74,76, Aga–75b, and Joh–78), though they are considerably reduced when the fixed-scatterer assumption is relaxed.

4.3d The Fixed-Scatterer Approximation

We must now return to a more detailed examination of the consequences of the closure assumption that we have used to eliminate the detailed features of nuclear structure, involving all the excited states and excitation spectrum explicitly, in favor of the nuclear correlation functions. This topic was discussed briefly at the end of Section 4.1. For this purpose we return to the original Lippmann-Schwinger equation (Eq. 45) of our multiple-scattering approach, written out for convenience here in terms of the

---

*Note that if the formalism of Eqs. 86–97, incorporating antisymmetrization, is used then the antisymmetrizing correction to the "single-nucleon" amplitude τ_i of Eq. 86 or τ_i' of Eq. 97 must be evaluated on the same footing as the binding correction, and may modify the latter appreciably by blocking the contribution of low-lying states through the Pauli principle.

In model calculations on the deuteron, numerical cancellations have been found (Fäl–77) between binding corrections and rescattering corrections. These may also enter in heavier nuclei to mitigate some higher-order effects, though the matter has yet to be tested there even in a three-body (projectile-active nucleon-core) model, except for zero energy (see Kol–77).

configuration space wave function:*

$$\Psi^{(+)}(\mathbf{r}_1,\ldots,\mathbf{r}_A;\mathbf{r}) = \Phi_0(\mathbf{r}_1,\ldots,\mathbf{r}_A)\phi_\mathbf{k}(\mathbf{r})$$

$$+ \frac{1}{E - H_N(\mathbf{r}_1,\ldots,\mathbf{r}_A) - K(\mathbf{r}) + i\eta} \sum_{i=1}^{A} v_i(\mathbf{r},\mathbf{r}_i)$$

$$\times \Psi^{(+)}(\mathbf{r}_1,\ldots,\mathbf{r}_A;\mathbf{r}). \qquad [175]$$

The equation is written in terms of the notation of Eqs. 44–46 in which we make explicit which operators involve nuclear coordinates $\{\mathbf{r}_i, i = 1,\ldots,A\}$ and which operators involve the projectile variable $\mathbf{r}$. The function Φ_0 refers to the nuclear ground state, and $\phi_\mathbf{k}$ to a projectile plane wave; since $\Psi^{(+)}$ involves the scattering projectile, we designate that we are using boundary conditions for an outgoing spherical wave for it. The interactions v_i in Eq. 175 are taken to be local.

Inserting the action of the nuclear Hamiltonian and the projectile kinetic energy operator we obtain (see, e.g., Eq. 146)

$$\Psi^{(+)}(\mathbf{r}_1,\ldots,\mathbf{r}_A;\mathbf{r}) = \Phi_0(\mathbf{r}_1,\ldots,\mathbf{r}_A)\phi_\mathbf{k}(\mathbf{r})$$

$$- \frac{2m}{4\pi} \sum_\alpha \Phi_\alpha(\mathbf{r}_1,\ldots,\mathbf{r}_A) \int d\mathbf{r}_1' \ldots d\mathbf{r}_A' \, d\mathbf{r}' \frac{e^{iK_\alpha|\mathbf{r}-\mathbf{r}'|}}{|\mathbf{r}-\mathbf{r}'|}$$

$$\times \Phi_\alpha^\dagger(\mathbf{r}_1',\ldots,\mathbf{r}_A') \sum_{i=1}^{A} v_i(\mathbf{r}',\mathbf{r}_i')\Psi^{(+)}(\mathbf{r}_1',\ldots,\mathbf{r}_A';\mathbf{r}'),$$

$$[176]$$

where

$$K_\alpha^2 = k^2 + 2m(E_0^{\text{nucl}} - E_\alpha^{\text{nucl}}). \qquad [177]$$

The assumption that allows a correct use of closure is that

$$k^2 \gg 2m(E_\alpha^{\text{nucl}} - E_0^{\text{nucl}}); \qquad [178]$$

that is, the projectile kinetic energy is much greater than the nuclear excitation energies typically encountered. Then we can replace $K_\alpha \rightarrow \bar{K}$ and use closure

$$\sum_\alpha \Phi_\alpha(\mathbf{r}_1,\ldots,\mathbf{r}_A)\Phi_\alpha^\dagger(\mathbf{r}_1',\ldots,\mathbf{r}_A') = \delta(\mathbf{r}_1 - \mathbf{r}_1')\ldots\delta(\mathbf{r}_A - \mathbf{r}_A') \qquad [179]$$

*The present approach to this problem is discussed in Fol–69.

to write

$$\Psi^{(+)}(\mathbf{r}_1,\ldots,\mathbf{r}_A;\mathbf{r}) = \Phi_0(\mathbf{r}_1,\ldots,\mathbf{r}_A)\phi_\mathbf{k}(\mathbf{r}) - \frac{2m}{4\pi}\int d\mathbf{r}'\,\frac{e^{i\bar{K}|\mathbf{r}-\mathbf{r}'|}}{|\mathbf{r}-\mathbf{r}'|}$$

$$\times \sum_{i=1}^{A} v_i(\mathbf{r}',\mathbf{r}_i)\Psi^{(+)}(\mathbf{r}_1,\ldots,\mathbf{r}_A;\mathbf{r}'). \qquad [\,180\,]$$

This is easily solved in terms of a nuclear ground state and distorted projectile wave function $\psi_\mathbf{k}^{(+)}$,

$$\Psi^{(+)}(\mathbf{r}_1,\ldots,\mathbf{r}_A;\mathbf{r}) = \Phi_0(\mathbf{r}_1,\ldots,\mathbf{r}_A)\psi_\mathbf{k}^{(+)}(\mathbf{r}), \qquad [\,181\,]$$

to give

$$\psi_\mathbf{k}^{(+)}(\mathbf{r}) = \phi_\mathbf{k}(\mathbf{r}) - \frac{2m}{4\pi}\int d\mathbf{r}'\,\frac{e^{i\bar{K}|\mathbf{r}-\mathbf{r}'|}}{|\mathbf{r}-\mathbf{r}'|}\sum_{i=1}^{A} v_i(\mathbf{r}',\mathbf{r}_i)\psi_\mathbf{k}^{(+)}(\mathbf{r}') \qquad [\,182\,]$$

from which all the nuclear physics complications have been removed. The nucleus is described in terms of A fixed scatterers at frozen positions $\mathbf{r}_1,\ldots,\mathbf{r}_A$ and acts with potentials $v_1,\ldots,v_A$ at these points. Ultimately, these positions are to be averaged with respect to the nuclear ground state wave function that contains all the correlational information about the nucleons. Thus, for example, the meson-nucleus scattering amplitude can be read off from Eq. 182 by taking $|\mathbf{r}-\mathbf{r}'|\to\infty$ and by using the definition of Eqs. 29–31 to give

$$\langle \mathbf{k}'|f|\mathbf{k}\rangle = -\frac{2m}{4\pi}\int d\mathbf{r}_1\ldots d\mathbf{r}_A\,\Phi_0^\dagger(\mathbf{r}_1,\ldots,\mathbf{r}_A)\int d\mathbf{r}\phi_\mathbf{k'}^\dagger(\mathbf{r})$$

$$\times \sum_{i=1}^{A} v_i(\mathbf{r},\mathbf{r}_i)\psi_\mathbf{k}^{(+)}(\mathbf{r})\Phi_0(\mathbf{r}_1,\ldots,\mathbf{r}_A). \qquad [\,183\,]$$

This separation of the problem into a calculation of scattering from frozen nucleons followed by an averaging over their ground state positions is, in practice, an enormous simplification, but it introduces the artificial feature that during the multiple-scattering phase of the discussion we do not yet have at hand the nuclear structure part of the problem, which enters only after we carry out the nuclear averaging. In particular, our expansion of the multiple-scattering series in Section 4.2 in terms of correlation functions is only possible in a fixed-scatterer formalism, though small recoil effects might possibly be expanded in nucleon momenta and therefore in gradient operators. The central approximation (Eq. 178) requires that at each stage of the multiple scattering the energy transfer be small; hence for *each* scattering the nucleon recoil energy must be much less than the projectile kinetic energy. Thus a necessary but not sufficient condition for the applicability of the frozen nucleon approximation is that the quasi-elastic process lie at much lower energy than that of the incident projectile,

$$\frac{(\mathbf{k}-\mathbf{k}')^2}{2M} \ll \frac{k^2}{2m}, \qquad [184]$$

or

$$\sin\frac{\theta}{2} \ll \tfrac{1}{2}\sqrt{M/m} ; \qquad [185]$$

that is, we may be severely restricted to forward directions. In this we are helped slightly in pion applications by the small pion mass which, at a given incident energy, implies smaller momenta than for a nucleon, say, and therefore smaller momentum transfers. (Of course, there remain other questionable features in applying the theory at larger angles, not least of which proves to be the need to consider increasingly higher orders of correlations.) We note that once the fixed-scatterer approximation is invoked there is no reference whatsoever in our formalism to the distinction between scattering on a free nucleon or on a nuclear one, and thus we do not have to appeal to the arguments surrounding Eqs. 164–174 to justify the impulse approximation. On the other hand, our original approach is in principle a complete one, which does not reject from the start the inclusion of nuclear structure effects, or at least the systematic, successive examination of each such effect. In contrast, it is intrinsic to the fixed-scatterer approximation that there is no systematic way to explore its validity short of detailed calculation without it.

   The same general estimates that we have made for elastic scattering apply also to the distorted wave impulse approximation, as discussed in Section 4.2c, for use with inelastic results. Of course, selection rules that may be operative in the inelastic case—as, for example, the particular excitation of nuclear spin or isospin degrees of freedom—may change the specifics of the estimates somewhat. At the very least we must expect that inelastic scattering will vanish in the forward direction unless there is some degree of coherence in the process; thus approximations that require small-angle scattering are more questionable in the inelastic situation. Therefore the applicability of the DWIA method to inelastic reactions requires specific testing for each case; we discuss this in the context of the particular instances treated in Chapter 5.

Before leaving the subject of the Watson approach to multiple scattering, a few words of overall assessment are in order. The theory in its general form provides a complete statement on the multiple-scattering problem, but in that form requires a thorough treatment of nuclear dynamics both in obtaining the projectile-nucleon* amplitudes in the presence of the nuclear medium and in dealing with projectile propagation and the use of these amplitudes in the multiple-scattering series. Since this full program is clearly difficult, recourse has often been made to two key approximations. The first of these is the impulse approximation, in which the projectile-nucleon amplitudes are replaced by the free ones. The second is that of the fixed-scatterer approximation, in which highly detailed nuclear structure information is eliminated in favor of nuclear correlation functions. These are still very formidable both in technical manipulation and in terms of our knowledge, or lack thereof, of nucleon correlations beyond the two-particle (or even density?) level. The great virtue of the Watson approach lies, however, in the fact that it provides a systematic and complete framework in which to explore corrections to the lowest-order approximation $\mathcal{V}_c = At\rho$, while even this very simple approximation contains in it enough of the crucial physical ingredients for the description of multiple-scattering to allow for meaningful, if tentative, comparison with data at medium energies. The higher-order corrections to this involve an expansion in correlations whose parameter of smallness contains the projectile-nucleon amplitude, the relevant correlation length, the density, and the reciprocal of the projectile momentum; these corrections also have to relate to deviations from the impulse approximation. Should the expansion in clusters fail to converge, then one is forced to consider regroupings of the scattering series (such as through the use of doorway states, discussed in Chapter 6) which may have to be explored, at least initially, in rather phenomenological ways. Likewise, severe breakdowns in the validity of the impulse approximation, or relatively circumscribed corrections to it, may require rearrangement of the multiple-scattering series to incorporate more physical effects in the t_i's at an earlier stage † and possibly in a somewhat phenomenological fashion.

Last, we note that nothing in our multiple-scattering formalism thus far deals in a substantive way with the problems of double counting; that is, our nuclear dynamics do not include explicit mesic degrees of freedom and

---

*One can also contemplate applications based on nucleon constituents (quarks?) or nucleon aggregates (α-clusters?); the former would, of course, tend to make the theory depart rather far from its origins in the Schroedinger equation.

† A case in point is the attempt to deal consistently with the internal energy dependence of the t_i's noted with respect to Eq. 40. This has been considered in Fra–56 and Bet–78.

therefore there is a risk of entangling the transfer of a pion in a multiple-scattering term with the exchange of a pion in producing the one-pion-exchange part of the nucleon-nucleon interaction. This deeper and more difficult problem is discussed in Chapters 6 and 7.

4.4 Glauber Theory

The multiple-scattering formalism we have developed thus far suffers from at least three practical difficulties (even before the problems of double counting in its mesic applications are reached): (1) It involves infinite series of dubious convergence properties. (2) The propagators that appear in it,

$$G_0 = \frac{1}{E - H_0 + i\eta} = \frac{P}{E - H_0} - \pi i\, \delta(E - H_0), \qquad [186]$$

contain off-shell propagation through the principal part piece and this requires knowledge of the basic projectile-nucleon amplitudes t_i off their energy shells; this information is at best hard to come by and uncertain. (3) The multiple-scattering approach depends, for its consistency, on the use of a potential in a Lippmann-Schwinger equation. (Indeed one is obliged to use a single-nucleon amplitude t_i in the multiple-scattering series that arises from such a projectile-nucleon potential, and not just any arbitrary model or phenomenological form.) The theory that we now develop, given originally by Glauber (see the references at the end of this chapter), avoids the first two difficulties; that is, it involves a finite scattering series and essentially only on-shell amplitudes—and may also transcend the need for a potential theory. In this last respect especially it may offer a considerable advantage over the multiple-scattering approach. Naturally, all of this does not come without its *quid pro quo*, in this case expressed in the fact that the Glauber theory is applicable only in the limit of very high energy (while still being basically a nonrelativistic theory!) and does not possess a formally exact dynamical development, as in multiple-scattering theory, which can allow for the systematic introduction of corrections and refinements. Nonetheless, many practical applications that are made of the multiple-scattering theory also require fairly high energies for the validity of the impulse approximation and for the speedy truncation of the expansion in clusters. Thus it is not self-evident before a particular application is studied in detail that there is an appreciable energy regime that is high enough for a low-order optical potential to work but still so low that the theory of Glauber is inaccurate.

4.4a Heuristic Derivation of the Glauber Series

We start from the standard expression for the partial wave analysis of the scattering amplitude for a spinless particle:

$$f(\theta) = \frac{1}{2ik} \sum_{l=0}^{\infty} (2l+1)(e^{2i\delta_l} - 1) P_l(\cos\theta), \qquad [187]$$

where δ_l is the phase shift in the partial wave of orbital angular momentum l and $P_l(\cos\theta)$ is the Legendre polynomial. We now imagine that we deal with very high energy scattering of a basically diffractive nature, so that $kR \gg 1$. Then many partial waves enter in the scattering, and the discrete variable l, with its unit increments, can be replaced by a continuous impact parameter variable

$$b = \frac{l}{k} \quad \text{with} \quad db = \frac{1}{k}\Delta l = \frac{1}{k}. \qquad [188]$$

The phase shift is identified correspondingly with the phase parameter

$$2\delta_l = \chi(b), \qquad [189]$$

and at high energies we expect predominantly small-angle scattering so that we can use the approximation* (Gra–65, p. 1003)

$$P_l(\cos\theta) = J_0\left((2l+1)\sin\frac{\theta}{2}\right) + \text{order}\left(\sin^2\frac{\theta}{2}\right). \qquad [190]$$

Inserting these into Eq. 187, we get

$$f(\theta) = -ik \int_0^\infty J_0\left(2kb\sin\frac{\theta}{2}\right)(e^{i\chi(b)} - 1)b\,db$$

$$= -ik \int_0^\infty J_0(\Delta b)(e^{i\chi(b)} - 1)b\,db, \qquad [191]$$

where in the last step we have inserted the momentum transfer $\Delta = |\mathbf{k} - \mathbf{k}'|$ $= 2k\sin\frac{1}{2}\theta$ for this elastic scattering situation. We note that the result of Eq. 191 is a natural generalization of that of Eq. 35.

*A vastly better numerical approximation is given by the Hilb formula (Hil–19)

$$P_l(\cos\theta) = \sqrt{\frac{\theta}{\sin\theta}}\ J_0\left[(l+\tfrac{1}{2})\theta\right] + \text{order}\ (l^{-3/2}),$$

which provides an excellent numerical result even for quite large values of the angle (Eri–78).

Now let us imagine that the scatterer is made up of many individual scattering elements. It is then natural to assume that the total phase shift, for going through the target, is equal to the sum of the individual phase shifts for the individual scatterers located at $\{\mathbf{r}_1; \ldots; \mathbf{r}_A\} = \{\mathbf{b}_1, z_1; \ldots; \mathbf{b}_A, z_A\}$. That is,

$$\chi(\mathbf{b}) = \sum_{i=1}^{A} \chi_i(\mathbf{b} - \mathbf{b}_i), \qquad [192]$$

which is known as the assumption of the additivity of phase parameters. We show below that for local potentials this assumption is exact (within the framework of the eikonal theory); the bid of the present theoretical approach to go beyond potential theory stems from the hope that Eq. 192 may apply much more generally. To systematize the insertion of Eq. 192 into Eq. 191, we introduce the profile function

$$\Gamma_i(\mathbf{b}) \equiv e^{i\chi_i(\mathbf{b})} - 1, \qquad [193]$$

in terms of which*

$$F(\theta) = -ik \int_0^{\infty} J_0(\Delta b)(e^{i\sum_{i=1}^{A}\chi_i(\mathbf{b}-\mathbf{b}_i)} - 1)b\,db$$

$$= -ik \int_0^{\infty} J_0(\Delta b)\left[\prod_{i=1}^{A}(\Gamma_i(\mathbf{b}-\mathbf{b}_i)+1) - 1\right]b\,db$$

$$= -ik \int_0^{\infty} J_0(\Delta b)\left[\sum_{i=1}^{A}\Gamma_i(\mathbf{b}-\mathbf{b}_i) + \sum_{i=1}^{A}\sum_{j<i}\Gamma_i(\mathbf{b}-\mathbf{b}_i)\Gamma_j(\mathbf{b}-\mathbf{b}_j)\right.$$

$$+ \sum_{i=1}^{A}\sum_{j<i}\sum_{k<j}\Gamma_i(\mathbf{b}-\mathbf{b}_i)\Gamma_j(\mathbf{b}-\mathbf{b}_j)\Gamma_k(\mathbf{b}-\mathbf{b}_k) + \cdots$$

$$\left. + \sum_{i=1}^{A}\sum_{j<i}\cdots\sum_{q<p}\Gamma_i(\mathbf{b}-\mathbf{b}_i)\cdots\Gamma_q(\mathbf{b}-\mathbf{b}_q)\right]b\,db. \qquad [194]$$

This result is a finite series in terms of single-, double-, ..., A-fold scatterings represented by the profile functions. The restrictions on the

---

*Note that if we consider spin and isospin degrees of freedom for the projectile the additivity of phase parameters is not valid, and instead one must order the profile functions in the sequence in which they are encountered by the projectile.

summations—which are *different* from those of our previous multiple-scattering approach—embody the physics of the high-energy situation, whereby the projectile progresses across the nucleus and encounters the nucleons in their spatial ordering, without revisiting any, whence also the restriction to A scatterings or less. The profile functions can be determined from Eqs. 191 and 193 (or through the chain involving Eq. 189 again) for the individual scatterer and the inverse Fourier-Bessel transform

$$\Gamma_i(\mathbf{b}) = \frac{i}{2\pi k} \int e^{-i\boldsymbol{\delta}\cdot\mathbf{b}} f_i(\boldsymbol{\delta}) d^2\boldsymbol{\delta}, \qquad [195]$$

where the scattering amplitude (Eq. 191) is expressed in terms of $\boldsymbol{\delta} = \boldsymbol{\Delta}$, on the assumption that at high energies the amplitude should depend only on the momentum transfer. Either from Eq. 195 or from its definition in terms of on-shell phase shifts, the profile function clearly requires only on-shell information for its determination.* Thus the Glauber series of Eq. 194 is a finite, on-shell series and may not require an underlying potential for its justification.

From our treatment here, and, more specifically, Eq. 195, it should be clear that we have assumed the validity of a high-energy, eikonal approximation for scattering on an *individual nucleon*, as well as on the nucleus. This rather stringent condition is required for consistency and must be kept clearly in mind. It is, for example, obviously violated for πN scattering at the 3,3 resonance, since a resonance in one partial wave manifestly violates the requirement that many partial waves should contribute equivalently. Indeed the differential πN cross section at the 3,3 resonance is nothing like diffractive, being in fact equal at backward directions to its forward value. In addition, the amplitude at resonance is purely imaginary, so the profile function of Eq. 195 is real, implying an imaginary phase parameter χ that contradicts the fact that at these energies there is no inelasticity in the πN amplitudes. Nonetheless, as we see in Chapter 5, Glauber theory can be applied to elastic πA scattering in the 3,3 region with some empirical success, presumably for more or less accidental reasons. Its applicability to inelastic processes in this kinematic domain is riskier (see, for example, Loh–77). It involves Eq. 194 averaged between initial and final nuclear states, and the off-diagonal feature makes the cross sections correspondingly less diffractive. In general, the virtue of the Watson approach over that of Glauber is that, at least in principle, it

---

*Note that Eq. 195 requires contributions from the unphysical region $\delta > 2k$ which complicates the practical applicability of these results, making them invalid unless those contributions are small (see Loh–77).

offers a complete theoretical framework into which higher-order corrections can be incorporated.*

The requirement that the diffractive condition $ka \gg 1$, where a is the relevant nucleon hadronic radius, apply for the target constituent as well as for the full nucleus $(kR \gg 1)$, automatically leads to the fulfillment, in practice, of the further implicit condition $ka^2/R \gg 1$. This is required in order that the eikonal wave function of Eq. 8, say, still be valid as the wave traverses from one nucleon to the next, which is not the case if the nucleons are far apart, since the eikonal form of the wave function breaks down at large distances from a scattering center. (For example, Eq. 8 lacks a straightforward interpretation as a plane wave plus outgoing spherical wave, unless we invoke further limiting assumptions.) Again in order to justify the validity of eikonal forms, we must also suppose that the projectile energy E is much greater than the potentials V which it experiences in the scattering, $E/|V| \gg 1$, a condition that is generally automatically satisfied in nucleon applications where we are in the diffractive region, but is not obviously fulfilled for pions in the 3,3 region, where the absolute value of the (imaginary) optical potential in the center of the nucleus reaches some 250 MeV while $E \sim 335$ MeV.

4.4b Relation to the Watson Series

In order to clarify further the nature of the approximations that enter in Glauber theory and to trace its relation to our previous multiple-scattering results, we now rederive the Glauber series starting from a Lippmann-Schwinger equation (Eq. 45) for a many-particle system with local potentials. We start with the general equation for a local potential

$$t = V + VG_0 t, \qquad\qquad [196]$$

with locality embodied in

$$\langle \mathbf{r}'|V|\mathbf{r} \rangle = V(\mathbf{r})\delta(\mathbf{r}' - \mathbf{r}). \qquad\qquad [197]$$

In the high-energy limit we invoke the same arguments that led to the linearized propagator of Eq. 148,

$$\langle \mathbf{r}'|G_0|\mathbf{r} \rangle \cong -iv^{-1}\delta^{(2)}(\mathbf{b}' - \mathbf{b})e^{ip(z'-z)}\theta(z'-z)$$
$$\text{(high-energy approximation),} \qquad\qquad [198]$$

---

*Methods of correcting Glauber theory systematically have been considered *inter alia* in Wal–73, a, b, c, and Won–75.

namely, that our predominant scattering at high energies is forward and nearly on shell. The rest of the development then follows without recourse to further assumptions. It is easy to verify that the solution to Eq. 196 using Eqs. 197 and 198 is

$$\langle \mathbf{r}' | t | \mathbf{r} \rangle = \delta^{(2)}(\mathbf{b}' - \mathbf{b}) e^{ip(z'-z)} \frac{d}{dz'} \theta(z'-z) e^{-iv^{-1} \int_z^{z'} V(\mathbf{b},\zeta) d\zeta} V(\mathbf{b},z)$$

$$= \delta^{(2)}(\mathbf{b}' - \mathbf{b}) e^{ip(z'-z)}$$

$$\times \left[\delta(z'-z) - iv^{-1} \theta(z'-z) V(\mathbf{b},z') e^{-iv^{-1} \int_z^{z'} V(\mathbf{b},\zeta) d\zeta} \right] V(\mathbf{b},z)$$

$$[199]$$

by exactly summing the iterated Lippmann-Schwinger series with Eqs. 197 and 198 or by substituting Eq. 199 into Eq. 196 on the right-hand side to get

$$\langle \mathbf{r}' | V + V G_0 t | \mathbf{r} \rangle = V(\mathbf{r}) \delta(\mathbf{r}' - \mathbf{r}) + \int d\mathbf{r}'' \, d\mathbf{r}''' \langle \mathbf{r}' | V | \mathbf{r}'' \rangle \langle \mathbf{r}'' | G_0 | \mathbf{r}''' \rangle \langle \mathbf{r}''' | t | \mathbf{r} \rangle$$

$$= V(\mathbf{b},z) \delta^{(2)}(\mathbf{b}' - \mathbf{b}) \delta(z' - z) - iv^{-1} \int d^2\mathbf{b}'' \, dz'' \, V(\mathbf{b}',z') \delta^{(2)}(\mathbf{b}' - \mathbf{b}'') e^{ip(z'-z'')}$$

$$\times \theta(z' - z'') \delta^{(2)}(\mathbf{b}'' - \mathbf{b}) e^{ip(z''-z)} \frac{d}{dz''} \theta(z'' - z) e^{-iv^{-1} \int_z^{z''} V(\mathbf{b},\zeta) d\zeta} V(\mathbf{b},z),$$

$$[200]$$

which through an integration by parts for z'' and use of

$$-\frac{d}{dz''} \theta(z' - z'') = \delta(z' - z'')$$

$$[201]$$

leads immediately to the second form for t in Eq. 199. Note that all of the results have a characteristic high-energy structure: they involve traversal of the nucleus while preserving the value of the impact parameter, propagation in the forward direction only, and a plane wave phase factor modulated by a phase involving an integration of the potential along the classical trajectory of the particle—this last phase leading to the effective modified propagation vector within the scattering region.

On the energy shell, that is for

$$(\mathbf{k})_z = (\mathbf{k}')_z = p,$$

$$[202]$$

the momentum space matrix element for the amplitude becomes especially simple, namely,

$$_{on}\langle \mathbf{k}'|t|\mathbf{k}\rangle_{on} = \int d^2\mathbf{b}\, dz\, d^2\mathbf{b}'\, dz'\, e^{-i\mathbf{k}'_\perp \cdot \mathbf{b}'} e^{-ik'_z z'}$$

$$\times \delta^{(2)}(\mathbf{b}' - \mathbf{b}) e^{ip(z'-z)} \frac{d}{dz'} \theta(z'-z)$$

$$\times e^{-iv^{-1}\int_z^{z'} V(\mathbf{b},\zeta)d\zeta} V(\mathbf{b},z) e^{i\mathbf{k}_\perp \cdot \mathbf{b}} e^{ik_z z}$$

$$= \int d^2\mathbf{b}\, e^{i(\mathbf{k}_\perp - \mathbf{k}'_\perp)\cdot \mathbf{b}} \int_{-\infty}^{\infty} dz\, e^{-iv^{-1}\int_z^{\infty} V(\mathbf{b},\zeta)d\zeta} V(\mathbf{b},z)$$

$$= -iv \int d^2\mathbf{b}\, e^{i(\mathbf{k}_\perp - \mathbf{k}'_\perp)\cdot \mathbf{b}} \int_{-\infty}^{\infty} dz\, \frac{d}{dz} e^{-iv^{-1}\int_z^{\infty} V(\mathbf{b},\zeta)d\zeta}$$

$$= iv \int d^2\mathbf{b}\, e^{i\Delta \cdot \mathbf{b}} \left[e^{-iv^{-1}\int_{-\infty}^{\infty} V(\mathbf{b},\zeta)d\zeta} - 1 \right]. \qquad [203]$$

This last, simple form results only for the on-shell case. Recalling that $t = -(4\pi/2m)f$, we can immediately relate Eq. 203 to the (azimuthally symmetric) result of Eq. 191 with the identification of the phase parameter there as

$$\chi(\mathbf{b}) = -v^{-1} \int_{-\infty}^{\infty} V(\mathbf{b},\zeta)\, d\zeta \qquad [204]$$

for the local potential case. In that case, it is trivially true that the phase parameters are additive, since the potential is additive for a many-particle target (note that our approach implicitly uses a fixed-scatterer approximation, with the amplitude ultimately averaged between nuclear ground states):

$$V = \sum_{i=1}^{A} v_i(\mathbf{b} - \mathbf{b}_i, z - z_i), \qquad [205]$$

whence, for on-shell scattering on a full nucleus,

$$_{on}\langle \mathbf{k}'|T|\mathbf{k}\rangle_{on} = iv \int d^2\mathbf{b}\, e^{i\Delta \cdot \mathbf{b}} \left[e^{-iv^{-1}\int_{-\infty}^{\infty} \sum_{i=1}^{A} v_i(\mathbf{b}-\mathbf{b}_i,\zeta - z_i)d\zeta} - 1 \right]$$

$$= iv \int d^2\mathbf{b}\, e^{i\Delta \cdot \mathbf{b}} \left[\sum_{i=1}^{A} (e^{-iv^{-1}\int_{-\infty}^{\infty} v_i d\zeta} - 1) \right.$$

$$\left. + \cdots + \sum_{i=1}^{A} \sum_{j<i} \cdots \sum_{q<p} (e^{-iv^{-1}\int_{-\infty}^{\infty} v_i d\zeta} - 1)\ldots(e^{-iv^{-1}\int_{-\infty}^{\infty} v_q d\zeta} - 1) \right],$$

$$[206]$$

just as in Eq. 194. Of course, Eq. 194, with the interpretation of the profile functions through Eq. 195 rather than

$$\Gamma_i(\mathbf{b} - \mathbf{b}_i) \to e^{-iv^{-1}\int_{-\infty}^{\infty} v_i(\mathbf{b} - \mathbf{b}_i, \zeta - z_i)d\zeta} - 1, \qquad [207]$$

as for the specific local potential case, offers an apparent greater generality in that it may not require an underlying potential theory.

The result of Eq. 194 or Eq. 206 is remarkable, as contrasted with the Watson series, for the fact that it is a finite, on-shell series. One is prompted to ask what became of an infinite number of terms in the Watson series and of the off-shell contributions there. The explicit answer (Har–69, Eis–72) can be generated, for example, by solving the defining relationship (Eq. 52) for an exact version of the Watson series in the high-energy limit, based on Eqs. 197 and 198. It then emerges, as we have seen indirectly here, that the off-shell contributions exactly cancel all the terms involving $(A + 1), (A + 2), \ldots$, scatterings along with the parts of the first A orders of scattering, which differentiate between the Watson summation prescription and the Glauber summation prescription,

$$\text{Watson} = \sum_{i=1}^{A} \sum_{j \neq i} \sum_{k \neq j} \ldots \quad \text{versus} \quad \text{Glauber} = \sum_{i=1}^{A} \sum_{j < i} \sum_{k < j} \ldots .$$

The expressions in Eqs. 194 or 206 are ultimately to be sandwiched between nuclear wave functions to obtain the projectile-nucleus scattering amplitude. This involves

$$F_{f0}(\Delta) = -\frac{ik}{2\pi} \int d^2 \mathbf{b} \, e^{i\Delta \cdot \mathbf{b}} \Phi_f^{\dagger}(\mathbf{r}_1, \ldots, \mathbf{r}_A)$$

$$\times \left[\sum_{i=1}^{A} \Gamma_i(\mathbf{b} - \mathbf{b}_i) + \cdots + \sum_{i=1}^{A} \sum_{j < i} \cdots \sum_{q < p} \Gamma_i(\mathbf{b} - \mathbf{b}_i) \right.$$

$$\left. \times \Gamma_j(\mathbf{b} - \mathbf{b}_j) \ldots \Gamma_q(\mathbf{b} - \mathbf{b}_q) \right] \Phi_0(\mathbf{r}_1, \ldots, \mathbf{r}_A)$$

$$\times \delta \left(\frac{1}{A} \sum_{i=1}^{A} \mathbf{r}_i \right) d\mathbf{r}_1 \ldots d\mathbf{r}_A, \qquad [208]$$

where we have inserted a δ-function to deal with the nuclear center-of-mass coordinate, here taken at the origin. The appearance of $\delta(\mathbf{R})$ strips off a factor $(2\pi)^3 \delta(\mathbf{K}' - \mathbf{K})$ that would otherwise be present to express conservation of total momentum. For elastic scattering, $\Phi_f = \Phi_0$. To incorporate the single-nucleon amplitudes explicitly, we can write

$$F_{f0}(\Delta) = -\frac{ik}{2\pi} \int d^2b \, e^{i\Delta \cdot b} \Phi_f^\dagger(r_1,\ldots,r_A)$$

$$\times \left[\sum_{i=1}^{A} \left(\frac{i}{2\pi k}\right) \int d^2\delta_i e^{-i\delta_i \cdot (b-b_i)} f_i(\delta_i) + \cdots \right.$$

$$+ \sum_{i=1}^{A} \sum_{j<i} \cdots \left(\frac{i}{2\pi k}\right) \int d^2\delta_i e^{-i\delta_i \cdot (b-b_i)} f_i(\delta_i) \ldots$$

$$\left. \times \left(\frac{i}{2\pi k}\right) \int d^2\delta_q e^{-i\delta_q \cdot (b-b_q)} f_q(\delta_q) \right] \Phi_0(r_1,\ldots,r_A)$$

$$\times \delta\left(\frac{1}{A}\sum_{i=1}^{A} r_i\right) dr_1 \ldots dr_A. \qquad [209]$$

For elastic scattering on the deuteron, this reduces, after insertion of relative and center-of-mass coordinates for proton and neutron of equal mass

$$r \equiv r_1 - r_2 = r_p - r_n, \quad R \equiv \tfrac{1}{2}(r_1 + r_2) = \tfrac{1}{2}(r_p + r_n), \qquad [210]$$

to

$$F(\Delta) = f_p(\Delta) \int e^{i\Delta \cdot s/2} |u(r)|^2 \, dr$$

$$+ f_n(\Delta) \int e^{-i\Delta \cdot s/2} |u(r)|^2 \, dr$$

$$+ \frac{i}{2\pi k} \int d^2\delta f_n(\Delta - \delta) f_p(\delta) \int e^{i(\delta - \Delta/2) \cdot s} |u(r)|^2 \, dr, \qquad [211]$$

where we have ignored spin-dependence in the amplitudes, labeled them as to whether they refer to proton or neutron, and introduced s, the perpendicular (impact parameter) component of the separation vector r. This result—already encountered in Eq. 3.22—shows once again that the single-scattering term involves the single-nucleon amplitude times the Fourier transform of the square of the spatial wave function, and the double scattering involves a folding integral of two amplitudes and that Fourier transform. This double-scattering term embodies the shadowing of one nucleon by another as well as optical absorptive effects and inter-ference effects in the scattering. Because of the high-energy, forward nature of the Glauber result, nothing higher than second order can appear for the deuteron. As a result of this structure, the double-scattering term can support higher momentum transfer than the single-scattering term by sharing it between the two scatterings, so that at some angle there will be a trade off between these, the single scattering giving way to the double scattering. For more scatterers, as in the general expression given in Eq. 209, double scattering eventually yields to triple scattering, and so on. This expression also makes explicit the fact that the Glauber approach em-bodies in it the full correlational information of the wave function in the successive orders of scattering. Thus, if we can calculate all these orders (and if we have a wave function sufficiently reliable to have in it an accurate description of all orders of correlations), the Glauber result will involve directly the same kind of information that emerges term-by-term in the Watson series for the optical potential.

4.4c The Optical Limit and Its Cluster Expansion

To establish more complete contact with our earlier optical approach, we now consider the Glauber results for an optical potential. This means that, having obtained the full elastic nuclear amplitude from Eq. 194 or Eq. 206 taken between nuclear ground states

$$F(\Delta) = -\frac{ik}{2\pi} \int d^2 b \, e^{i\Delta \cdot \mathbf{b}} \Phi_0^\dagger(\mathbf{r}_1, \dots, \mathbf{r}_A)$$

$$\times \left\{ \prod_{i=1}^{A} \left[\Gamma_i(\mathbf{b} - \mathbf{b}_i) + 1 \right] - 1 \right\} \Phi_0(\mathbf{r}_1, \dots, \mathbf{r}_A) \, d\mathbf{r}_1 \dots d\mathbf{r}_A,$$

$$[212]$$

we define an equivalent optical phase parameter $\chi_{\text{opt}}(\mathbf{b})$ that will yield the same amplitude,

$$F(\Delta) = -\frac{ik}{2\pi} \int d^2 b \, e^{i\Delta \cdot \mathbf{b}} (e^{i\chi_{\text{opt}}(\mathbf{b})} - 1). \qquad [213]$$

From this we can identify

$$\chi_{\text{opt}}(\mathbf{b}) = -i \log \int \Phi_0^\dagger(\mathbf{r}_1, \dots, \mathbf{r}_A) \prod_{i=1}^{A} \left[\Gamma_i(\mathbf{b} - \mathbf{b}_i) + 1 \right] \Phi_0(\mathbf{r}_1, \dots, \mathbf{r}_A)$$

$$\times d\mathbf{r}_1 \dots d\mathbf{r}_A; \qquad [214]$$

the corresponding local optical potential can be inferred from Eq. 204:

$$\chi_{opt}(\mathbf{b}) = -v^{-1}\int_{-\infty}^{\infty}\mathcal{V}(\mathbf{b},\zeta)\,d\zeta. \qquad [215]$$

For our present purposes, we explore the consequences of this form for a simple situation parallel to that of our opening optics discussion; namely, we ignore nucleon spin and isospin, and content ourselves with the lowest two orders for the optical potential. This last step we do by expanding the logarithm in Eq. 214 about the leading term in its argument, that is, about unity. To second order in the profile functions, which, as we see in a moment, is also second order in true nucleon correlations, this gives

$$\chi_{opt}(\mathbf{b}) \cong -i\log\int d\mathbf{r}_1\dots d\mathbf{r}_A |\Phi_0(\mathbf{r}_1,\dots,\mathbf{r}_A)|^2$$

$$\times\left[1 + \sum_{i=1}^{A}\Gamma_i(\mathbf{b}-\mathbf{b}_i) + \tfrac{1}{2}\sum_{i=1}^{A}\sum_{j\neq i}\Gamma_i(\mathbf{b}-\mathbf{b}_i)\Gamma_j(\mathbf{b}-\mathbf{b}_j)\right]$$

$$\cong -i\int d\mathbf{r}_1\dots d\mathbf{r}_A |\Phi_0(\mathbf{r}_1,\dots,\mathbf{r}_A)|^2\sum_{i=1}^{A}\Gamma_i(\mathbf{b}-\mathbf{b}_i)$$

$$-\tfrac{1}{2}i\left\{\int d\mathbf{r}_1\dots d\mathbf{r}_A |\Phi_0(\mathbf{r}_1,\dots,\mathbf{r}_A)|^2\sum_{i=1}^{A}\sum_{j\neq i}\Gamma_i(\mathbf{b}-\mathbf{b}_i)\Gamma_j(\mathbf{b}-\mathbf{b}_j)\right.$$

$$\left.-\left[\int d\mathbf{r}_1\dots d\mathbf{r}_A |\Phi_0(\mathbf{r}_1,\dots,\mathbf{r}_A)|^2\sum_{i=1}^{A}\Gamma_i(\mathbf{b}-\mathbf{b}_i)\right]^2\right\}, \qquad [216]$$

where in the first step we have chosen to write the summation over the profile functions in the form we have used for the Watson multiple-scattering approach (with a compensating factor of $\tfrac{1}{2}$), and in the second step we have expanded systematically to Γ^2. Using the definitions of the nuclear (spin/isospin scalar) single-particle density, Eqs. 126 or 141, and the two-particle correlation function, Eq. 140, we have

$$\chi_{opt}(\mathbf{b}) \cong -iA\int d^2\mathbf{b}'\,\Gamma(\mathbf{b}-\mathbf{b}')\int_{-\infty}^{\infty}dz'\,\rho(\mathbf{b}',z')$$

$$-\tfrac{1}{2}i\left[A(A-1)\int d^2\mathbf{b}'\,d^2\mathbf{b}''\,\Gamma(\mathbf{b}-\mathbf{b}')\Gamma(\mathbf{b}-\mathbf{b}'')\right.$$

$$\times\int_{-\infty}^{\infty}dz'\,dz''\,P^{(2)}(\mathbf{b}',z';\mathbf{b}'',z'')$$

$$-A^2\int d^2\mathbf{b}'\,d^2\mathbf{b}''\,\Gamma(\mathbf{b}-\mathbf{b}')\Gamma(\mathbf{b}-\mathbf{b}'')$$

$$\left.\times\int_{-\infty}^{\infty}dz'\,dz''\,\rho(\mathbf{b}',z')\rho(\mathbf{b}'',z'')\right], \qquad [217]$$

where the expression in brackets relates to *true* second-order correlations. From this, Eq. 215 together with Eqs. 191 and 193 gives the optical potential in the limit of large A and for small nucleon interaction range (i.e., $\Gamma(\mathbf{b} - \mathbf{b}')$ strongly peaked at $\mathbf{b}' \approx \mathbf{b}$) as

$$\mathcal{V}(\mathbf{r}) \cong -\frac{4\pi}{2m} f_0 \left\{ A\rho(\mathbf{r}) + \frac{\pi i}{k} f_0 A(A-1) \int_{-\infty}^{\infty} dz'' \left[P^{(2)}(\mathbf{b}, z; \mathbf{b}, z'') \right. \right.$$

$$\left. \left. - \rho(\mathbf{b}, z)\rho(\mathbf{b}, z'') \right] \right\}, \qquad [218]$$

where f_0 is the forward projectile-nucleon scattering amplitude. Clearly, had we retained higher terms in Γ, we would obtain triple and higher-order correlations. Inserting the definition of the correlation function $C(\mathbf{r}', \mathbf{r})$, Eq. 143, with the large nucleus approximation of Eq. 151 and noting that

$$\int_{-\infty}^{\infty} C(\rho) \, d\rho = 2R_{\text{corr}} \qquad [219]$$

from Eq. 153, we obtain from Eq. 218

$$\mathcal{V}(\mathbf{r}) = -\frac{4\pi}{2m} f_0 A \rho(\mathbf{r}) \left[1 + \frac{2\pi i}{k} f_0 (A-1)\rho(r) R_{\text{corr}} \right]. \qquad [220]$$

This is exactly the result of Eq. 154, which was obtained as the high-energy eikonal limit for the leading two terms of the Watson series for the optical potential.

This result closes the ring on our various approaches to a description of multiple-scattering. All of them have as their essential ingredient an expansion in the basic scattering amplitude and in successively higher orders of correlation within the scattering medium; in the latter lies our main hope for a relatively speedy truncation of the series, at least at high energies. These series express the basic physics of optical absorption, in the sense of removal of the system from the elastic channel (not true absorption in which the projectile is annihilated), and diffraction as well as much more detailed aspects of multiple-scattering in the medium. At very high energies, eikonal results pertain, which considerably simplify the formal description and lead to truly diffractive scattering; at lower energies the Watson series offers, at least in principle, a systematic method for developing a complete theory of multiple scattering.

References

The development of the approach of Watson to multiple-scattering theory, on which Sections 4.1–4.3 are strongly based, was given by him and his collaborators in a series of papers: Wat–53, Fra–53a, b, Fra–56, Wat–57, Wat–58. A unified treatment of these is presented in Gol–64 and Fet–65. The very convenient exploitation of the antisymmetry of the nuclear wave function to simplify the multiple-scattering formalism was introduced in Ker–59, and a detailed comparison between this approach and that of Watson is given in Aus–77. The approach of Glauber was developed by him in Gla–55 and Gla–59, and subsequent discussions were given in Fra–66, Gla–67, and Gla–70. More recent summaries are in Czy–71 and Czy–75.

The applications of multiple-scattering theories to pion-nucleus scattering are discussed in the review articles by Kol–69 and Huf–75a and in the compendium in Gib–73.

5 Meson Scattering by Nuclei – Applications of Multiple Scattering Theory

It is our purpose in the present chapter to survey the main physical features that make their appearance when multiple-scattering theory is applied to meson-nucleus scattering reactions. We focus primarily on the major qualitative aspects of the subject, on the reliability of the more detailed descriptions and their approximation, and on the question of what can be learned from such reactions about nuclear structure, the reaction mechanisms, or the elementary particles themselves. We do not attempt to deal with detailed, technical aspects of the calculations, nor do we analyze with any thoroughness the experimental situation. There are several reasons for this: These topics would tend to obscure the main lines of theoretical development that we intend to convey here. The situation with regard to them is now in a great state of flux, and it is our present purpose to set forth the basic physical aspects that we believe likely to persist in the future, more complete theoretical and experimental analyses. Last, any effort to catalog the details of recent advances would vastly expand the size of this book; such material has its more natural place in review articles (of which several exist: Kol–69, Huf–75a, Ste–74), in the series of biennial conferences on high energy physics and nuclear structure (Cer–63, Ale–67, Dev–70, Dzh–72, Tib–74, Nag–75, Loc–77) and in seasonal schools and topical conferences devoted to meson-nucleus interactions (Eri–71a, Bec–71, Gib–73, War–76, Bar–76), to which the reader is referred.

190

5.1 Elastic Scattering of Pions

5.1a Basic Features of the Optical Potential

We begin our discussion with the consideration of elastic scattering, which is not only a natural point of departure and experimentally the most accessible scattering process, but also supplies needed input for a subsequent treatment (e.g., in DWIA) of inelastic events. We have seen in Chapter 2 that from zero pion lab kinetic energy to above 300 MeV the overwhelmingly dominant feature of pion-nucleon scattering is the resonance in the $J = T = \frac{3}{2}$ channel. It is reasonable to expect this feature to have important consequences for pion-nucleus scattering in the first few hundred MeV above scattering threshold that we wish to explore, and the fact that the physics of this resonance is relatively well known may aid us in this endeavor.

We first focus on the resonance purely in terms of its ability to provide a very large πN cross section at a particular energy, and we thus temporarily ignore special features that arise, for example, from its p-wave nature. Then taking a strict forward approximation for the amplitude producing the lowest-order optical potential in Eq. 4.131, we have

$$\mathcal{V}_c(\mathbf{r}) \cong t(\theta = 0°)A\rho(\mathbf{r}) = -\frac{4\pi}{2m}f_0 A\rho(\mathbf{r}). \qquad [1]$$

The justification for the forward approximation is expected to lie in the relatively large size of the nucleus and the consequent forward-peaking in the Fourier transform of the nuclear density, which cuts down contributions to the lowest-order optical potential at large angles. We have seen in Chapter 2 that the on-shell πN amplitude, in the range from 0–300 MeV, can be parametrized by a form involving s- and p-wave terms, namely,

$$\langle \mathbf{k}'|t(E_\pi)|\mathbf{k}\rangle = b(E_\pi) + c(E_\pi)\mathbf{k}'\cdot\mathbf{k} + id(E_\pi)\boldsymbol{\sigma}\cdot\mathbf{k}'\times\mathbf{k}, \qquad [2]$$

where $\mathbf{k}$ and $\mathbf{k}'$ are the initial and final center-of-mass momenta for the πN system and $\boldsymbol{\sigma}$ is the nucleon Pauli spin matrix. The operator of Eq. 2 is intended for use between nucleon spinors. The energy-dependent coefficients $b(E_\pi)$, $c(E_\pi)$, and $d(E_\pi)$ are related to the s- and p-wave channel amplitudes or phase shifts in Chapter 2; b, as the coefficient of the s-wave part, has no resonant behavior in this energy range, while c and d resonate at the position of the 3,3 resonance. All of these coefficients involve both isoscalar and isovector parts, but we often restrict ourselves at least preliminarily to light nuclei with nearly equal numbers of protons and neutrons so that the isovector part of the amplitude averages to a contribution much smaller than that of the isoscalar piece in the linear, lowest-order average* for t. Similarly, the nuclear targets tend to be spin saturated, that is, to have a spin down nucleon to counterbalance each spin up nucleon; hence the spin vector term with coefficient $d(E_\pi)$ is ignored. Of course, it is no problem in principle to include these terms.

In the forward direction this then yields for the approximate optical potential

$$\mathcal{V}_c(\mathbf{r}) \cong -\frac{4\pi}{2m} f_0 A \rho(\mathbf{r}) = \left[b(E_\pi) + k^2 c(E_\pi) \right] A \rho(\mathbf{r}). \qquad [3]$$

As we approach resonance, $c(E_\pi)$ grows and becomes predominantly imaginary, the imaginary part of the forward amplitude being given by the optical theorem as

$$\mathrm{Im} f_0 = \frac{k}{4\pi} \langle \sigma_{\pi N}^{\mathrm{tot}}(E_\pi) \rangle_{T \cong 0}, \qquad [4]$$

where we have indicated an average of the total πN cross section over (approximately) equal numbers of protons and neutrons. At resonance, the pertinent value is $\frac{1}{2}(\sigma_{\pi^\pm p}^{\mathrm{tot}} + \sigma_{\pi^\pm n}^{\mathrm{tot}}) \cong 135$ mb, which, with $k \cong 304$ MeV$/c$, yields[†] $f_0 \sim (1.7$ fm$)i$, whereas well above the region of πN resonances, say for $E_\pi > 3$ GeV, the total cross section settles down to a relatively smooth and constant value of about 30 mb. Because of the resonance, which through Eqs. 3 and 4 makes the imaginary part of the optical potential very large, the scattering even at that rather low energy value is highly diffractive in the sense that it may be well approximated by scattering from a black, completely absorptive sphere. This is dramatically illustrated in Fig. 5.1, which shows a cross section for $\pi^- - {}^{12}C$ elastic scattering at 180 MeV calculated from a model based on Eqs. 4.37 and 4.38 for a completely absorptive, sharp-edged nucleus with a further factor incorporated to diffuse the edge slightly. The remarkable agreement with experiment obtained in this very simple model underlines the fact that for a highly absorptive situation, the main result is dictated largely by geometry. The same point is also reflected (Huf–75b) in a considerable insensitivity to certain of the parameters that go into defining the p-wave term in the Kisslinger potential (cf. Eq. 8). So long as we look at elastic scattering at

---

*For higher-order terms, the isospin and spin vector terms in the optical potential have been considered, for example, in Lam–73.
†Note that this value does not incorporate spin/isospin (coherent) saturation, nor does it restrict itself to the 3,3 channel, as did the estimate $f_{\mathrm{res}} \sim \frac{4}{3} i / k_{\mathrm{res}}$ under Eqs. 4.155 and 4.173. The present estimate seems more appropriate for $\mathrm{Im}\,\mathcal{V}$, which involves knockout processes.

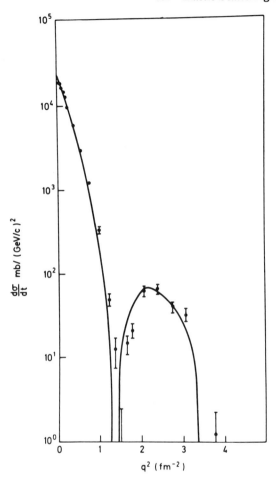

Fig. 5.1 The differential cross section for $\pi^- - {}^{12}\text{C}$ elastic scattering at 180 MeV compared with a "trivial" model calculation using merely a black sphere nucleus (Wil–71). The data are from Bin–70.

other than large angles and very near to the 3,3 region, the dominant physical parameter is the nuclear hadronic radius (and the πN forward scattering amplitude). To get away from this constraint, we must consider other processes, examine larger angles, or work above or below the resonance. Each of these has its own hazards: The study of processes other than elastic scattering is generally substantially more complicated and often requires input information that should be first tested for the elastic case. Larger angle scattering, as we see, introduces many uncertainties, chief among them the unknown features of nucleon correlations. Working well above the $\Delta(1232)$ resonance, while feasible and interesting, is experimentally more difficult, encounters other resonances, and raises the specter of increasingly severe problems with the relativistic aspects of the problem. Thus, in some ways, the most immediately attractive prospect is work in the region of 50–100 MeV, which, while requiring great attention to the nuclear structure corrections in the impulse approximation, Fermi averaging, and cluster expansion of the optical potential, may offer opportunities to examine more revealing aspects of the πA system (Eis–76a). Moreover, in this energy range a further interesting feature enters in the form of appreciable interference between the s- and p-wave πN components near the forward direction. This tends to weaken the optical potential of Eq. 3 somewhat, and also to produce great sensitivity to calculational details.

5.1b   A Closer Study of the Optical Potential

In order to treat the energy region below and at the $3,3$ resonance more fully, it is clear that we can no longer rely on the forward approximation of Eqs. 1 and 3 and must generally deal more thoroughly with the p-wave nature of the resonance. This in turn requires that we reexamine Eq. 2 in greater detail. The form of the amplitude given there assumes implicitly that we are contemplating an on-energy-shell situation; hence the magnitudes k and k' ($= k$ in the center-of-mass system for elastic πN scattering, as here) can always be compensated for in the coefficients $c(E_\pi)$ and $d(E_\pi)$. Then the term $c(E_\pi)\mathbf{k}'\cdot\mathbf{k}$, for example, is simply a convenient way of expressing $c'(E_\pi)P_1(\cos\theta)$, the $l=1$ partial wave in the scattering angle θ, with $c' = k^2 c$. Of course, we could as well have taken $c'\hat{\mathbf{k}}'\cdot\hat{\mathbf{k}} = c'\cos\theta$ with unit vectors $\hat{\mathbf{k}}$ and $\hat{\mathbf{k}}'$, which in the on-energy-shell case would make no difference. In our nuclear applications, however, we require the πN amplitude in the presence of other nucleons so that off-shell effects are always present and our spin-scalar s- and p-wave representation for the scattering transition matrix should really be written more generally as

$$\langle \mathbf{k}'|t(E_\pi)|\mathbf{k}\rangle = b(E_\pi;k',k) + c(E_\pi;k',k)\mathbf{k}'\cdot\mathbf{k}. \qquad [\,5a\,]$$

Common experience with local potentials tends to suggest that the dependence of b and c on the off-energy-shell momenta is only important at fairly high values of k and k', the p-wave aspect being correctly represented near threshold by $\mathbf{k}'\cdot\mathbf{k}$ and the s-wave by a factor with no momentum-dependence, and so, preliminarily, we consider a simplified parametrization already introduced in Eq. 2 with the further restriction of dropping the spin term so that

$$\langle \mathbf{k}'|t(E_\pi)|\mathbf{k}\rangle = b(E_\pi) + c(E_\pi)\mathbf{k}'\cdot\mathbf{k}. \qquad [\,5b\,]$$

The form of Eq. 5b is now to be used for the construction of an optical potential, as in Eq. 4.124, in momentum space, for instance, after which the Schroedinger equation with that potential is solved for the pion scattering problem.* From Eq. 4.120 we know that we must transform the πN amplitude into the appropriate kinematical frame for the nucleon under consideration, to be averaged over the momentum distribution of the nucleons. For the moment, however, we ignore this complication on the grounds that the nucleon has a much higher rest mass energy than the pion energies we are considering; thus the πN center-of-mass frame is close to the lab frame, which in turn is not very different—certainly on the average—from the various individual frames of the nucleons with their Fermi motion. Then we take the lowest-order optical potential in momentum space in the approximate form (Eq. 4.130)

$$\langle \mathbf{k}'|\mathcal{V}_c|\mathbf{k}\rangle = A\langle \mathbf{k}'|t|\mathbf{k}\rangle \int e^{i(\mathbf{k}-\mathbf{k}')\cdot\mathbf{r}}\rho(\mathbf{r})\,d\mathbf{r}, \qquad [6]$$

where $\rho(\mathbf{r})$ is the spin/isospin-scalar nuclear density, or, for $S=T=0$ nuclei, just the ground state density.

It is clear in Eq. 6 that, unless the amplitude $\langle \mathbf{k}'|t|\mathbf{k}\rangle$ depends on the momentum *difference* $\mathbf{k}-\mathbf{k}'$, the optical potential will not be local in configuration space. In principle this need not be a source of great

---

*We note that in much of the region of interest to us the pion is relativistic; for example, near the 3,3 resonance its velocity is ~ 0.9. Thus, although the Watson multiple-scattering approach on which we are primarily focusing is intrinsically nonrelativistic, we anticipate making at least "kinematic" modifications for the relativistic nature of the projectile. We do not generally use the fully relativistic Klein-Gordon equation for the pion, since it is not clear what transformation properties should be attributed to our nonrelativistic potential, nor are we prepared to go beyond a one-particle approximation for the pion field. As one alternative (Gol-64, pp. 24-25), we may modify the Schroedinger equation for a relativistic spinless particle in a potential $\mathcal{V}$,

$$(\omega+\mathcal{V})\psi = E\psi, \quad \omega = \sqrt{p^2+m^2}\,,$$

multiplying by $(\omega-\mathcal{V})$ and using the equation itself to get

$$(\omega^2+[\omega,\mathcal{V}])\psi = (E-\mathcal{V})^2\psi$$

which apart from the commutator $[\omega,\mathcal{V}]$ is the Klein-Gordon equation for an interaction $\mathcal{V}$ which is the fourth component of a four-vector. For a potential that is small compared to the meson mass, and varies slowly, the solutions of this equation are essentially the same as those of the Klein-Gordon equation. (See also Section 6.1, where the relation of the Klein-Gordon form to crossing is discussed.)

concern, but in practice it is an appreciable complication, since calculations in configuration space with local or "essentially" local potentials are much easier (in the modern age this means that they require far less computer time) than more general calculations in momentum space, say. There has therefore been a strong tendency to start the theoretical program of pion-nucleus scattering with a reasonably simple configuration-space potential. Kisslinger (Kis–55) accomplished this for the approximate lowest-order optical potential of Eqs. 5b and 6 by noting that it could be written as

$$\langle \mathbf{k}' | \mathcal{V}_c | \mathbf{k} \rangle = Ab(E_\pi) \int e^{-i\mathbf{k}'\cdot\mathbf{r}} \rho(\mathbf{r}) e^{i\mathbf{k}\cdot\mathbf{r}} d\mathbf{r} + Ac(E_\pi) \int (\nabla e^{-i\mathbf{k}'\cdot\mathbf{r}}) \cdot \rho(\mathbf{r}) (\nabla e^{i\mathbf{k}\cdot\mathbf{r}}) d\mathbf{r}$$

$$= Ab(E_\pi) \int e^{-i\mathbf{k}'\cdot\mathbf{r}} \rho(\mathbf{r}) e^{i\mathbf{k}\cdot\mathbf{r}} d\mathbf{r} - Ac(E_\pi) \int e^{-i\mathbf{k}'\cdot\mathbf{r}} [\nabla \cdot \rho(\mathbf{r}) \nabla] e^{i\mathbf{k}\cdot\mathbf{r}} d\mathbf{r},$$

$$[7]$$

where in the second term we have integrated by parts, and each gradient operator is understood to act on all functions of $\mathbf{r}$ to the right of it. The configuration space optical potential can now immediately be read off from Eq. 7 by stripping away the plane wave factors in the integrand

$$\mathcal{V}_c(\mathbf{r}) = Ab(E_\pi)\rho(\mathbf{r}) - Ac(E_\pi)\nabla \cdot \rho(\mathbf{r})\nabla.$$

$$[8]$$

The s-wave result here has been obvious from the start. In the second, p-wave part the nonlocality of the potential is expressed in the fact that the gradients will act on the pion wave function when this potential is used in the Schroedinger equation. This nonlocality complication is not, however, of an essential nature for practical purposes, since the Schroedinger equation may be transformed into the local form by means of a substitution for the pion wave function $\psi(\mathbf{r})$ in the form*

$$\psi(\mathbf{r}) = \phi(\mathbf{r}) / \sqrt{1 + 2mAc(E_\pi)\rho(\mathbf{r})} .$$

$$[9]$$

The resulting "potential" in the second-order equation satisfied by $\phi(\mathbf{r})$—itself not, of course, a true wave function—is messy but local; thus standard and rapid methods of solution can be applied. This makes the Kisslinger potential, Eq. 8, useful for relatively simple and speedy calculations of pion scattering.

Equation 8 exhibits the first in a series of nonlocal features that we find within general πA optical potentials. The Kisslinger potential "just misses"

---

*This transformation also points up the presence of a singularity at $1 + 2mAc(E_\pi)\rho(\mathbf{r}) = 0$ which is only avoided in nuclear applications at low energies by the fact that $\mathrm{Im}\, c(E_\pi) \neq 0$.

being local, involving as it does the behavior of the wave function at a point r and in its immediate vicinity through derivatives of the wave function. In arriving at this form, both the historical development and our presentation here have been guided by latent prejudices as to the "natural" manifestation in a potential of the usual k^{2l} threshold behavior in the amplitude. This feeling is strengthened somewhat by the successful treatment by Chew and Low of the πN 3,3 resonance assuming a separable off-shell dependence on the momenta variables of this sort (see Chapter 2). We could also have exploited the unknown off-shell behavior to produce a different and local optical potential. We would then have replaced the *on*-shell πN amplitude of Eq. 5b by

$$\langle \mathbf{k}'|t(E_\pi)|\mathbf{k}\rangle = \left[b(E_\pi) + k^2 c(E_\pi) \right] - \tfrac{1}{2} c(E_\pi)\mathbf{q}^2, \qquad [10]$$

where $\mathbf{q}=\mathbf{k}-\mathbf{k}'$, and we have lumped the terms in square brackets together, taking k^2 there with its on-shell value. The last term, being a function of $\mathbf{k}-\mathbf{k}'$, immediately yields a local optical potential

$$\mathcal{V}_c^{\mathrm{loc}}(\mathbf{r}) = A\left[b(E_\pi) + k^2 c(E_\pi) \right]\rho(\mathbf{r}) + \tfrac{1}{2}Ac(E_\pi)\nabla^2\rho(\mathbf{r}) \qquad [11]$$

with the laplacian acting only on the density, which can be seen as in Eq. 7. For lowest-order treatment of elastic or nearly elastic processes Eqs. 8 and 11 will be nearly equivalent, but where highly off-energy-shell momenta are involved they will not be. Both incorporate p-wave effects, such as a minimum at 90°, but induce differences at large angles. The local form of Eq. 11, incidentally, is also useful for applications to kaon-nucleus interactions, since D-state K^-N resonances are there important and lead to terms of the form $\nabla^4\rho(\mathbf{r})$, here taken as local. Such terms will strongly emphasize surface interactions (Kis–76b). They may also appear in general amplitudes at higher energy if these are parametrized in terms of the momentum transfer, $f(\mathbf{q}^2)=f_0+f_1\mathbf{q}^2+f_2\mathbf{q}^4+\ldots$, thus leading to local pieces in the optical potential involving derivatives of the density.

Both the nonlocal and the local potentials of Eqs. 8 and 11 are deficient in their failure to deal correctly with the damping of high, off-energy-shell momentum components. This should correctly be done through the introduction of a vertex form factor for the πN scattering amplitudes, which will be the source of still more nonlocality, and will require treatment of the problem in momentum space. The p-wave part of the πN amplitude might then be parametrized, under the influence of the theory of Chew and Low say, in a separable form (Eq. 2.100)

$$t_{\pi N}^{p-\mathrm{wave}} = c(E_\pi)g(k')g(k)\mathbf{k}'\cdot\mathbf{k}, \qquad [12]$$

with $g(k_0) = 1$, where k_0 is the on-shell momentum associated with E_π, and $g(k) \underset{k \to \infty}{\to} 0$. This is often parametrized as $g(k) = [(k_0^2 + \Lambda_n^2)/(k^2 + \Lambda_n^2)]^n$, with $n = 1$ or 2. The rate of fall-off of the vertex factor Λ_n reflects, in part, the difficult physics of higher-energy inelastic channels coupled to the πN system and is therefore rather uncertain (see Section 2.4c), though not without appreciable influence on a variety of calculational results.

Both the potentials of Eqs. 8 and 11 have as their main effect a modification of the cross section at large angle from what would be obtained without the inclusion of the specifically p-wave feature of the πN scattering. In a qualitative way one can see from Eqs. 6 and 10 that the momentum space optical potential involves in its p-wave part

$$\langle \mathbf{k}' | \mathcal{V}_c^{\text{loc}} | \mathbf{k} \rangle \cong A b \rho(\mathbf{q}) + A c k^2 \left(1 - \frac{q^2}{2k^2} \right) \left(1 - \frac{q^2 R^2}{6} + \cdots \right), \quad [13]$$

where in the last factor we have expanded the Fourier transform of a uniform nuclear density of radius R to second order in qR. The presence of the p-wave factor $(1 - q^2/2k^2)$ could be incorporated into a conventional calculation to this order, without density derivatives, by using the effective radius from Eq. 13,

$$R^{*2} = R^2 + \frac{3}{k^2} . \qquad [14]$$

Such an approach even meets with reasonable quantitative success (Ste-74).

One effect induced by the nonlocal Kisslinger potential appears even in the forward direction, and arises from the particular way in which it assigns off-energy-shell momentum dependence within the πN amplitude. This in turn determines the internal pion momentum for propagation within the medium, and may lead to an important qualitative effect in pion-nucleus scattering. We find the possibility of a large shift in the position of the resonance in the πA, as opposed to the πN, system (Eri-70a). To see this we may start from Eq. (4.40) for the internal pion momentum, which was the dispersion relation

$$K^2 = k^2 + 4\pi \rho f(k; K). \qquad [15]$$

The effect is, of course, just equivalent to using the first order optical potential in the Schroedinger equation for a medium of constant density ρ (nuclear matter). We stress that Eq. 15 naturally requires that we make a specific, model-dependent assignment of the off-energy-shell behavior of the amplitude. We parametrize the forward scattering amplitude f as in Eq. 5b, taking for convenience the limit $M \gg m$ so that the center-of-mass and lab systems coincide and $f = (-2m/4\pi)t$. We drop the s-wave piece as being immaterial to the present issue and in any event very small near the p-wave resonance, which we incorporate through

$$c(E_\pi) = -\frac{4\pi}{2m} \frac{C}{E_\pi - E_{\text{res}} + \frac{1}{2} i \Gamma(E_\pi)}, \qquad E_\pi = \sqrt{k^2 + m^2} \ . \quad [16]$$

Then Eq. 15 becomes

$$K^2 = k^2 + 4\pi\rho \frac{CK^2}{E_\pi - E_{\text{res}} + \frac{1}{2} i \Gamma(E_\pi)}, \qquad [17]$$

or

$$K^2 = \frac{E_\pi - E_{\text{res}} + \frac{1}{2} i \Gamma(E_\pi)}{E_\pi - E_{\text{res}} - 4\pi\rho C + \frac{1}{2} i \Gamma(E_\pi)} k^2 = n^2 k^2, \qquad [18]$$

where n is the index of refraction, Eq. 4.4. From Eq. 4.13 the optical potential, in terms of the external pion kinematical variables, is

$$\mathcal{V} = -\frac{k^2}{2m}(n^2 - 1) = -\frac{4\pi}{2m}\rho \frac{Ck^2}{E_\pi - E_{\text{res}} - 4\pi\rho C + \frac{1}{2} i \Gamma(E_\pi)} \ ;$$

$$[19]$$

that is, the position of the resonance effect *in the optical potential* would be shifted downward by the amount $4\pi\rho C \sim 40$ MeV for central densities ($\rho \sim 0.17$ fm^{-3}, $f_{\text{res}} \sim 0.9$ fm) to the extent that the assumptions of this rather oversimplified model are correct. This is a considerable overestimate, not only because the central density is not the operative quantity—the nucleus being highly absorptive there—but also because near the nuclear resonance Eq. 18 indicates that K will inevitably be very large even though the external momentum may not be. Thus the fact that the πN interaction has a nonzero range enters the picture by suppressing the contribution from large internal momenta K. The effect of this is to reduce the shift* in the πA peak position from the $4\pi\rho C \sim 40$ MeV indicated in

---

*An alternative, related discussion of the shift of the πA cross section peak from the πN resonance position is given in Loc-71, while recent re-analyses relating to Eri-70a are provided in McV-77 and Sed-77. There are also effects that produce an effective shift of the resonance peak through kinematic factors whose consequences change if the resonance is broadened, for example, Lan-73.

Eq. 19 to perhaps about half that value for plausible choices of the off-shell behavior in the πNN vertex, depending rather sensitively on the particular choice made in that vertex (Mon–73). We note that if we were to use the "local" prescription of Eq. 11 in place of the Kisslinger potential then the dispersion relation Eq. 15 would yield

$$K^2 = k^2 + 4\pi\rho \frac{Ck^2}{E_\pi - E_{\text{res}} + \frac{1}{2}i\Gamma(E_\pi)} \qquad [20]$$

instead of Eqs. 17 and 18, or

$$\mathcal{V} = -\frac{4\pi}{2m}\rho \frac{Ck^2}{E_\pi - E_{\text{res}} + \frac{1}{2}i\Gamma(E_\pi)}, \qquad [21]$$

whence there would be no shift effect.

Equation 19, when juxtaposed with Eq. 16, underscores the different manifestations of the 3,3 resonance in the basic πN system and in the nuclear case. In the former instance, the resonance form appears directly in the relevant amplitude and involves the channel phase shift $\delta_{3,3}$ going through the value $\pi/2$, in this case in an elastic fashion, $\eta = 1$. This was shown in the Argand diagrams of Fig. 2.4. The resonance in the *nucleus* is characterized here by the imaginary part of the *optical potential* going through a maximum (while the real part vanishes), that is, a maximal absorptivity in the system which may be shifted, through a number of different mechanisms, from the resonance position in the basic system. We have already remarked on the consequences of this extreme absorptivity in Fig. 5.1. It is also illustrated by the Argand plots for the πA case of Fig. 5.2. Because of the large nuclear size, the resonance effect—originally p-wave in the πN system—is spread over several partial waves, all of which "resonate" at *roughly* the same energy in a highly inelastic way (i.e. the complex amplitude f_L passes counterclockwise through a purely imaginary value, with $\delta_L = \pi/2, \eta_L \ll 1$; see Eq. 2.13a and Fig. 2.4a). The "resonance" in the nuclear case has an increase in cross section that represents the increased probability for scattering on each constituent nucleon. This in turn raises the probability of inelastic events (nucleon knock-out, for instance) which represents optical absorption in the many-nucleon system. Even for a system as light as the deuteron this situation exists (Hoe–74), that is, the 3,3 πN resonance in the compound system produces enhanced scattering from each constituent with a corresponding increase in the elastic scattering and the absorption, as discussed in Section 4.1, but no resonance between the pion and the full nucleus in the same sense as between the pion and the single nucleon.

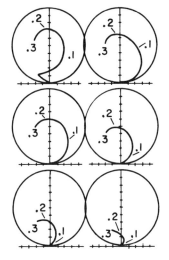

Fig. 5.2 Argand diagrams for pion-carbon partial waves as a function of energy (from Sek-73). Energy units are GeV and, reading left to right from top, $L = 0, 1, \ldots, 5$.

Thus far we have ignored totally the motion of the nucleons as it effects the properties of the optical potential. In point of fact, this Fermi motion will have a "smearing" effect on the basic amplitude $t_{\pi N}$, which must appear in a form such that an average over the nuclear momenta has been performed. Once again this requires us to know the off-shell behavior of the amplitude so that we can identify the momentum dependence over which to average. Even within the confines of the potential of Eq. 8, that is, the neglect of detailed off-shell momentum behavior, a complication must be addressed in the form of corrections due to the transformation of the πN amplitude from the center-of-mass frame, for which Eq. 5b, say, is an appropriate parametrization, to the actual frame of scattering for a pion on a particular nucleon in the nucleus. This transformation is easily treated if we construct the amplitudes in a manifestly covariant way, which we do, for the moment, in a nonrelativistic form, replacing Eq. 5b with

$$\langle \mathbf{k}'_\pi \mathbf{p}'_N | t(E_\pi) | \mathbf{k}_\pi \mathbf{p}_N \rangle = b(E_\pi)$$

$$+ c(E_\pi) \left[\frac{M}{M+m} \mathbf{k}'_\pi - \frac{m}{M+m} \mathbf{p}'_N \right] \cdot \left[\frac{M}{M+m} \mathbf{k}_\pi - \frac{m}{M+m} \mathbf{p}_N \right].$$

$$[22]$$

In a nonrelativistic approach, the t-matrix element is Galilei invariant*.

---

*For instance, to lowest order t is the matrix element of a (nonrelativistically) invariant potential. The scattering amplitude f, on the other hand, transforms as $(\cos\theta)^{-1/2}$ as can be seen from $d\sigma/(d\cos\theta\,d\phi) = |f|^2$ and the invariance of the total cross section. In the center-of-mass frame *only*, the relationship between them reduces to simply $t = (-2m_{\text{red}}/4\pi)f$, where m_{red} is the reduced mass. For forward angles, the invariance of the total cross section and the optical theorem can be used to establish $f_{\text{lab}}/k_{\text{lab}} = f_{\text{c.m.}}/k_{\text{c.m.}}$.

Moreover, the vector combination in the brackets, involving the pion and nucleon momenta $\mathbf{k}_\pi$ and $\mathbf{p}_N$ in a general frame and ratios of their masses m and M, are invariant since they are constructed so as to be proportional to the difference of the pion and nucleon velocities $\mathbf{v}_\pi - \mathbf{v}_N$, whence the effects of a Galilei transformation to a frame moving with velocity $\mathbf{V}$ cancel:

$$\mathbf{v}_\pi - \mathbf{v}_N \rightarrow \mathbf{v}'_\pi - \mathbf{v}'_N = (\mathbf{v}_\pi - \mathbf{V}) - (\mathbf{v}_N - \mathbf{V}) = \mathbf{v}_\pi - \mathbf{v}_N. \qquad [23]$$

In the center-of-mass frame, $\mathbf{k}_\pi = \mathbf{k}$, $\mathbf{p}_N = -\mathbf{k}$ and Eq. 22 reduces to Eq. 5b, but it is also a valid description in any other frame and basically embodies automatically the transformations of the cosine of the scattering angle and the magnitudes of the momenta from frame to frame.

In the laboratory frame, $\mathbf{p}_N = 0, \mathbf{p}'_N = \mathbf{k}_\pi - \mathbf{k}'_\pi$, the amplitude becomes

$$\langle \mathbf{k}'_\pi, \mathbf{k}_\pi - \mathbf{k}'_\pi | t(E_\pi) | \mathbf{k}_\pi, 0 \rangle = \left[b(E_\pi) - c(E_\pi) \frac{mM}{(M+m)^2} k_\pi^2 \right]$$

$$+ c(E_\pi) \frac{M}{M+m} \mathbf{k}'_\pi \cdot \mathbf{k}_\pi. \qquad [24]$$

The effect of the transformation is to modify the coefficients of the s- and p-wave parts in a manner involving changes of order* $m/M \sim 1/6.7$. Because of the relative smallness of this ratio, the nucleon laboratory frame is also a good approximation to the various individual frames for pion-nucleus scattering in the region around the 3,3 resonance, since the nucleon Fermi velocity is much smaller than the pion velocity (though the Fermi momentum is close to the pion momentum at the resonance). Of course, Eq. 24 is not yet a solution to the more specialized problem of producing a potential in configuration space which will generalize the Kisslinger potential, Eq. 8, since we must determine how to take $\mathbf{k}_\pi$ off the energy shell. (This feature is discussed in Mil–74a and the more complete kinematic averaging in Mac–72, 73.)

There is, of course, no problem in carrying out these kinematic transformations for on-shell amplitudes in the relativistic case.[†] A serious problem arises in the relativistic case, however, in that there is no well-defined meaning to a Lorentz transformation of an off-energy-shell t-amplitude (Hel–76). For example, since the intermediate total energy in the off-shell amplitude need not generally be conserved, the Lorentz transformed total

---

*In reality, relativistic effects enter appreciably near the 3,3 region, say, and the relevant construct to produce Eq. 23 involves the pion total relativistic energy ω instead of the mass m; near the 3,3 resonance $\omega/M \sim \frac{1}{3}$; that is, the numerical magnitudes of the effect are somewhat larger than implied by a purely nonrelativistic approach.

three-momentum, though conserved in one frame, need not be in another (where the four components are mixed). For the fully off-energy-shell amplitude, one would indeed need to consider separate Lorentz transformations for the initial and final states as well (Hel–76). Thus what is a straightforward kinematical problem in the nonrelativistic case becomes a more intricate problem in dynamics for the relativistic situation.

5.1c A Brief Survey of Calculations with the Lowest-Order Optical Potential

The early precision measurement (Bin–70) of π^- scattering on ^{12}C between 100 and 300 MeV, followed by a similar study (Bin–75; see also Cro–69) on ^4He, spurred a considerable number of calculations of these processes in the kinematic region of the 3,3 resonance, based on multiple-scattering theories. As a characterization of the degree of success with which such efforts met, we show in Fig. 5.3 one of the first of them (Ste–70; see also Kre–70 and Mai–71) for $\pi^- - {}^{12}C$ elastic scattering. These calculations are based on the lowest-order Kisslinger optical potential, Eq. 8, with coefficients $b(E_\pi)$ and $c(E_\pi)$ taken from πN data and Fermi averaged, and give a remarkably good representation of the data, especially at forward angles. If the b and c coefficients are varied the fit is somewhat improved, though generally at larger angles, where higher-order corrections to the optical potential must be expected to enter, and to which we turn shortly. The general agreement with the data may be seen as a partial vindication of the approximations that have led us to the πA optical potential, bearing in mind the relative insensitivity to detailed features of the theory which was already noted with regard to Fig. 5.1. As we have mentioned, the data below resonance may be more sensitive to detailed nuclear features, especially to the motion of the nucleons. A remarkably

---

†The Galilei invariant vectors in the square brackets of Eq. 22 can, for example, be replaced by corresponding constructs

$$k_\mu \equiv (k_\pi - k_N)_\mu - \frac{(k_\pi - k_N)_\nu (k_\pi + k_N)_\nu}{(k_\pi + k_N)_\lambda (k_\pi + k_N)_\lambda} (k_\pi + k_N)_\mu,$$

where all the quantities subscripted with Greek indices are four-vectors, and repeated indices are to be summed. In the center-of-mass system

$$k_\pi - k_N = (2\mathbf{k}, \sqrt{m^2 + \mathbf{k}^2} - \sqrt{M^2 + \mathbf{k}^2}) \quad \text{and} \quad k_\pi + k_N = (\mathbf{0}, \sqrt{m^2 + \mathbf{k}^2} + \sqrt{M^2 + \mathbf{k}^2})$$

so that $k_\mu = (2\mathbf{k}, 0)$ and the Lorentz invariant $\frac{1}{4} k'_\mu k_\mu$ is $\mathbf{k}' \cdot \mathbf{k}$ in that system; it is easily evaluated in any other system, and, as in Eq. 24 leads to a modification of the partial wave decomposition in the various frames.

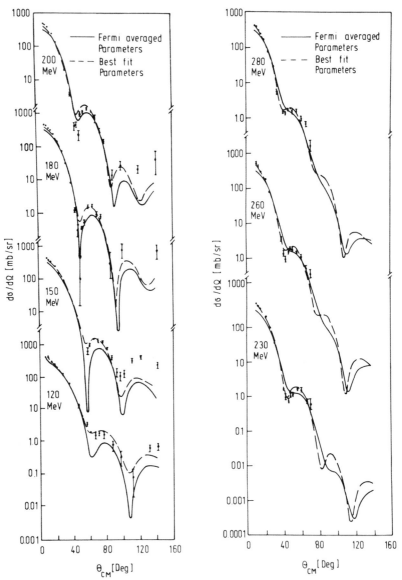

Fig. 5.3 Differential cross sections for $\pi^- - {}^{12}C$ elastic scattering at various energies throughout the 3,3 resonance region, as calculated from a lowest-order optical potential derived either from Fermi-averaged πN parameters (solid line) or best-fit parameters (Ste–70). The data are from Bin–70.

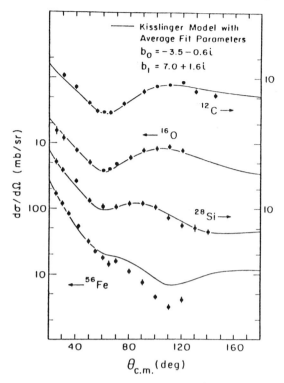

Fig. 5.4 Fits of the Kisslinger potential to 50 MeV π^+ elastic scattering data for various nuclei (from Dyt–79).

good fit of the Kisslinger potential to 50 MeV π^+ data for various nuclei is shown in Fig. 5.4. This fit requires considerable adjustment of the optical potential parameters with respect to their free $t_{\pi N}$ values, presumably representing such effects as Fermi motion, nucleon binding, true pion absorption, the angle transformation of Eq. 24, and correlation effects to which we now turn. These last effects are also expected to be pertinent for interpreting scattering in the resonance region at angles beyond the first diffraction minimum.

5.1d   Correlations and Other Higher-Order Effects in the Optical Potential

The basic theoretical approach to two-nucleon correlation terms as they enter in the cluster expansion of the optical potential has been to use formalism 1 to second order, Eq. 4.75, with auxiliary assumptions to simplify this innocent-looking, but in reality rather formidable result. The main lines have followed the original investigations of these terms in pA scattering by Feshbach and his coworkers (Fes–70, 71, Lam–73, Ull–74, Bir–75, Gar–78). The preference for using formalism 1 arises from the large effect that propagation in the medium has on the projectile—here the pion near the 3,3 region with its short mean free path—and the consequent expectation that formalism 1, by summing over much of this effect from the start, may lead to a more rapidly convergent series. The second-order optical potential is written as (cf. Eq. 4.75 and Fes–71)

$$\mathcal{V}_c^{(2,1)} \cong \langle 0| \sum_{i=1}^{A} t_i |0\rangle$$

$$+ \sum_{\nu \neq 0} \langle 0| \sum_{i=1}^{A} t_i |\nu\rangle \frac{1}{E - E_\nu^{\text{nucl}} - K - \langle \nu| \sum t |\nu\rangle} \langle \nu| \sum_{j=1}^{A} t_j |0\rangle,$$

$$[25]$$

in which we now make the first major assumption of closure for the sum over nuclear states, similar to that discussed in connection with Eqs. 4.137–4.144 or Eqs. 4.175–4.180. We define an average nuclear excitation energy $\overline{E}^{\text{nucl}}$, independent of nuclear state ν, which is to replace E_ν^{nucl} in the propagation denominator, and assume that the projectile propagation takes place predominantly at sufficiently high energy that we do not need to know $\overline{E}^{\text{nucl}}$ too precisely, and can approximate it by some fairly low value. We next assume that the state-dependent lowest-order optical potential in the denominator is much the same for the nuclear states of interest, and is well-approximated by the ground-state result $\mathcal{V}_c^{(1)} = \langle 0| \Sigma t_i |0\rangle$. Closure may then be used to remove the nuclear sum and introduce the true correlation function as in Eqs. 4.139–4.143. The presence of the lowest-order optical potential in the propagator of formalism 1 is still troublesome, however, since it prevents this quantity from being diagonal in projectile space, and therefore fairly restrictive assumptions have usually been made at this stage (e.g., Lee–77a) that this potential may be approximated by a constant value $\langle \mathbf{k}| \Sigma t_i |\mathbf{k}\rangle$ where $\mathbf{k}$ is the incident momentum. The problem then requires evaluating a form much like that in Eq. 4.139,

$$\langle \mathbf{k}'| \mathcal{V}_c^{(2,1)} |\mathbf{k}\rangle \approx A \langle \mathbf{k}'| t |\mathbf{k}\rangle \rho(\mathbf{k} - \mathbf{k}')$$

$$+ A(A-1) \int \langle \mathbf{k}'| t |\mathbf{k}''\rangle \frac{C(\mathbf{k}', \mathbf{k}; \mathbf{k}'')}{E - \overline{E}^{\text{nucl}} - E_\pi(\mathbf{k}'') - \mathcal{V}_c^{(1)}} \langle \mathbf{k}''| t |\mathbf{k}\rangle \frac{d\mathbf{k}''}{(2\pi)^3},$$

$$[26]$$

where

$$\rho(\mathbf{k} - \mathbf{k}') \equiv \int e^{i(\mathbf{k} - \mathbf{k}') \cdot \mathbf{r}} \rho(\mathbf{r}) \, d\mathbf{r} \qquad [27]$$

is the Fourier transform of the nuclear density, and

$$C(\mathbf{k}', \mathbf{k}; \mathbf{k}'') \equiv \int e^{i(\mathbf{k}'' - \mathbf{k}') \cdot \mathbf{r}'} \left[P^{(2)}(\mathbf{r}', \mathbf{r}) - \rho(\mathbf{r}')\rho(\mathbf{r}) \right] e^{i(\mathbf{k} - \mathbf{k}'') \cdot \mathbf{r}} \, d\mathbf{r} \, d\mathbf{r}' \qquad [28]$$

is the transform of the true correlation function. In Eq. 26, of course, more elaborate integration over the nuclear motion may be included, and, to some degree, more satisfying approximations for $\mathcal{V}_c^{(1)}$ in the denominator may be studied (Bir–75). The correlations generally included for the evaluation of Eqs. 28 and 26 involve Pauli and hard-core effects; relatively little consideration of attractive, intermediate-range tensor correlations has thus far been included. The former are simply the correlations induced by antisymmetrizing the nuclear wave function, and the hard-core effects arise from the suppression of the two-nucleon wave function for small values of their separation distance due to the strong nucleon-nucleon repulsion there. The consequences of the short-range correlation are most dramatic for very light nuclei, for example, ^{4}He. These have smaller dimensions, a fact which tends to enhance short-range effects, and, in the case of ^{4}He taken in a shell model with $(1s)^4$ configuration, the Pauli correlation vanishes. (On the other hand, for light nuclei the presence of an A-particle correlation, induced by the requirement that the center of mass $\mathbf{R} = A^{-1} \times (\mathbf{r}_1 + \mathbf{r}_2 + \cdots + \mathbf{r}_A)$ be fixed is more important.) Results* (Lee–77a) of including the second-order two-particle correlation effect for $\pi^- - {}^4$He elastic scattering are shown in Fig. 5.5. At low energies on this very small nucleus the theory performs poorly, while at higher energies the main effect of the correlation taken in the form of Eq. 4.151 with $C(\rho) = -\exp[-\rho^2/2l_c^2]$, $l_c = 0.4$ fm, enters at and beyond the second maximum. By the time the third maximum is encountered one must anticipate the need to include also three-nucleon correlations (Ull–74) and so on. As we have seen, there are nonnegligible technical difficulties† in carrying out even the second-order calculations. As a consequence, both the reliability and sensitivity of these calculations to various parameters will be established only after much further effort is expended. This leaves unresolved for the time being such fascinating questions as the study of the effects of possible α-clusters in nuclei on medium-energy elastic scattering.

---

*For ^{12}C such results have been obtained in Sch–72a and Bir–75.

†For very light nuclei, Monte Carlo methods may be exploited to deal directly with the full multiple-scattering series for the T-matrix. The case of ^{4}He, for example, is treated in this way with considerable success in Gib–76.

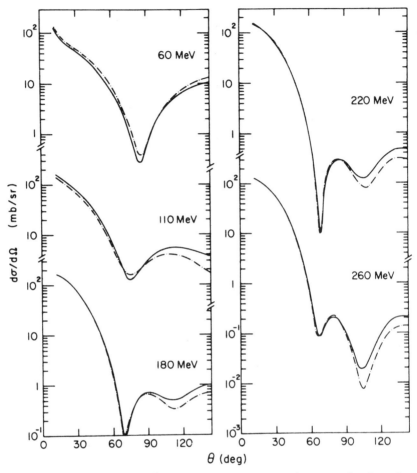

Fig. 5.5 A calculation of $\pi^- - {}^4\mathrm{He}$ elastic scattering at various energies throughout the 3,3 region using a second-order optical potential (solid line) with repulsive short-range correlations, and a first-order optical potential (dashed line). The theoretical curves are from Lee-77a.

Before leaving the subject of second-order effects, it is well to note that these may also arise from other features of the theory that we have noted. For example, the binding corrections, which differentiate between scattering of the pion on the bound nucleon or on a free nucleon as in Eq. 4.58, also involve second-order effects. This is the case also for efforts to generate a theory that evaluates the various internal scattering amplitudes $t_{\pi N}(E_\pi)$ at a self-consistent scattering energy E_π through dispersion relations, such as Eq. 15. These various effects generate corrections to the lowest-order optical potential that are nonlinear in the density, just as for the correlation effects, and naturally enter on a similar footing in analyzing elastic scattering data.

Thus far in our discussion of the application of multiple-scattering theory to the meson-nucleus system we have not paid direct attention to the special features of the system—in particular the πN resonance—except insofar as the relevant physical parameters were taken for the situation of πN $3,3$ resonance region and the p-wave nature of that resonance has appeared through derivative terms. In Section 4.3 we saw from our estimate of the validity of the impulse approximation for this case that the presence of a resonance may require more detailed consideration of its effects on the multiple-scattering theory. Indeed, even in the lowest-order optical potential, the fact that the pion may produce a Δ resonance with a given nucleon, which Δ then propagates through the nucleus until its natural decay time returns it to the πN channel, introduces significant corrections to the fixed scatterer approximation and a nonlocality in $\mathcal{V}_c^{(1)}$. For the purpose of examining this point (Len–75, Hir–77a), let us take the free πN amplitude in the $3,3$ channel for a nucleus with spin saturation in the form (see for example, Eq. 16)

$$\langle \mathbf{k}'_\pi \mathbf{p}'_N | t^{\text{free}}(E_\pi) | \mathbf{k}_\pi \mathbf{p}_N \rangle \propto (2\pi)^3 \delta(\mathbf{k}'_\pi + \mathbf{p}'_N - \mathbf{k}_\pi - \mathbf{p}_N) t_0 \frac{\Gamma(E_\pi)}{2}$$

$$\times \frac{1}{E_\pi - E_{\text{res}} + \frac{1}{2} i \Gamma(E_\pi)}, \qquad [29]$$

where we have put in evidence the three-dimensional delta function of momentum conservation but ignored the explicit momentum-dependence in the numerator arising from the p-wave aspect of the problem and the vertex form factor. The normalization factor t_0 in Eq. 29 is fixed by the optical theorem at resonance to be

$$t_0 = \frac{|\mathbf{v}_\pi - \mathbf{v}_N|}{2\kappa^2} \sigma_{\text{total}} \bigg|_{E_\pi = E_{\text{res}}}, \qquad [30]$$

where $\kappa = (M\mathbf{k}_\pi - m\mathbf{p}_N)/(M + m)$. In using Eq. 29 in the impulse approximation for the first-order optical potential, as in Eq. 4.117 and Section 4.3 generally, we fix the pion energy at which t^{free} is to be evaluated as

$$E_\pi = E - H_N(A - 1) - \frac{(\mathbf{k}_\pi + \mathbf{p}_N)^2}{2M^*}, \quad M^* = M + m, \qquad [31]$$

as in Eqs. 4.164–4.168. Here we have separated the part of the nuclear

Hamiltonian $H_N(A-1)$ referring to the $A-1$ passive spectator nucleons and the part involving the active nucleon. In the latter we have ignored the binding part, which can be treated subsequently in terms of the interaction between the resonating πN system (a Δ particle) and the residual nucleus (Len–75, Hir–77a) and have retained the kinetic energy of the πN subsystem in which we now are primarily interested.

We characterize the nuclear ground state through a shell model description, denoting the single-particle states by ϕ_n, with corresponding removal energies

$$\langle \phi_n^{-1} | H_N(A-1) | \phi_n^{-1} \rangle = -\varepsilon_n, \qquad [32]$$

where ϕ_n^{-1} refers to the hole state remaining after the removal of a nucleon from the state ϕ_n. Then the first-order optical potential of Eq. 4.117 is

$$\langle \mathbf{k}'_\pi | \mathcal{V}_c^{(1)}(E) | \mathbf{k}_\pi \rangle \propto \tfrac{1}{2} t_0 \Gamma$$

$$\times \sum_{n=1}^{A} \int \frac{d\mathbf{p}_N}{(2\pi)^3} \left[\phi_n^*(\mathbf{p}'_N) \frac{1}{E + \varepsilon_n - E_{\text{res}} - \dfrac{(\mathbf{k}_\pi + \mathbf{p}_N)^2}{2M^*} + \dfrac{i\Gamma}{2}} \phi_n(\mathbf{p}_N) \right] \qquad [33]$$

where we have taken a constant width $\Gamma \approx 110$ MeV. In order to see the physical consequence of retaining the πN (resonating) subsystem recoil energy in Eq. 33, we carry out the integral there. The optical potential in configuration space is then

$$\langle \mathbf{r}' | \mathcal{V}_c^{(1)} | \mathbf{r} \rangle = -\frac{2M^*}{4\pi} \left(\tfrac{1}{2} \bar{\kappa}^2 t_0 \Gamma \right) \sum_{n=1}^{A} \phi_n^*(\mathbf{r}') \frac{e^{iK_n |\mathbf{r}' - \mathbf{r}|}}{|\mathbf{r}' - \mathbf{r}|} \phi_n(\mathbf{r}), \qquad [34]$$

with

$$K_n^2 = 2M^* \left(E + \varepsilon_n - E_{\text{res}} + \frac{i\Gamma}{2} \right). \qquad [35]$$

The nonlocality in the optical potential of Eq. 34 is manifest. For a very short-lived Δ resonance, $\Gamma \to \infty$ and the "propagation" of the nonlocality goes over to a delta function,

$$\frac{1}{4\pi i} M^* \Gamma \frac{e^{iK_n |\mathbf{r}' - \mathbf{r}|}}{|\mathbf{r}' - \mathbf{r}|} \xrightarrow[\Gamma \to \infty]{} \delta(\mathbf{r}' - \mathbf{r}), \qquad [36]$$

so that the first-order optical potential becomes local:

$$\langle \mathbf{r'}|\mathcal{V}_c^{(1)}|\mathbf{r}\rangle \underset{\Gamma\to\infty}{\longrightarrow} -i\bar{\kappa}^2 t_0 \sum_{n=1}^{A} |\phi_n(\mathbf{r})|^2 = -i\frac{v}{2}\sigma_{total}\rho(\mathbf{r}) \qquad [37]$$

where the last step follows from Eq. 30, and its result is the same as Eq. 4.15 for the first-order optical potential in this limiting situation.

The nonlocality is induced here by the recoil of the resonance (and the energy-dependence of the transition amplitude) and the nonvanishing lifetime of the active resonance. (Indeed, kinematic effects introduced by the energy behavior of the amplitude, as through the coefficients $b(E_\pi)$ and $c(E_\pi)$ of Eq. 22, suffice to produce nonlocalities.) We have already seen other sources of nonlocality in the optical potential: These may come about due to a particular momentum-dependence in the basic meson-nucleon interaction, perhaps arising from the dominance of a particular partial wave, as in the case of the Kisslinger potential. They also enter inevitably—except in the limit of very high energies—from the higher-order correlation corrections to the optical potential. We discuss yet another aspect of nonlocality in the context of the doorway state approach to πA scattering in Section 6.4.

The nonlocality effect in the nonstatic optical potential of Eq. 35, which allows baryon recoil, is measured by

$$l_\Delta \sim (2\,\mathrm{Im}\,K_n)^{-1} \sim \frac{p}{M^*\Gamma} \sim 0.5 \text{ fm}, \qquad [38]$$

which is an appreciable length compared to the others in our system. The systematic inclusion of these nonlocality effects, as well as those of the interaction between the Δ subsystem and the residual nucleus, can be carried out (Hir–77a) for ^{4}He, for instance, and gives fine agreement with the data provided a phenomenological "spreading potential" is introduced to account for strong coupling of the Δ/residual-nucleus channel to other channels that are not explicitly included in the theory. Results that include the effects of this potential as fit to the total and elastic π-^{4}He cross sections are shown in Fig. 5.6 for π-^{4}He elastic differential cross sections at energies all through the 3,3 region (see also Mon–76, Dil–77, Kli–78, Mon–78.)

In the course of considering the consequences of special features of the πN 3,3 resonance, or Δ formation, on πA scattering, one may also make use of other formulations concerning deviations from the impulse approximation, $t_i = t_i^{free}$, which arise from the production of the resonance in the nuclear medium (Saw–72, Dov–73a, Bar–73a, Bet–73, Eis–73a,

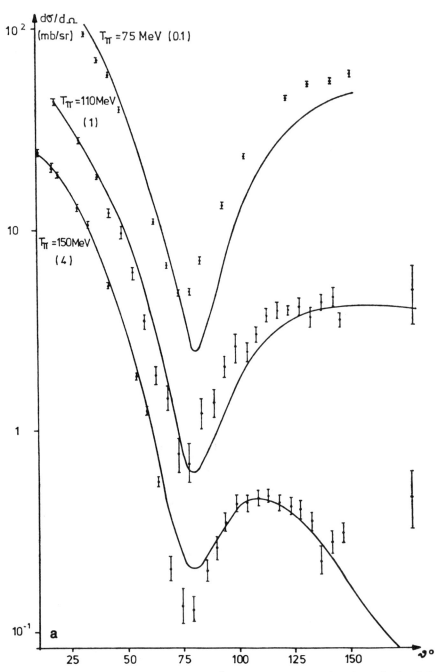

Fig. 5.6 Elastic scattering of pions on ^{4}He at various energies through the 3,3 resonance region compared with a calculation that includes dynamic effects of the Δ that is formed when the pion resonates with a target nucleon (Hir–77a). The numbers in parentheses are the factors by which the corresponding cases have been divided so that all may be shown in the same figure.

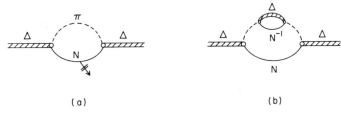

(a)   (b)

Fig. 5.7 Effects of the nuclear medium in modifying properties of the 3,3 resonance. In (*a*) is shown the Pauli blocking of the intermediate nucleon in the Δ
self-energy diagram, while (*b*) gives a quenching effect for the intermediate pion,
which may resonate with another nucleon and remove the πA system from the
elastic channel.

Nat–73b, Eis–74, Dov–76a), as opposed to production in free space. These
points can be examined, for instance, with reference to the model of the
3,3 resonance of Chew and Low (Che–56) discussed in Chapter 2.
Although this model is not derived from a Lippmann-Schwinger equation
with a conventional potential, results obtained using it may give guidance
with regard to other situations as well, including those closer to forms
needed for the multiple-scattering theories we have discussed so far. Using
this model, one may consider modifications arising from Pauli exclusion
principle effects (Fig. 5.7a) on the active nucleon and quenching effects
(Fig. 5.7b) on the transferred pion as it plows through the nuclear medium
while trying to resonate with the active nucleon; binding effects for the
active nucleon require some more considerable generalization of the Chew-
Low model which in its primeval form involves a static nucleon and
therefore cannot easily encompass its interaction with other nucleons. Both
the blocking and the quenching effects are estimated to be very large at
nuclear matter densities, but vastly smaller and essentially negligible at the
peripheral densities that are actually seen in πA scattering near the 3,3
region, where the blackness of the more central volume renders it without
detailed influence on the elastic scattering. Thus we are faced with the
situation that the second-order effects in elastic scattering require high
densities to be large enough to be appreciable, but at high densities
absorption is so great as to mask those regions and their effects. At large
angles many of these effects enter in some degree, though, of course, they
are there very entangled with each other.

5.1e The Lorentz-Lorenz Effect: Another Aspect of Higher-Order Scattering for Pions

Quite early in the development of optical potentials for use with low-
energy πA systems it was realized by Kroll (Kro–61) that a strong analogy

exists between pion-nucleus scattering near threshold and an optical phenomenon that occurs when long wavelength photons traverse matter. This is known as the Lorentz-Lorenz effect (Jac–62). The analogy arises from the realization that in both cases the scattering on an individual constituent is dominated by p-waves (in the case of the photon because the dipole interaction prevails for long wavelength, and in the case of the pion because of the dominant 3,3 resonance), and the constituents are held apart from each other (in the nucleus because of repulsive correlations and in matter because of such effects or because of lattice arrangements). As a consequence, there arises in both systems a nonlinear dependence of the scattering on the medium density. In the case of electromagnetism this is expressed by a nonlinear relationship between the index of refraction and the medium density, which is the dynamic analog of the static, zero-frequency Clausius-Massotti equation (Jac–62) between the dielectric constant and the density or the molecular polarizability. From our discussion in Chapter 4, we know that we can take over results for the index of refraction for use with the optical potential (see, for example, Eq. 4.13). The Lorentz-Lorenz effect is another instance of the insights afforded by the optical analogy. The consequences of this effect for pionic atoms were explored extensively by the Ericsons (Eri–66a, Kre–69).

Although the Lorentz-Lorenz effect as a zero-energy phenomenon arises most naturally in the context of pionic atoms, it in fact plays a central role in a number of other situations as well, which is our main reason for discussing it at length here. As we see, it offers a somewhat unique model in which the multiple-scattering series can be summed exactly. The resulting form, in addition to being pertinent for pionic atoms, may have some validity at higher energies as well, perhaps even into the 3,3 region (Gar–78). It may also be relevant for pions that are far from their mass shell, as in pion condensation (see Section 8.2), where it may be important in raising the critical density for condensation by a factor of two or so from what would otherwise be obtained.

There are a variety of ways in which the Lorentz-Lorenz effect for the πA system may be derived, based on the optics analogy (Eri–66b, Sch–71, Bay–75), on alternative multiple-scattering formalisms (Eri–66a), on nuclear matter formulations involving excitations of the Δ isobar (Bro–75), and on the Watson multiple-scattering series (Eis–73b). We choose the last approach as being in keeping with the spirit of our development here, and because it offers a unique opportunity to perform an exact summation of the complete series in this instance, which is not without some surprises. We choose to start with the complete, exact expansion of the optical potential in formalism 0, Eq. 4.84,

$$
\mathcal{V}_c = -\frac{4\pi}{2m} \langle 0| \sum_{i=1}^{A} f_i + \sum_{i=1}^{A} \sum_{j\neq i} f_i G_0(ij)(1-P_0)f_j
$$

$$
+ \sum_{i=1}^{A} \sum_{j\neq i} \sum_{k\neq j} f_i G_0(ij)(1-P_0)f_j G_0(jk)(1-P_0)f_k + \cdots |0\rangle,
$$

$$[39]$$

where P_0 is the nuclear ground state projection operator and we envision using the impulse approximation so that $f_i = -(2m/4\pi)t_i$ is the free πN scattering amplitude. Because we are using formalism 0 the propagator between nucleon i and j in configuration space, say, is simply

$$
G_0(ij) = -\frac{4\pi}{2m}\frac{e^{ik_0|\mathbf{r}_i - \mathbf{r}_j|}}{|\mathbf{r}_i - \mathbf{r}_j|}, \qquad k_0 = \sqrt{2mE} . \qquad [40]
$$

The use of formalisms $1,2,\ldots$ of Chapter 4 would, of course, involve a much more complicated form for the propagator, and is in any event unnecessary since we intend to sum the full series exactly (see also War–78a).

As we have seen in Chapter 4, a more explicit form for the first few terms of the optical potential in momentum space, but in the limit of zero pion momentum, $\mathbf{k}\to 0$, $\mathbf{k}'\to 0$, is

$$
\mathcal{V}_c = \mathcal{V}_c^{(1)} + \mathcal{V}_c^{(2)} + \mathcal{V}_{c,\text{three-body}}^{(3)} + \mathcal{V}_{c,\text{reflection}}^{(3)} + \cdots, \qquad [41]
$$

with the first-order term

$$
\mathcal{V}_c^{(1)} = -\frac{4\pi}{2m} A \int f_1 \rho(\mathbf{r}_1)\, d\mathbf{r}_1 \qquad [42]
$$

for nuclear density ρ normalized such that $\int \rho\, d\mathbf{r} = 1$, and with the second-order term

$$
\mathcal{V}_c^{(2)} = -\frac{4\pi}{2m} A(A-1) \int f_1 G_0(1,2) f_2 \big[P^{(2)}(\mathbf{r}_1,\mathbf{r}_2) - \rho(\mathbf{r}_1)\rho(\mathbf{r}_2) \big]\, d\mathbf{r}_1\, d\mathbf{r}_2
$$

$$[43]$$

for the two-nucleon correlation function $P^{(2)}(\mathbf{r}_1,\mathbf{r}_2)$ of Eq. 4.140. The third-order term divides into a three-particle correlation part,

$$
\begin{aligned}
\mathcal{V}^{(3)}_{c,\,\text{three-body}} = &-\frac{4\pi}{2m} A(A-1)(A-2) \int f_1 G_0(1,2) f_2 G_0(2,3) f_3 \\
&\times \big[P^{(3)}(\mathbf{r}_1,\mathbf{r}_2,\mathbf{r}_3) - \rho(\mathbf{r}_1) P^{(2)}(\mathbf{r}_2,\mathbf{r}_3) \\
&- P^{(2)}(\mathbf{r}_1,\mathbf{r}_2)\rho(\mathbf{r}_3) + \rho(\mathbf{r}_1)\rho(\mathbf{r}_2)\rho(\mathbf{r}_3) \big]\, d\mathbf{r}_1\, d\mathbf{r}_2\, d\mathbf{r}_3,
\end{aligned}
$$

$$[44]$$

and a two-nucleon piece corresponding to reflections, or rescatterings between a pair of nucleons (j and $i = k$ in Eq. 39),

$$
\begin{aligned}
8 f V^{(3)}_{c,\,\text{reflection}} = &-\frac{4\pi}{2m} A(A-1) \int f_1 G_0(1,2) f_2 G_0(2,1') f_{1'} \\
&\times \big[P^{(2)}(\mathbf{r}_1,\mathbf{r}_2)\,\delta(\mathbf{r}_1-\mathbf{r}_1') - \rho(\mathbf{r}_1) P^{(2)}(\mathbf{r}_2,\mathbf{r}_1') \\
&- P^{(2)}(\mathbf{r}_1,\mathbf{r}_2)\rho(\mathbf{r}_1') + \rho(\mathbf{r}_1)\rho(\mathbf{r}_2)\rho(\mathbf{r}_1') \big]\, d\mathbf{r}_1\, d\mathbf{r}_2\, d\mathbf{r}_1' \quad [45]
\end{aligned}
$$

(cf. Eqs. 4.156–4.158).

We now restrict our attention to the p-wave part of the πN amplitude, parametrizing it as

$$
\langle \mathbf{k}' | f_{\pi N}(E) | \mathbf{k} \rangle = c_0(E) \mathbf{k}' \cdot \mathbf{k}. \qquad [46]
$$

As we have seen earlier in this section, in configuration space the momenta are to be replaced by gradient operators, here made to act on the propagators, and in the limit of zero pion kinetic energy ($k_0 \to 0$) we see from Eq. 40 that we require

$$
\nabla_l \nabla_m \frac{1}{r} = -\frac{4\pi}{3}\,\delta_{lm}\,\delta(\mathbf{r}) + 3\,\frac{r_l r_m - \frac{1}{3}\delta_{lm} r^2}{r^5}. \qquad [47]
$$

We assume that the correlation functions in Eqs. 43–45 support only a negligible contribution* from the quadrupole part of Eq. 47, and the effect of the delta function there is then to evaluate the correlation functions at the same point; that is, we require $P^{(2)}(\mathbf{r},\mathbf{r}), P^{(3)}(\mathbf{r},\mathbf{r},\mathbf{r}), \ldots$, all of which vanish if we assume a repulsive hard core correlation so that there is zero probability to find two nucleons at the same point. Then we have in

---

*The quadrupole contributions have been estimated (War–78a) to be 10% corrections, or less, to the Lorentz-Lorenz effects.

configuration space

$$\mathcal{V}_c(r) = \mathcal{V}_c^{(1)} + \mathcal{V}_c^{(2)} + \mathcal{V}_{c,\,\text{three-body}}^{(3)} + \mathcal{V}_{c,\,\text{reflection}}^{(3)} + \cdots$$

$$= -\frac{4\pi}{2m} A c_0 \nabla \cdot \left\{ \rho(\mathbf{r}) - \frac{4\pi}{3}(A-1)c_0\rho^2(\mathbf{r}) \right.$$

$$\left. + \left(\frac{4\pi}{3}c_0\right)^2 (A-1)(A-2)\rho^3(\mathbf{r}) + \left(\frac{4\pi}{3}c_0\right)^2 (A-1)\rho^3(\mathbf{r}) + \cdots \right\} \nabla$$

$$= -\frac{4\pi}{2m} A c_0 \nabla \cdot \frac{\rho(\mathbf{r})}{1 + \dfrac{4\pi}{3} c_0(A-1)\rho(\mathbf{r})} \nabla. \qquad [48]$$

This is the Lorentz-Lorenz modification to the Kisslinger potential that makes it dependent on the density in a nonlinear way.

The effect in Eq. 48 is numerically rather large, since $4\pi c_0(E{\to}0){\sim}6$ fm^3 and central nuclear densities are about $(A-1)\rho(\mathbf{r}=0){\sim}0.17$ fm^{-3}, so that Eq. 48 implies 33% reductions in the p-wave part of the optical potential. In practice, however, the Lorentz-Lorenz feature is masked by uncertainties in the other parameters of the optical potential as well as by the question: if we see a quadratic density effect how do we know it does not arise from some other higher-order correction, say to the impulse approximation as used in the optical potential? The result of Eq. 48 raises disconcerting formal questions as well. Unlike our estimate in Section 4.3, it does not involve an expansion in correlation lengths, and the effect does not vanish for very-short-range anticorrelations, $r_c{\to}0$. It also tends to suggest that one can probe a presumably fairly short-range correlation at zero external pion momentum, and not at $k{\sim}1/r_{\text{corr}}$.

It is clear that the feature of the problem that has thus far been omitted is the finite extent of the scatterer as seen by the projectile. Clearly, if the πN interaction range is such that the pion sees the nucleons as if they were all overlapping then the fact that the nucleon centers may be held slightly apart by a small correlation core is of no great significance. We can make an estimate of this effect—though in doing so we can no longer sum the optical potential series exactly—by incorporating a vertex form factor of the πN amplitude off the energy shell. This we do in the form

$$\langle \mathbf{k}'| f_{\pi N}(E)|\mathbf{k}\rangle = c_0(E)\mathbf{k}' \cdot \mathbf{k}\, h(k')h(k), \qquad h(0)=1, \qquad [49]$$

where a simple and convenient choice for the vertex function is

$$h(k) = (1 + k^2/\alpha^2)^{-1}. \qquad [50]$$

A rough estimate of the consequent reduction in the Lorentz-Lorenz effect is given by arguing that even for a pion of zero external momentum, $k_0 = 0$, the internal momentum involved in the exchange between two correlated nucleons is $\sim 1/r_{corr}$ so that in the denominator of Eq. 48 we have

$$c_0 \to \xi c_0 \to h^2 (1/r_{corr}) c_0. \qquad [51]$$

A more refined estimate for ξ can be made by assuming that the correlations are of Jastrow type,

$$P^{(n)}(\mathbf{r}_1, \mathbf{r}_2, \ldots, \mathbf{r}_n) = \rho(\mathbf{r}_1) \rho(\mathbf{r}_2) \cdots \rho(\mathbf{r}_n) \prod_{i<j} \left[1 + g(|\mathbf{r}_i - \mathbf{r}_j|) \right]. \qquad [52]$$

Then one arrives at

$$\xi = - \int g(r) \Delta(\mathbf{r}) \, d\mathbf{r}, \qquad [53]$$

where

$$\Delta(\mathbf{r}) = \int h^2(k) e^{i\mathbf{k}\cdot\mathbf{r}} \frac{d\mathbf{k}}{(2\pi)^3} \xrightarrow[h \to 1]{} \delta(\mathbf{r}). \qquad [54]$$

Using the vertex function of Eq. 50 and a correlation function $g(r) = -\theta(r_{corr} - r)$, we get

$$\xi = 1 - \left[1 + \alpha r_{corr} + \tfrac{1}{2}(\alpha r_{corr})^2 \right] e^{-\alpha r_{corr}} \approx \tfrac{1}{6}(\alpha r_{corr})^3. \qquad [55]$$

For a rough estimate of the correlation length we take $r_{corr} \sim 0.5$ fm, while for the poorly known πN range parameter we estimate (Myh–73b) $\alpha \sim 350$ MeV$/c$, whence $\xi \leqslant 0.1$; that is, the Lorentz-Lorenz effect is greatly reduced by the relatively long range ($\alpha^{-1} \sim 0.6$ fm) of the πN interaction.* One might ascribe the correlation in the Lorentz-Lorenz effect to the Pauli principle (Dov–71, Del–76) in which case they have a substantially longer range, but the potential must then be carefully averaged over spin and isospin states for the nucleons, since only the $S = T = 0$ or $S = T = 1$ pairs will experience an exclusion from $\mathbf{r} = 0$ separation distance.

We have noted in the context of nonlocal and local versions of the Kisslinger potential, Eqs. 8 and 11, that when correlations are introduced

---

*A similar situation may frequently pertain in the usual optical case as well (Gue–64). We note further that sizable effects in the s-wave part of the pionic atom optical potential due to finite πN range also occur (Iac–74).

that bring about a nonoverlapping situation for the nucleons ($\alpha^{-1} \ll r_{corr}$, $\xi \approx 1$) the πA amplitude must no longer depend on off-energy-shell features of the πN amplitude, so that the two versions of the potential must then be equivalent. The Lorentz-Lorenz mechanism is present, at zero energies, in just that situation, and indeed its inclusion (Sch–72b; see also Eri–73a) brings the results for the nonlocal and local calculations for pionic atoms into line. This is, in fact, accomplished by reducing the zero-energy πA elastic scattering amplitude to its lowest-order form $T_c = \langle 0|\Sigma t_i|0\rangle$ *exactly*, as may be seen from Eqs. 39–47. The point is that the series for the optical potential, Eq. 39, becomes that for the elastic amplitude if we simply remove the projection operator P_0 there. (This operator projects out the ground state for $\mathcal{V}_c$ to prevent its double counting when $\mathcal{V}_c$ is iterated to produce T_c.) In Eqs. 43–45, the higher-order terms $T_c^{(n)}$ then just involve the nth order correlations $P^{(n)}(\mathbf{r}_1, \mathbf{r}_2, \ldots, \mathbf{r}_n)$—not the true correlations—so that no density factors appear. When the first term of Eq. 47 is put into play it annihilates all of these since $P^{(n)}(\mathbf{r}, \mathbf{r}, \ldots, \mathbf{r}) = 0$. Thus only the lowest-order term, analogous to Eq. 42 survives, and it only involves the on-shell πN amplitude.

It has also been pointed out (Bay–75) that the exchange of ρ mesons can produce an effect very much like that of Lorentz and Lorenz. Again, it is easy to see how this comes about in our multiple-scattering formalism provided we generalize it slightly to include "excitations" of the projectile —in this case a meson that in its entrance channel is a pion but can also convert to a ρ-meson.* The ground state projections considered hitherto must then be thought of as pertaining to both the nucleus and the projectile so that $P_0|\text{nucleus; meson}\rangle = 0$ if the nucleus is not in its ground state and/or the meson is not a pion. Then, to the terms already shown in Eqs. 41–45 must be added

$$\mathcal{V}_c^{(2\rho)} = -\frac{4\pi}{2m} A(A-1) \int f_1^{\pi \leftarrow \rho} G_0^\rho(1,2) f_2^{\rho \leftarrow \pi} P^{(2)}(\mathbf{r}_1, \mathbf{r}_2), \qquad [56]$$

and so forth, where only $P^{(n)}(\mathbf{r}_1, \mathbf{r}_2, \ldots, \mathbf{r}_n)$ enters for exclusively ρ exchanges since $P_0|\rho\rangle = 0$; the action of P_0 must be retained for intermediate π states, of course. The conversion amplitudes $f^{\pi \leftarrow \rho}$ or $f^{\rho \leftarrow \pi}$ can be determined by considering the presumably dominant magnetization or tensor part of the nonrelativistic ρNN coupling (Bro–75, Bay–75, Lev–77) in the form $(f_\rho/m_\rho)\boldsymbol{\varepsilon} \cdot \boldsymbol{\sigma} \times \mathbf{k}_\rho$ for ρ coupling constant f_ρ, mass m_ρ, momentum $\mathbf{k}_\rho$, polarization vector $\boldsymbol{\varepsilon}$, and nucleon spin $\boldsymbol{\sigma}$. Taken with the πNN vertex

---

*This generalization is treated at length in Lon–75. Other aspects of the coupling of additional channels are considered in Chapter 6.

$(f_\pi/m_\pi)\boldsymbol{\sigma}\cdot\mathbf{k}_\pi$, and averaging over spin, the amplitude has the form

$$\langle\mathbf{k}'|f^{\rho\leftarrow\pi}(E)|\mathbf{k}\rangle = c'(E)\boldsymbol{\varepsilon}\cdot\mathbf{k}'\times\mathbf{k}. \qquad [57]$$

We now insert this into $\mathcal{V}_c$, sum over ρ polarization, and proceed as for the pion case; that is, we replace the momenta by gradient operators acting on the ρ propagator

$$G_0^\rho(ij) = -\frac{4\pi}{2m_\rho}\frac{e^{-\kappa|\mathbf{r}_i-\mathbf{r}_j|}}{|\mathbf{r}_i-\mathbf{r}_j|}, \qquad \kappa=\sqrt{m_\rho^2-m_\pi^2}, \qquad [58]$$

which is a strongly decaying exponential, since we are considering a situation near π threshold and therefore far below ρ threshold. Then

$$\nabla_i\nabla_j\frac{e^{-\kappa r}}{r} = \frac{e^{-\kappa r}}{r^3}\left\{\tfrac{1}{3}\delta_{ij}\kappa^2 r^2 + \left(\frac{3r_ir_j-r^2\delta_{ij}}{r^2}\right)\left(1+\kappa r+\tfrac{1}{3}\kappa^2 r^2\right)\right\} - \frac{4\pi}{3}\delta_{ij}\delta(\mathbf{r}),$$
$$[59]$$

where we will once again drop the quadrupole term; hence

$$\mathcal{V}_c^{(2\rho)} = -\frac{4\pi}{2m}A(A-1)\left(\frac{4\pi}{3}\cdot 2\right)c'^2\mathbf{k}'\cdot\mathbf{k}$$
$$\times\int\left[\frac{\kappa^2}{4\pi}\frac{e^{-\kappa|\mathbf{r}_1-\mathbf{r}_2|}}{|\mathbf{r}_1-\mathbf{r}_2|}-\delta(\mathbf{r}_1-\mathbf{r}_2)\right]P^{(2)}(\mathbf{r}_1,\mathbf{r}_2)\,d\mathbf{r}_1\,d\mathbf{r}_2, \qquad [60]$$

and similarly for the higher-order terms. The factor of two in comparison with the π case arises from the vector nature of the ρ. We note that for infinite nuclear matter, where only forward elastic scattering is possible, the construct $f^{\pi\leftarrow\rho}\sim\boldsymbol{\varepsilon}\cdot\mathbf{k}_\pi\times\mathbf{k}_\rho$ must vanish; this is still implicit in Eq. 60, which vanishes for no correlations and constant density, $P^{(2)}(\mathbf{r}_1,\mathbf{r}_2)=\rho(\mathbf{r}_1)\rho(\mathbf{r}_2)=\rho_0^2$. Thus in the ρ Lorentz-Lorenz effect, as in that for pions, the correlations play a crucial role. If we assume, as before, that hard-core anticorrelations are present, $P^{(2)}(\mathbf{r},\mathbf{r})=0$ so we drop the *second* term in brackets in Eq. 60, quite unlike the pion case. The ρ Lorentz-Lorenz effect then comes from arguing that in the *first* term there $\kappa\sim m_\rho\sim 770$ MeV$/c\sim$ $(0.3$ fm$)^{-1}$ so that the ρ propagation is very short ranged and the slowly-varying density factors ρ of $P^{(2)}(\mathbf{r}_1,\mathbf{r}_2)=\rho(\mathbf{r}_1)\rho(\mathbf{r}_2)[1+g(|\mathbf{r}_1-\mathbf{r}_2|)]$ can be evaluated at the same point. A result analogous to the Lorentz-Lorenz one is then obtained, but with a damping factor ξ_ρ due to the more rapidly varying correlation piece,

$$\mathcal{V}_c^{(2\rho)} = -\frac{4\pi}{2m}A(A-1)\left(\frac{4\pi}{3}\cdot 2\right)c'^2\mathbf{k}'\cdot\mathbf{k}\rho^2(\mathbf{r})\xi_\rho, \qquad [61]$$

with

$$\xi_\rho = \int \frac{\kappa^2 e^{-\kappa R}}{4\pi R} [1 + g(R)] \, d\mathbf{R}. \qquad [62]$$

A rough parametrization of $g(R) = -e^{-R/r_{corr}}$ yields

$$\xi_\rho = \frac{1 + 2\kappa r_{corr}}{(1 + \kappa r_{corr})^2} \sim 0.6 \qquad [63]$$

for $\kappa = \sqrt{m_\rho^2 - m_\pi^2} = 760$ MeV and $r_{corr} \sim 0.5$ fm; that is, there is a suppression of the ρ-exchange Lorentz-Lorenz effect for actual nuclear parameters, which is made considerably stronger if a finite-range ρNN vertex function is introduced so that there is a subtractive contribution from the modification of the delta function of Eq. 60. The higher-order terms in the optical potential, including the mixed π/ρ exchanges, can then be treated in a similar way to yield the full Lorentz-Lorenz form. The precise magnitude of the effect depends on the poorly known ρNN coupling constant (as well as on the correlation length and the ρNN interaction range). If it is large it is apt to have important consequences not only for pionic atoms but also for pion condensation, where it raises the critical density for the onset of condensation (Bro–76a; see also Section 8.2). Its role in pion scattering, $E > 0$, is somewhat more obscure, since we have seen for example in Section 4.3 that we must then consider an expansion for the optical potential involving Fourier transforms of the correlation functions with potentially nonnegligible sensitivity to the momentum values.

5.1f   Glauber Theory and Other More Specialized Techniques

In order to incorporate all orders of nuclear correlations in the multiple-scattering theory, as well as to work with a finite, on-energy-shell series, it is tempting to apply the Glauber series to pion-nucleus scattering. This is without fundamental justification in the region of the 3,3 resonance, since the applicability of the theory requires that the scattering on the *individual* nucleon be diffractive, as well as that on the entire nucleus, and this criterion is clearly not met by scattering which is overwhelmingly dominated by one partial wave, the p-wave in the 3,3 case. Nonetheless, the Glauber theory has been applied to π-^{12}C scattering with considerable success (Sch–70, Wil–70), as can be seen from Fig. 5.8, even down to energies as low as 120 MeV pion lab kinetic energy (Hes–73). The reasons for this may again lie in the blackness of the nucleus near resonance and

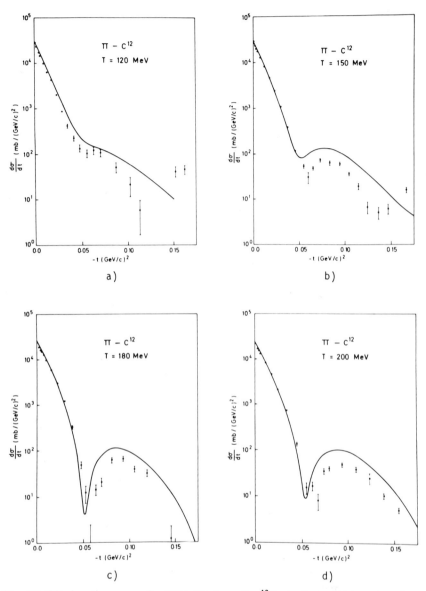

Fig. 5.8 Glauber theory results (Wil–70) for $\pi^- - {}^{12}C$ elastic scattering at various energies through the 3,3 region, showing striking agreement with the data (Bin–70).

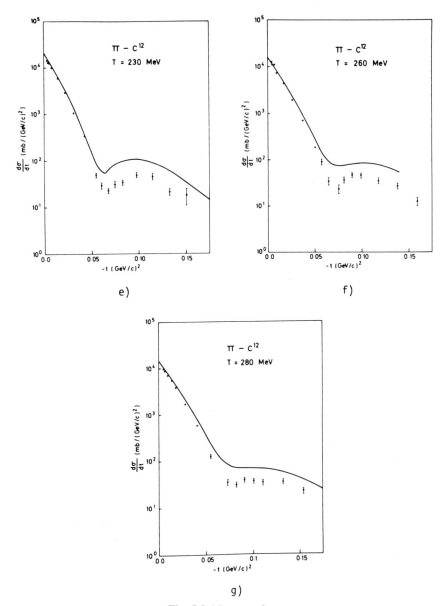

Fig. 5.8 (*Continued*)

223

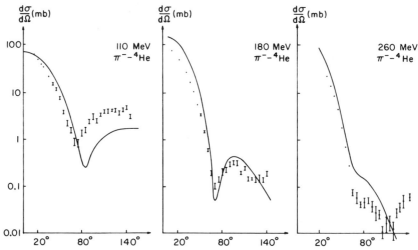

Fig. 5.9 A relatively unsuccessful application of Glauber theory to $\pi^- - {}^4$He scattering in the 3,3 region (Bin–71).

the general diffractive features of the scattering on a nucleus of reasonable size $(kR \gg 1)$ which may be adequately accounted for once a reasonably correct πN amplitude is used with proper nuclear geometry. For a very small nucleus, such as ^{4}He, the Glauber calculation fails (Bin–71b, Hes–73), as shown in Fig. 5.9, where by "failure" we mean the inability to give the correct forward fall-off and the position of the first diffraction minimum. This comes about presumably because the nucleus is too small to be diffractive.

Various other specialized techniques have been advanced for solving the nuclear multiple-scattering problem in special model situations. One of these is based on the use of separable πN potentials that allow, in principle, for an exact solution of the entire multiple-scattering problem (Fol–69). As a matter of practice, this formal solution is of no vast help for $A > 2$, and the separable potential assumption is used in the context of a conventional cluster expansion of the optical potential to establish the off-energy-shell feature of the πN scattering.*

Another specialized technique in multiple-scattering theory has been the exploitation of the approximation of nonoverlapping scatterers. If the basic projectile-nucleon potentials have finite ranges and are sufficiently separated by repulsive correlations so as to be disjoint, that is, to have regions of zero potential between each pair of scattering centers, then the projectile

*The application of separable interactions to πA scattering is reviewed in Wal–73d; see also Tab–76 and Eis–76a and the discussion comments following the former.

propagates within these regions according to the dynamic equation for a free particle. Although the uncertainty principle *allows* violation of momentum conservation, so that off-shell propagation may occur, the free propagation guarantees that only on-energy-shell information concerning the basic projectile-nucleon amplitude is needed (Bèg–61, Fol–69, Aga–73, Huf–73); this result is, of course, valid only in potential theories such that inelastic processes involving particle production are excluded. As a consequence of the restriction to on-shell πN information, it is clear that the ambiguities relating to the treatment of off-shell behavior noted in connection with Eqs. 5–8 and 10 and 11 must disappear in the nonoverlapping limit; thus it is not fully consistent to address them without going at least to higher orders in the optical potential where the effects of correlations that may hold the nucleons apart can first be felt. In general, the nonoverlapping limit can serve as a test of consistency for theoretical treatments of multiple-scattering, as well as allowing for the summation (Aga–75a, b) of so-called reflections—terms in which the pion runs back and forth an arbitrary number of times between a particular pair of nucleons during its propagation (see the discussion at the end of Section 4.3c)—and forming a convenient point of departure for approximative calculational schemes (Gur–75, 76).

5.1g The Hadronic Matter Distribution as Seen in πA Scattering

In the course of studying the distribution of hadronic matter within the nucleus by pion scattering it is tempting to analyze the data so as to extract nuclear neutron radii for comparison with the proton radii measured in elastic electron scattering. For example, one may work (All–72, All–73) in the region around 700 MeV pion lab kinetic energy, where the 3, 3 resonance has died and indeed the $\pi^+ p$ total cross section has a minimum, while that of the $\pi^- p$ has maxima corresponding to some $T = \frac{1}{2}$ resonances. The ratio of cross sections at that energy is

$$\frac{\sigma_{\text{total}}(\pi^- p)}{\sigma_{\text{total}}(\pi^+ p)} = 2.6 = \frac{\sigma_{\text{total}}(\pi^+ n)}{\sigma_{\text{total}}(\pi^- n)}, \qquad [64]$$

where the last part of the equality follows from charge symmetry. In the central nuclear region, the pion is strongly absorbed, while at the surface the π^- scattering is especially sensitive to the proton distribution and the π^+ to neutrons (the opposite of the situation for the $T = \frac{3}{2}$ resonance), though the ratio of Eq. 64 is reduced by the effects of Fermi averaging. The most complete analysis for neutron radii using pions has been carried out by exploiting the nuclear reaction cross sections (to be touched on again in the next section), which are closely related to $\mathrm{Im}\,\mathcal{V}_c$, as in Eq. 4.21. Measurements (All–72, All–73, Clo–74) of the ratio $\sigma_{abs}(\pi^- A)/\sigma_{abs}(\pi^+ A)$ to better than about 1% can then yield meaningful information on the neutron distribution through detailed optical calculations* and indeed suggest that for ^{208}Pb, for example, the neutron and proton root-mean-square radii are equal to within about 1% or 2%. For a wide variety of nuclei (C, Al, Ca, Ni, Sn, Ho, Pb) it is found (All–73) that $|r_n^{rms} - r_p^{rms}| < 0.1$ fm.

Efforts to pin down greater detail in the hadronic matter distribution than the neutron radius, and perhaps surface thickness, have particularly focused on two-particle correlations, as they enter in the second-order optical potential, and on α-particle clusters within the nucleus. Theoretical capabilities cannot yet distinguish the latter and, as we have seen earlier in this section, are not presently able to determine the former conclusively. The most interesting part of the two-nucleon correlations, namely, that which arises from the strong, short-range repulsion in the NN interaction, is especially elusive. It is intrinsically a manifestation of all the unknowns in the high-energy, short-distance behavior of the two-nucleon system. As such, the places where it might strongly influence the multiple-scattering process generally require a highly reliable treatment of many higher-order corrections, including those that are outside of the framework of multiple-scattering theories that omit coupling to particle production channels and are very imperfect in their handling even of simpler relativistic effects. We return to these points in the next chapter, where we also are in a position to explore the consequences of true absorption for the scattering processes. In a general way, it is clear that the great majority of the early theoretical work has focused in some detail on approximations that yield local (or nearly local, as in the case of the Kisslinger form) optical potentials; this work has been based on the fixed-scatterer assumption (or nonoverlapping potentials). Nonlocal effects and improvements on the frozen nucleus and impulse approximations have yet to be explored fully. Indeed, we are not yet in a position to say with much confidence how these will interact with each other or which will lead to qualitative new features or important numerical consequences. What is clear, however, is that the lowest-order optical potential represents a reasonable point of departure for further analysis of elastic scattering. It agrees reasonably with the experimental data out to the first minimum in the differential cross section and theoretical estimates of corrections to it have not seriously impaired its validity. Thus, in addition to its usefulness in calculating differential and total

---

*The original analysis was based on the lowest-order optical potential. Higher-order terms may mask the dependence on the neutron radius, and thus should ultimately be included.

elastic cross sections and absorption cross sections, we may expect that the wave functions that it yields will provide a reasonable tool for use in DWIA approaches to particular inelastic processes.

5.2 Total Cross Sections and Forward Scattering Amplitudes*

As we have seen, the ratio of πA absorption or reaction cross sections for π^+ and π^- scattering can yield information on the neutron distribution in nuclei. The systematic measurement (Wil–73, Car–76) of total cross sections near the 3,3 region can be analyzed (Dov–76b, McV–77) in terms of generalizations of the heuristic absorption model of Section 4.1b, Eqs. 4.20–4.25, and of the πA optical potential. The total cross section is made up of two pieces, $\sigma_{tot} = \sigma_{el} + \sigma_{react}$, namely, the elastic and reaction cross sections. The latter includes both inelastic scattering, in which the nucleus is excited but the pion is not annihilated, and true absorption, in which the pion is destroyed and its rest mass energy transferred to the nuclear system (or the electromagnetic field). In principle, the optical potential determines both the elastic and reaction cross sections, with the imaginary part of the potential providing for the loss of flux from the elastic to the reaction channels. Then the reaction cross section corresponds to what was previously called (in Chapter 4.1) the optical absorption. However, the analysis of σ_{tot} is complicated by the fact that $\sigma_{true\,abs}$ is not usually incorporated in optical potential analyses; we return to this point in Chapter 6. As we have seen in Chapter 4 (cf. Eqs. 4.21, 4.22, and 4.132 for example), once we omit $\sigma_{true\,abs}$ from the discussion, the reaction cross section gives direct information on the imaginary part of the optical potential, which in lowest order is $\mathcal{V}_c^{(1)} = At\rho$ so that this information may be related to the nuclear density ρ. With this omission, $\text{Im}\,\mathcal{V}_c^{(1)}$ relates to the reaction cross section $\sigma_{opt\,abs}$ which is then essentially the cross section for nucleon knockout (see Section 5.6). If true absorption is incorporated in the optical potential, then, of course, its imaginary part will pertain to the total loss of pion flux in traversing the target.

In carrying out such an analysis of the total cross section using $\mathcal{V}_c^{(1)}$, the simple model of Section 4.1 must be enlarged to encompass the possibility of a tail in the nuclear density in place of the sharp cutoff assumed in Section 4.1b, and then the interplay between the large basic πN total cross section and this density tail can even cause the πA total cross section to

---

*This section is a minor digression from the main line of development and is not needed for the understanding of subsequent material.

exceed the geometric limit $\sigma_{total} = 2\pi(R + \lambda)^2$ discussed in connection with Eqs. 4.24, 4.25, and 4.39. Detailed analysis (Dov-76b) in terms of partial wave total cross sections then motivates a relatively simple parametrization in terms of k, R, and A-dependent resonance position and width, which proves quite successful and offers a way of studying these basic parameters in the nuclear medium. To complete such a description of the total cross section, more reliable information is needed both from experiment and theory on the *true* absorption component of the reaction cross section.

The measured total cross sections can also be used as input for the evaluation of dispersion relations for forward elastic pion-nucleus scattering amplitudes. In applying dispersion relations to πA experiments we have a different technique from that of multiple-scattering theory for analyzing and systematizing data. The dispersion relations can be used in a relatively model-independent way with the measured total πA cross section as input to determine the real part of the forward amplitude. Then, for example, the vanishing of $\text{Re} f(\omega)$ is an indication of the resonance position in the nucleus, while the residue in the pole term gives the effective πA coupling constant relatively directly. We can compare with experimental information on $\text{Re} f(0°)$ obtained from the rather difficult phase shift analysis of πA scattering or from the interference between the hadronic and Coulomb amplitude (Mut-75, Coo-76, 77). The point of departure for the dispersion relation analysis (Eri-70b, Dub-75) of πA scattering is the usual assumption that the forward elastic scattering amplitude $f(\omega)$, as a function of complex energy ω, is analytic apart from certain singularities that must be specified in each case and depend on the masses (including those of excitations!) of the scattering particles which enter. In particular, it is assumed analytic for $\text{Im} \omega > 0$. This last feature can be expected intuitively on the grounds of a causality argument (in this version so oversimplified as hardly to deserve being called an argument). The forward scattering amplitude is the Fourier transform of the nuclear response

$$f(\omega) = \int_{-\infty}^{\infty} R(t)e^{i\omega t}\,dt = \int_0^{\infty} R(t)e^{i\omega t}\,dt, \qquad [65]$$

where the second version of the transform, in which the range of integration is restricted to $t \geq 0$, results from a causality assumption that there can be no response until the arrival of the initial wave at some moment taken as $t = 0$. Since the integral of Eq. 65 exists for real ω, it will surely be more convergent for $\text{Im} \omega > 0$, that is, in the upper-half-plane for ω.

We then apply Cauchy's theorem to $f(\omega)$ analytic in the upper-half-plane,

$$f(\omega + i\eta) = \frac{1}{2\pi i} \int_C \frac{f(z)\,dz}{z - \omega - i\eta}, \qquad \eta \to 0^+, \qquad [66]$$

where the contour C runs along the real axis and is then closed with an infinite semicircle in the upper-half-plane, on which, we assume, $f(z)$ is sufficiently small to give negligible contribution. Noting the usual

$$\frac{1}{z-\omega-i\eta} = \frac{P}{z-\omega} + i\pi\,\delta(z-\omega), \qquad [67]$$

where P denotes a principal part integration, we have

$$\text{Re} f(\omega) = \frac{1}{\pi} P \int_{-\infty}^{\infty} \frac{\text{Im} f(\omega')}{\omega'-\omega}\,d\omega'. \qquad [68]$$

To aid in the convergence of this integral for large $|\omega'|$, we may, if necessary, subtract its value at some convenient point ω_0 to write

$$\text{Re} f(\omega) - \text{Re} f(\omega_0) = \frac{\omega-\omega_0}{\pi} P \int_{-\infty}^{\infty} \frac{\text{Im} f(\omega')}{(\omega'-\omega_0)(\omega'-\omega)}\,d\omega', \qquad [69]$$

which is more convergent by one power of ω' in the denominator. This process can be repeated as much as needed, though at the cost of introducing each time a new (generally unknown) constant $\text{Re} f(\omega_i)$ for the ith subtraction point ω_i.

We also wish to invoke crossing (see Chapters 2 and 6) to relate the scattering amplitude $f(\omega)$ to that for the antiparticle* on the same target $\bar{f}(\omega)$:

$$f(-\omega) = \bar{f}^*(\omega), \qquad [70]$$

where spins are also to be reversed; this enables us to extract information on the otherwise inaccessible negative frequencies. For the positive frequencies—with positive kinetic energy as well, that is, situations that are above scattering threshold—the integrand on the right-hand side can be evaluated using the optical theorem,

$$\text{Im} f(\omega) = \frac{k}{4\pi}\sigma(\omega), \qquad \omega \geqslant m, \ k = \sqrt{\omega^2 - m^2}. \qquad [71]$$

Correspondingly, for $\omega \leqslant -m$, we can use Eq. 70 to write

$$\text{Im} f(\omega) = -\frac{k}{4\pi}\bar{\sigma}(-\omega), \qquad \omega \leqslant -m, \qquad [72]$$

---

*Note that the π^+ and π^- are each other's antiparticles so that isospin invariance is of great use in exploiting the crossing property. See also Section 6.1b, Eq. 6.6.

where $\bar{\sigma}$ is the cross section for antiparticle scattering at energy $\bar{\omega}$, $\bar{\sigma}(-\omega) = \bar{\sigma}(\bar{\omega})$. In the unphysical region, $-m < \omega < m$, there may occur poles in $f(\omega)$ for intermediate systems that obey all the relevant selection rules for the scattering and have just one particle of given mass; these may be introduced into the right-hand side of Eq. 68 or Eq. 69 with a term in $\text{Im} f(\omega)$ proportional to $\delta(\omega - \omega_{\text{pole}})$. There may also occur unphysical cuts (the physical cut is that along the real axis above scattering threshold, where real scattering actually takes place), if the allowable intermediate states involve more than one particle and hence have a continuous energy distribution. Such cuts may even start below the ω-value dictated by considering zero internal energy for the intermediate state in question, and are then termed anomalous cuts. These are often encountered for weakly bound systems, but are relatively innocuous for forward scattering.

For purposes of πA forward scattering, we must first consider uncrossed and crossed pole graphs, as in Fig. 5.10. These are the analogs of the single-nucleon Born graphs discussed in Chapter 2, and contribute poles

$$\frac{2r_i}{\omega - \omega_i} \quad \text{and} \quad \frac{-2r_i}{\omega + \omega_i} \qquad [73]$$

in precise analogy to the situation there. The pole positions are

$$\omega_i = \mathfrak{M}_i - \mathfrak{M} - \frac{m^2}{2\mathfrak{M}}, \qquad [74]$$

where $\mathfrak{M}$ is the mass of nucleus A and $\mathfrak{M}_i$ is the mass of the intermediate state i, and their residues are r_i, proportional to the square of the coupling constant of the $\pi A A_i$ vertex. The amplitudes are divided in the conventional way into even and odd charge parts* (removing Coulomb effects),

$$f^{(\pm)}(\omega) \equiv \tfrac{1}{2} [f_{\pi^+ A}(\omega) \pm f_{\pi^- A}(\omega)], \qquad [75]$$

where, under crossing,

$$f^{\pm}(-\omega) = \pm f^{(\pm)*}(\omega). \qquad [76]$$

For $T=0$ nuclei, only $f^{(+)}(\omega) = f_{\pi^\pm A}(\omega)$ enters, while for $T = \tfrac{1}{2}$,

$$f(\omega) = f^{(+)}(\omega) - 2(\vec{I} \cdot \vec{T}) f^{(-)}(\omega), \qquad [77]$$

where $\vec{I}$ and $\vec{T}$ are the pion and nucleus isospin operators. For $T > \tfrac{1}{2}$, there are isotensor terms in the amplitude but these may be ignored and Eq. 77

*Compare Eq. 2.16b.

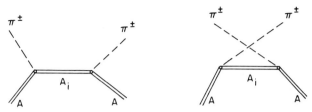

Fig. 5.10 Pole terms in the application of dispersion relations to πA scattering, the poles arising for the various possible bound-states A_i of the nucleus A.

may be retained as a reasonable approximation, since double charge exchange cross sections are known to be small.

The dispersion relation for the even amplitude, from the basic Eqs. 69–71 and the poles and crossing in Eqs. 73–76 with the subtraction point $\omega_0 = m$, is $(k^2 = \omega^2 - m^2)$:

$$\mathrm{Re}f^{(+)}(\omega) - \mathrm{Re}f^{(+)}(m) = \sum_i \frac{2\omega_i r_i}{\omega^2 - \omega_i^2}\frac{k^2}{k_i^2} + \frac{2k^2}{\pi}P\int_{\omega_L}^{\infty}\frac{\omega'\,\mathrm{Im}f(\omega')}{k'^2(\omega'^2 - \omega^2)}\,d\omega',$$

$$[78]$$

where the beginning of the unphysical region is at $\omega_L \approx 0$ (arising, e.g., from absorption with ejection of a nucleon or more) and extends up to $\omega = m$. For the odd integral no subtraction is necessary, since the combination of total cross sections that appears, $\sigma^{(-)}(\omega) = \frac{1}{2}[\sigma_{\pi^+A}(\omega) - \sigma_{\pi^-A}(\omega)]$, vanishes as ω becomes infinite assuming the validity of Pomeranchuk's theorem, which states that at infinite energy the cross section for scattering of particles by a given target is equal to that for scattering of the antiparticles by the same target (Gas–66). Thus

$$\mathrm{Re}f^{(-)}(\omega) = \sum_i \frac{2\omega r_i}{\omega^2 - \omega_i^2} + \frac{2\omega}{\pi}P\int_{\omega_L}^{\infty}\frac{\mathrm{Im}f^{(-)}(\omega')}{\omega'^2 - \omega^2}\,d\omega'. \qquad [79]$$

The contribution of the pole terms to the scattering in the physical region is here substantially larger than for $f^{(+)}(\omega)$, since $\omega \geqslant m \gg \omega_i$.

The main obstacle to using the dispersion relations of Eqs. 78 and 79 is the unknown region $\omega_L < \omega < m$, dominated by all manner of absorption processes. For $^4\mathrm{He}$ or $^{12}\mathrm{C}$, say, these would involve

$$\pi + {}^4\mathrm{He} \longmapsto \mathrm{d} + 2\mathrm{N}$$
$$\longmapsto {}^3\mathrm{He} + \mathrm{N}$$
$$\longmapsto 4\mathrm{N} \quad \text{and so on}$$
$$\pi + {}^{12}\mathrm{C} \longmapsto {}^{10}\mathrm{B} + 2\mathrm{N}$$
$$\longmapsto {}^{10}\mathrm{B}^* + 2\mathrm{N}$$
$$\longmapsto {}^9\mathrm{Be} + 3\mathrm{N} \quad \text{and so on}$$

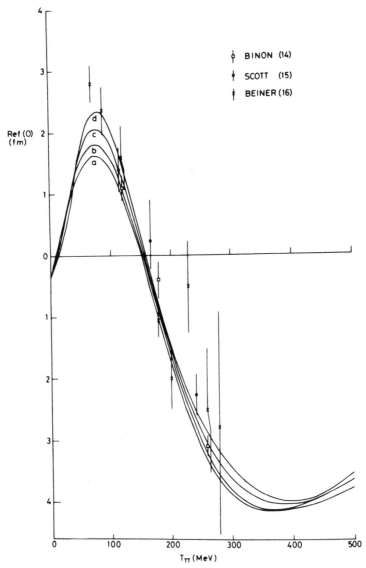

Fig. 5.11 The real part of the forward scattering amplitude for the $\pi^- - {}^{12}C$ system as a function of pion kinetic energy in the lab. The experimental results are extracted from Coulomb interference (Bin–71a, Sco–72) or phase shift analysis (Bei–73). The four calculated curves are taken from the dispersion relation and experimental information on $\pi^- - {}^{12}C$ total cross sections and differ in that curves a, b, c and d match the unphysical to the physical regions at $\omega = m_\pi$, $\omega = m_\pi + 10$ MeV, $\omega = m_\pi + 20$ MeV and $\omega = m_\pi + 30$ MeV, respectively (from Bat–73).

232

but these would appear empirically to produce only minor effects in the dispersion relations (Eri-70b), the contribution of the 3,3 region to $\mathrm{Im} f(\omega)$ $= k\sigma(\omega)/4\pi$ being substantially larger than that of the threshold region, at least away from $\omega \approx m$. This not only reduces worries about the unphysical contribution but also makes appropriate the use of the threshold subtraction point $\omega_0 = m$. The pole contributions arise from the region near $\omega = 0$, since the ω_i are small (compared to m), and for $f^{(+)}$ are negligible, while for $f^{(-)}$ a single-nucleon excess should give a pole contribution appropriate to one nucleon, $f^{(-)} = 2 \times 0.08/\omega$, for πNN coupling constant (squared) 0.08. Applications, to $^{12}\mathrm{C}$ for instance, have been made based on experimental total $\pi - ^{12}\mathrm{C}$ cross sections, using Glauber theory for the high-energy limit and pionic atom data for the threshold subtraction. The below-threshold information is introduced by fitting $\mathrm{Im} f(\omega)$ to a polynomial in k, which is then replaced by $k \rightarrow ik$ below threshold. Results are then compared (Bat-73), as in Fig. 5.11, with $\mathrm{Re} f(\omega)$ as extracted from Coulomb interference measurements (Bin-71a, Sco-72) or phase shift analysis (Bei-73). For $T = \frac{1}{2}$ nuclei, a similar analysis allows the determination of the pole residue, or coupling constant, which arises from taking the various poles that are grouped near $\omega = 0$ as if they were at that single point. Thus, for example, for $\pi^{\pm} - ^9\mathrm{Be}$ scattering, measurements (Wil-73, Clo-74) exist of $\Delta\sigma = \sigma_{\pi^+} - \sigma_{\pi^-}$ in the energy range 100 MeV–800 MeV, and higher energies can be dealt with using Glauber theory, while below 100 MeV an optical potential can be used. The result (Eri-75) for the effective coupling constant is 0.06, to be compared with 0.08 for a free nucleon, thus implying a perceptible change of the coupling in the nuclear situation.

5.3   Mesic Atoms

A useful tool for the study of interactions between mesons and nuclei has been the analysis of mesic atoms. These are formed when a negative pion or kaon is slowed down sufficiently to be trapped by the Coulomb attraction of the nucleus. An electromagnetic cascade then takes place until the meson reaches the lower Bohr orbits of the mesic atom. On the one hand, these are well within the lowest electron orbit, so that the meson feels the full Coulomb field. On the other hand, as lower levels are reached the mesic atom wave function involves appreciable probability for the meson to be within the nuclear volume, where strong interaction effects will play an important role. These will have two main consequences: a shift in the position of the otherwise hydrogenic level position due to the hadronic attraction or repulsion between the meson and the nucleus, and a broadening of the atomic level arising from the fact that it can now decay through strong interaction absorption of the meson by the nucleus in addition to further electromagnetic transitions. Eventually the probability for strong interaction absorption dominates and transitions to still lower atomic levels are not seen since the meson has little likelihood of penetrating further once strong absorption comes into play appreciably. The measurement (see, for example, Tau–75 for a review of these) of the energy shifts and widths of the levels in the mesic atom then offers a way of studying the meson-nucleus interaction, essentially equivalent to scattering and absorption reactions performed at zero energy. The energy shift then relates to the scattering length and the width to the rate of true pion absorption.

5.3a Multiple-Scattering at Zero Energy: The Optical Potential

The theoretical analysis of mesic atom level shifts and widths has generally* been carried out using the optical potential discussed formally in Chapter 4, and more concretely in Section 5.1. For mesic atoms the basic meson-nucleon amplitude must be considered at zero energy, and, in the case of the pion, one must expect πN s- and p-waves to dominate. For pionic atoms,⁻ the potential is thus taken, preliminarily, in the form (Eri–66a, Kre–69)

$$V = -\frac{4\pi}{2m_{red}}\left\{ b_0\rho(\mathbf{r}) + b_1\left[\rho_n(\mathbf{r}) - \rho_p(\mathbf{r})\right] + i\,\mathrm{Im}\,B_0\rho^2(\mathbf{r}) \right.$$
$$\left. - \nabla\cdot\left[c_0\rho(\mathbf{r}) + i\,\mathrm{Im}\,C_0\rho^2(\mathbf{r})\right]\nabla \right\} \qquad [80]$$

for use in a Klein-Gordon equation† with a Coulomb potential $V_c(r)$ present,

$$\left(\nabla^2 + \left\{\left[E - V_c(r)\right]^2 - m_{red}^2\right\}\right)\psi = 2m_{red}V\psi. \qquad [81]$$

Here

$$m_{red} = \frac{m\,\mathfrak{M}}{\mathfrak{M} + m} \approx m \qquad [82]$$

---

*An exception to this is a calculation (Ure–66) for kaonic atoms that generated a KN *potential* and summed it over the nucleons for the full KA interaction; this may serve as a reminder that there are other routes for such problems than that of the optical potential.
†See also the footnote on p. 195.

is the πA reduced mass for a nucleus of mass $\mathfrak{M}, \rho_n(\mathbf{r})$ and $\rho_p(\mathbf{r})$ are the neutron and proton densities, with $\rho(\mathbf{r}) = \rho_n(\mathbf{r}) + \rho_p(\mathbf{r})$ the total nuclear density, and normalization $\int \rho_n \, d\mathbf{r} = N, \int \rho_p \, d\mathbf{r} = Z, \int \rho \, d\mathbf{r} = A$. The quantities b_0, b_1, c_0, B_0, and C_0 are complex coefficients related to πN scattering and pion absorption in the nucleus. In particular, if we assume the optical potential to be given by its lowest-order approximation, the s-wave isoscalar and isovector coefficients are

$$b_0 = \tfrac{1}{3}(\alpha_1 + 2\alpha_3), \quad b_1 = \tfrac{1}{3}(\alpha_3 - \alpha_1), \qquad [83]$$

in terms of the zero-energy limit of the s-wave πN channel amplitudes for total πN isospin $T = \tfrac{1}{2}, \tfrac{3}{2}$ (see Eqs. 2.13 and 2.14a)

$$\alpha_{2T} = \frac{1}{k} e^{i\delta_T} \sin \delta_T. \qquad [84]$$

The isovector part $b_1(\rho_n - \rho_p)$ has been taken to be diagonal and is therefore proportional to the neutron-proton density difference. In the p-wave part the isovector piece is dropped as unimportant, and we have, again for the lowest-order result,

$$c_0 = \tfrac{1}{3}(4\alpha_{33} + 2\alpha_{31} + 2\alpha_{13} + \alpha_{11}), \qquad [85]$$

with p-wave πN amplitudes in the channels $J = \tfrac{1}{2}, \tfrac{3}{2}$ and $T = \tfrac{1}{2}, \tfrac{3}{2}$, to be taken in the zero-energy limit (Eq. 2.14b):

$$\alpha_{2T2J} = \frac{1}{k^3} e^{i\delta_{TJ}} \sin \delta_{TJ}. \qquad [86]$$

A minor kinematic modification to Eqs. 83 and 85 is usually introduced to incorporate the transformation from the πN center-of-mass system, in which α_{2T} and α_{2T2J} of Eqs. 84 and 86 are parametrized, to the πN laboratory system which is an appropriate approximation for the πA application. This can be incorporated approximately* by recalling from the optical theorem that $\mathrm{Im} f(0°) = k\sigma_{total}/4\pi$, while σ_{total} is invariant, so that $\mathrm{Im} f_{c.m.}(0°)/k_{c.m.} = \mathrm{Im} f_{lab}(0°)/k_{lab}$. We may assume that the transformation that is correct for the imaginary part of f in the forward direction is approximately correct for the amplitude in general, while $k_{c.m.}(M + m) = k_{lab} M$, so that factors of $M/(M + m) \approx 0.87$ must be appropriately introduced.

---

*The exact transformation can, of course, be derived from a consideration of the general dynamical definition of the scattering amplitude or from the transformation properties of the differential cross section. (See also the footnote on p. 201.) For our present purposes, the forward, approximate result suffices.

The parameter $b_0 = \frac{1}{3}(\alpha_1 + 2\alpha_3) = \frac{1}{2}(a_n + a_p)$, for proton and neutron scattering lengths a_n and a_p, is an order of magnitude smaller than b_1, and it therefore proves necessary to incorporate the next order in the optical potential from the start. For this zero energy case, this is

$$b_0 = \frac{1}{2}(a_n + a_p) - \langle \frac{1}{r} \rangle_{\text{corr}} \cdot \frac{1}{4} \left[2(a_n^2 + a_p^2) + (a_n - a_p)^2 \right] \quad [87]$$

where $\langle 1/r \rangle_{\text{corr}}$ is the zero-energy limit of the propagator e^{ikr}/r averaged for nucleon correlations. If these are taken as Pauli correlations for two identical nucleons of parallel spin, the result is (Kre–69), for Fermi momentum p_F, $\langle 1/r \rangle_{\text{corr}} = 3p_F/2\pi$.

The terms in the pionic atom optical potential of Eq. 80 involving $i \operatorname{Im} B_0 \rho^2(\mathbf{r})$ and $i \operatorname{Im} C_0 \rho^2(\mathbf{r})$ are intended to include the effects of true absorption. They are taken as quadratic in the density on the assumption that the main absorption reaction will lead to the ejection of two nucleons in order to facilitate energy-momentum conservation. (If one nucleon is emitted it must take up all 140 MeV of the pion rest mass energy and hence emerge with about 500 MeV/c momentum, which is far greater than that supplied by the Fermi momentum of $\lesssim 270$ MeV/c, the pion bringing in essentially no momentum.) The fact that two nucleons are probably involved in the absorption thus motivates the $\rho^2(\mathbf{r})$ choice. The real parts of B_0 and C_0 exist and represent dispersive effects through which the absorption mechanism can change the real part of the optical potential. They are difficult to calculate (Hac–78) and have generally been ignored. The coefficients B_0 and C_0 are far less easily related to the fundamental $\pi d \leftrightarrow 2N$ process, say, than is the case for b_0, b_1, c_0, and $\pi N \to \pi N$. This is because the absorption process may well be appreciably modified by the presence of the nuclear medium in which it takes place. For instance, the residual nucleus has a profound effect on the mutual off-shell scattering of the outgoing nucleons (Mor–73, Gar–74). For shifts of the $1s$ level, for example, as measured in $2p$-to-$1s$ transitions, good fits to the data are obtained when the B_0 and C_0 terms are ignored, as seen in Fig. 5.12; for other transitions, and for pionic atom widths, taking B_0 and C_0 as constant parameters to be fixed from experiment then yields good agreement over a range of nuclei, and the resulting values for B_0 and C_0 are about a factor of two larger than those predicted from the $\pi d \leftrightarrow 2N$ data (Kre–69). More detailed discussion of the role of true absorption is given in Chapters 6 and 7.

A basic problem in the application of the optical potential approach to pionic atoms lies in the convergence of the Watson series for it. Our estimates of Chapter 4 have implied that we cannot easily guarantee such convergence, or the negligibility of corrections to the impulse approximation, unless we are at fairly high energies. Of course, this is not to say that

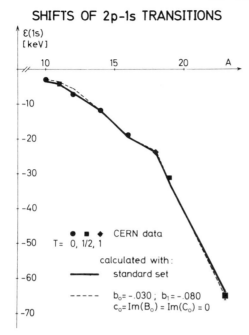

Fig. 5.12 Energy shifts due to strong interactions in $2p \to 1s$ transitions in pionic atoms for various nuclei with nucleon number A, and, as shown, isospin $T = 0$ (circle), $T = \frac{1}{2}$ (square), $T = 1$ (lozenge). The theoretical results are based on a pionic atom optical potential with standard parameters (solid line) or modified ones (dashed line) with the values shown (Kre–69).

in any other application the series will diverge; only detailed calculation of the higher-order terms can decide this. In the particular case of pionic atoms, the zero-energy scattering amplitudes tend to be quite small—on the order of 0.1 fm and hence much smaller than internucleon separation distances—and this would appear to aid convergence greatly. There is no guarantee, however, that even the condition $\langle f_{\pi N} \rangle / \langle r \rangle \ll 1$ assures success since the smallness of πN scattering *on*-energy-shell does not necessarily promise small off-shell scattering. Nonetheless, the results of optical potential calculations for pionic atoms seem to support the existence of rapid convergence. The validity of the impulse approximation at zero energy has been examined directly in a three-body formulation involving the pion, a nucleon, and the remaining nucleons taken as a fixed residual nucleus (Moy–69, Kol–77). This calculation suggests that the binding of the "active" nucleon modifies the equivalent πA scattering length by about 10%; the results of using a more recent πN analysis there are to give much poorer agreement with the pionic atom experimental data.

In a general way, pionic atom levels with $l \geq 1$ seem to be reasonably described by the lowest-order optical potential with input parameters from free, zero-energy πN scattering and the simple assumptions about true absorption noted above. For $l = 0$, the input πN information is less well known and we require at least the second-order potential as in Eq. 87. This, with a simple form for B_0, works, though not as well as for $l \geq 1$. There remain the problems of higher orders and off-shell effects such as that of binding and πN range (with respect to the latter, see Kwo–78), as well as the precise role of the dispersive corrections $\operatorname{Re} B_0$ and $\operatorname{Re} C_0$. Lastly we note that the Lorentz-Lorenz effect of Section 5.1e, when applied to its most natural physical situation—namely, the pionic atom—remains unproved in the sense that its influence can generally be vitiated by making changes in the other parameters of the optical potential. In all likelihood, it should be treated on the same footing as all other corrections to the lowest-order potential; that is, only the $\rho^2(\mathbf{r})$ term should be retained and its coefficient not regarded as given definitively by the formal arguments presented in Section 5.1e.

5.3b Kaonic Atoms

The relative success of the lowest-order optical potential in its application —with suitable higher-order corrections—to pionic atoms is not reproduced for kaonic atoms. There, rather, the strength of the meson-nucleon interaction is much greater and the optical potential has much stronger nonlocal features making the problem more difficult. Indeed, even the sign of the shifts of the levels (Bac–72) is not given correctly by $\mathcal{V}_c^{\overline{K}A}$ $= -(4\pi/2m_K^{red})f_{\overline{K}N}(E_K \to 0)A\rho(\mathbf{r})$. The reasons for this are fairly clear (Wyc–71, Bar–71) and arise from the presence of a resonance—the $\Lambda(1405)$—in the channel with the quantum numbers of the K^-p system, located at an energy slightly below (by about 27 MeV) K^-p threshold. When we consider the effective energy of *relative* motion at which K^-p scattering takes place for a kaon bound in an atomic orbital and a proton bound in a nucleus, we must subtract that part of the energy that goes into the center-of-mass motion of the K^-p system:

$$E_{rel}(K^-p) = M_p + m_K - \epsilon_p - P^2/2(M_p + m_K), \qquad [88]$$

where P is the momentum of the total K^-p system, ϵ_p is the mean proton binding energy, and we have ignored the small binding energy of the kaon in the atom.

For a rough estimate, we can suppose that the bulk of the K^-p kinetic energy resides in the more tightly confined proton, and approximate the magnitude of the last term in Eq. 88 by 10 MeV. The mean proton binding energy contributes about another 10 MeV reduction in the internal energy available for the K^-p system. In addition, there is a small dynamical effect that arises from the near-resonant interaction of the pion from the virtual decay of the $\Lambda(1405)[\to\Sigma+\pi(148 \text{ MeV}/c)]$ with nucleons in the nucleus. This effect raises the position of the resonance (Eis–76b) by some 5—10 MeV (as well as broadening the resonance by $\sim25\%$), so that the net effect is to close the 27 MeV gap between the $\Lambda(1405)$ resonance position and K^-p threshold ($M_p+m_K=938 \text{ MeV}+494 \text{ MeV}=1432 \text{ MeV}$) quite completely. Since the real part of the K^-p amplitude changes sign at the resonance, its effective introduction into the relevant kinematic region for K^-p scattering in the kaonic atom can lead to a reversal of the sign of the shift produced by the optical potential—an indication of how important it can be to include kinematic averaging of the impulse approximation and corrections to it.

To include the kinematic effects and dynamic effects more completely, one must address the full three-body problem of K^-, active proton and residual nucleus, with channel coupling to permit $\overline{K}N\to\Sigma\pi$, as well as $K^-p\to\overline{K}^0n$ (Alb–76). The input for the two-particle interaction uses separable interactions, and it is supposed that the $T=0$ channel is dominated by the $\Lambda(1405)$ resonance. The resulting optical potential is still of lowest-order in the sense of the Watson expansion; that is, $\mathcal{V}_c\propto t_{\overline{K}N}$. However, the $\overline{K}N$ t-matrix here is not given by the free amplitude (impulse approximation) but rather as the result of a calculation; the higher-order corrections to $\mathcal{V}_c$ coming from nuclear correlations are still neglected. The kaon optical potential $\mathcal{V}_K(\mathbf{r},\mathbf{r}')$ is also nonlocal, and to increase its utility various approximate local equivalents are constructed. These include an equivalent local potential

$$\mathcal{V}_{el}(\mathbf{r})\equiv\frac{1}{\phi_K(\mathbf{r})}\int\mathcal{V}_K(\mathbf{r},\mathbf{r}')\phi_K(\mathbf{r}')\,d\mathbf{r}' \qquad [89]$$

where $\phi_K(\mathbf{r})$ is the undistorted kaon atomic wave function, a local potential

$$\mathcal{V}_l(\mathbf{r})\equiv\int\mathcal{V}_K(\mathbf{r},\mathbf{r}')\,d\mathbf{r}', \qquad [90]$$

and a local density approximation. As seen in Fig. 5.13, none of these follows the nuclear density, as the lowest-order, local optical potential would. Moreover, their real parts change sign at the nuclear surface, as required by experiment.

Both the pionic and kaonic atoms, as well as other hadronic atoms such as those built with $\overline{p}$ or Σ^-, can exhibit dynamic mixing effects (Haf–70, Leo–76). In these, the electromagnetic interaction (or, possibly in heavier

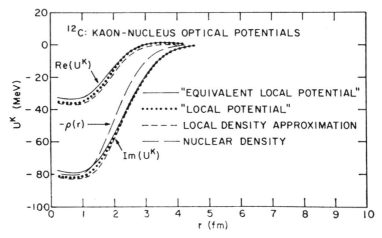

Fig. 5.13 Various local approximations to the nonlocal kaon-nucleus optical potential for use in $K^- - {}^{12}C$ kaonic atoms (Alb–76). The localizing prescriptions are given in and around Eqs. 89 and 90. The real parts of the "equivalent local" and "local" potentials change sign near the nuclear surface; that is, they are not just proportional to the nuclear density $\rho(\mathbf{r})$. This feature is seen in experiment.

nuclei, the strong interaction) admixes nuclear excited states with the nuclear ground state in the mesic atom, at the same time mixing the atomic states in which the meson finds itself. Such effects are particularly large if the nuclear excitation energy and the atomic transition energy in question are nearly degenerate, a situation which is met with for medium-weight nuclei. There the nuclear transitions from the 0^+ ground state to a low-lying 2^+ level may have energies comparable to those of the atomic $n=5 \rightarrow n=4$ or $n=4 \rightarrow n=3$ decays and the effects of the dynamic mixing may be seen by comparing mesic atom lines in neighboring isotopes. A specific instance occurs (Leo–76) in ${}^{112}Cd$, for example, where the atomic pion $5g \rightarrow 3d$ transition has an energy difference of 618.8 keV, as compared with a $0^+ \rightarrow 2^+$ nuclear excitation requiring 617.4 ± 0.3 keV.

5.4 Scattering with Exchange of Quantum Numbers

The class of scattering processes in which the main effect is to transfer a diagonal quantum number, such as charge or strangeness (or both), is especially interesting since one expects it to lead to nuclear states that are very closely related to the ground state. This holds out the hope of being able to study these interesting nuclear circumstances with special care, and of exploiting what may be a relatively simple nuclear situation for furthering our understanding of the meson-nucleus interaction. One can envision carrying out the following experiments: (1) Pion single charge exchange $A(\pi^{\pm}, \pi^0)A'$ in which the final nucleus A' is excited to a resonance that is the isobaric analog of the original nuclear ground state A, that is, is a rotation in isospin space of that state; this of course requires a nucleus with isospin $T \geqslant \frac{1}{2}$. Such reactions are generally done with π^+ beams (which are more copious than π^- beams) and are then parallel to (p, n) experiments. The π^0 is especially difficult to observe since it decays quickly to two photons, necessitating special techniques for the detection of this process (Sha–76), such as activation methods or special photon spectrometers (Als–79). (2) Pion double charge exchange $A(\pi^{\pm}, \pi^{\mp})A''$ in which the final nucleus A'' is the "double" analog of A, requiring $T \geqslant 1$, and two nucleons have of necessity been active in the process. Cross sections for this prove to be very small (Mar–77b), increasing the difficulty of its observation. (3) Strangeness exchange reactions (K, π), producing nuclear levels in which one nucleon has become a hyperon. The relatively low intensity of kaon beams makes these experiments (Pov–76) especially challenging. (4) One can also imagine single charge exchange using kaons, such as (K^-, K^0), as a sheer demonstration of virtuosity, or double charge exchange $(K^{\mp}, K^{\pm})$ which also requires double strangeness exchange. This last would clearly be extremely interesting, but its detailed discussion can probably safely await future editions of this work.

5.4a Pion Charge Exchange

Let us first consider the single charge exchange to the isobaric analog state or resonance (denoted by iar), which we shall take in the form $A(\pi^+, \pi^0)A'_{iar}$. A unique feature of this transition to the analog state is that for nuclei with $T = \frac{1}{2}$ a theory that predicts elastic scattering for both charged pions $\pi^{\pm}$ is completely constrained by isospin invariance with respect to its prediction for excitation of the analog (see also Eqs. 104 and 105 below). In order to illustrate the salient points of a dynamical calculation here, we describe the process using the distorted wave impulse approximation (DWIA), discussed at the end of Section 4.2, in an eikonal form. For the lowest-order DWIA result (see Eq. 4.116), we therefore require

$$\langle \pi^0; a | T | \pi^+; 0 \rangle = \int \chi_{\mathbf{k}'}^{(-)*}(\mathbf{r}) \langle a | \sum_{i=1}^{A} \left(t_s + t_v \vec{I} \cdot \vec{\tau}_i \right) \delta(\mathbf{r} - \mathbf{r}_i) | 0 \rangle \psi_{\mathbf{k}}^{(+)}(\mathbf{r}) \, d\mathbf{r},$$

$$[91]$$

where $\chi_{\mathbf{k}}^{(-)}(\mathbf{r})$ is the wave function for the outgoing π^0 distorted in the lowest-order optical potential, and $\psi_{\mathbf{k}}^{(+)}(\mathbf{r})$ is the wave function of the incident π^+ distorted in the full optical potential. The nuclear ground state $|0\rangle$ has as its analog here

$$|a\rangle = \frac{1}{\sqrt{N-Z}} T_+ |0\rangle = \frac{1}{\sqrt{N-Z}} \cdot \frac{1}{2} \sum_{j=1}^{A} (\tau_1 + i\tau_2)_j |0\rangle, \quad [92]$$

for

$$\tfrac{1}{2}(\tau_1 + i\tau_2)|n\rangle = |p\rangle, \quad \tfrac{1}{2}(\tau_1 + i\tau_2)|p\rangle = 0, \quad [93a]$$

$$\tau_3 |p\rangle = |p\rangle, \quad \tau_3 |n\rangle = -|n\rangle, \quad [93b]$$

whence, noting that here $T_- |0\rangle = 0$,

$$
\begin{aligned}
\langle a | a \rangle &= \frac{\langle 0 | T_- T_+ | 0 \rangle}{(N-Z)} = \frac{\langle 0 | [\, T_-, T_+ \,] | 0 \rangle}{(N-Z)} \\
&= \frac{1}{4(N-Z)} \langle 0 | \sum_{j,k} \left[(\tau_1 - i\tau_2)_j, (\tau_1 + i\tau_2)_k \right] | 0 \rangle \\
&= \frac{1}{4(N-Z)} \langle 0 | 4 \sum_j (-\tau_3)_j | 0 \rangle = \langle 0 | 0 \rangle = 1. \quad [94]
\end{aligned}
$$

The analog state as defined in Eq. 92 is to be taken as a good approximation to the actual nuclear state reached in the final nucleus; that is, we assume that $|a\rangle$ is very nearly an eigenstate of the nuclear Hamiltonian so that it is narrow and relatively unmixed* (Aue−72). This state carries coherence with respect to the ground state in the sense that it has good overlap with the ground state when they sandwich the isospin raising operator, so that near the forward direction the cross section in lowest-order does not vanish as for most inelastic processes but is proportional to $(N-Z)$. The πN amplitude has been decomposed into an isoscalar and isovector part, t_s and t_v, the latter carrying the nucleon and pion isospin operations $\vec{\tau}$ and $\vec{I}$. Then using Eq. 92 and

$$\langle \pi^0 | \vec{I} \cdot \vec{\tau} | \pi^+ \rangle = \sqrt{2} \cdot \tfrac{1}{2}(\tau_1 + i\tau_2), \quad [95]$$

Eq. 91 gives the general, lowest-order DWIA result

$$\langle \pi^0; a | T | \pi^+; 0 \rangle = \sqrt{2}\, t_v \int \chi_{\mathbf{k}'}^{(-)*}(\mathbf{r}) \frac{1}{\sqrt{N-Z}} \left(N\rho_n(\mathbf{r}) - Z\rho_p(\mathbf{r}) \right) \psi_{\mathbf{k}}^{(+)}(\mathbf{r})\, d\mathbf{r}.$$

$$[96]$$

---

*This will be much less the case for the strangeness analog state introduced below in Eq. 108.

This shows vividly the hoped-for simplicity of the quantum-number-exchange process (though still only in lowest order!): The amplitude involves directly the nuclear neutron and proton densities ρ_n and ρ_p [with $\int \rho_{n,p}(\mathbf{r})d\mathbf{r}=1$ here], and the distorted pion waves. Naturally, only the isovector part of the πN amplitude contributes.

We now approximate both the pion waves by those involving the lowest-order optical potential, thus $\psi_k^{(+)}(\mathbf{r}) \approx \chi_k^{(+)}(\mathbf{r})$, and take those in eikonal form,

$$\chi_k^{(\pm)}(\mathbf{r}) \cong e^{i\mathbf{k}\cdot\mathbf{r}} \exp - iv^{-1} \int_{\mp\infty}^{z} \mathcal{V}_c^{(\pm)}(\mathbf{b},\zeta)\,d\zeta, \qquad [97]$$

where the optical potential is that for the appropriate pion in the initial or final states,

$$\mathcal{V}_c^{(+)} = -\frac{4\pi}{2m}A\rho(\mathbf{r})\bar{f}_0(\pi^+)$$

$$\mathcal{V}_c^{(-)} = -\frac{4\pi}{2m}A\rho(\mathbf{r})\bar{f}_0^*(\pi^0). \qquad [98]$$

This approximate eikonal result for distorted pion waves is of general applicability and is used again several times in the following sections for estimates of distortive effects; the validity of the approximation is restricted to forward scattering situations.

In Eq. 98 $\rho(\mathbf{r})=(N\rho_n + Z\rho_p)/A$ is the nuclear density normalized to unity, and $\bar{f}_0$ is the forward πN amplitude, averaged over nucleon species. This last we parametrize using the optical theorem as

$$\bar{f}_0 = \frac{ik}{4\pi}\frac{1}{A}\left[N\sigma_{\pi n}(1-i\alpha_{\pi n})+Z\sigma_{\pi p}(1-i\alpha_{\pi p})\right], \qquad [99]$$

where $\sigma_{\pi N}$ is the total πN cross section ($N=n,p$) and $\alpha_{\pi N}$ is the ratio of real to imaginary part for the forward amplitude. We insert Eqs. 97–99 into Eq. 96 and approximate the eikonal classical path integral of Eq. 97 as if the pion traverses the nucleus in the forward direction, and take t_v in the forward direction as well, which is not unreasonable, since for our process there is a degree of coherence and the πA amplitude is forward peaked. Then

$$\langle \pi^0; a|T|\pi^+;0\rangle \cong \sqrt{2}\, t_v(0)\int e^{i(\mathbf{k}-\mathbf{k}')\cdot\mathbf{r}}\frac{1}{\sqrt{N-Z}}\left[N\rho_n(\mathbf{r}) - Z\rho_p(\mathbf{r})\right]$$

$$\times e^{-\bar{\sigma}(1-i\bar{\alpha})A/2\int_{-\infty}^{\infty}\rho(\mathbf{b},\zeta)\,d\zeta}\,d\mathbf{r}, \qquad [100]$$

where $\bar{\sigma}$ and $\bar{\alpha}$ have been averaged over N and Z, π^+ and π^0, so that

$\bar{\sigma} \approx \frac{2}{3}\sigma(\pi^+ p \to \pi^+ p)$. For the angle-integrated cross section the oscillatory part arising from $\bar{\alpha}$ plays no great role. This is because at the high energies appropriate to our eikonal estimate,

$$\sigma \equiv \left(\frac{\omega}{2\pi}\right)^2 \int |T|^2 d\Omega = \left(\frac{\omega}{2\pi k}\right)^2 \int_{\Delta < 2k} |T|^2 d^2\Delta, \qquad [101]$$

where $\Delta = (\mathbf{k} - \mathbf{k'})_\perp$, $\Delta \cong 2p \sin \frac{1}{2}\theta$, is the perpendicular momentum transfer. Thus

$$\sigma \cong \left(\frac{\omega}{2\pi}\right)^2 |\sqrt{2}\ t_v(0)|^2 \frac{1}{k^2(N-Z)} \int d^2\Delta \int d^2\mathbf{b}\, dz\, e^{i\Delta \cdot \mathbf{b}}\left[N\rho_n(\mathbf{r}) - Z\rho_p(\mathbf{r})\right]$$

$$\times e^{-1/2\bar{\sigma}(1-i\bar{\alpha})A \int_{-\infty}^{\infty} \rho(\mathbf{b},\zeta)d\zeta} \int d^2\mathbf{b'}\, dz'\, e^{-i\Delta \cdot \mathbf{b'}}\left[N\rho_n(\mathbf{r'}) - Z\rho_p(\mathbf{r'})\right]$$

$$\times e^{-1/2\bar{\sigma}(1+i\bar{\alpha})A \int_{-\infty}^{\infty} \rho(\mathbf{b'},\zeta)d\zeta}$$

$$\cong \sigma_0(\pi^+ n \to \pi^0 p)\left(\frac{2\pi}{k}\right)^2 \int d^2\mathbf{b}\, e^{-\bar{\sigma}A \int_{-\infty}^{\infty} \rho(\mathbf{b},\zeta)d\zeta}$$

$$\times \frac{1}{N-Z}\left|\int_{-\infty}^{\infty} dz\,(N\rho_n(\mathbf{b},z) - Z\rho_p(\mathbf{b},z))\right|^2 \qquad [102]$$

in which the oscillatory part has canceled; the manipulation performed here is again one of fairly widespread usefulness at high energies and forward directions.

Equation 102 shows explicitly the main physical ingredients of the lowest-order DWIA result for $A(\pi^+,\pi^0)A'_{iar}$. In the region of 100–300 MeV, the angle-integrated cross section will have a part that rises in response to the 3,3 resonance, namely, $\sigma_0(\pi^+ n \to \pi^0 p) = \sigma_0(\pi^- p \to \pi^0 n)$, the forward πN charge exchange cross section. But this feature is negated —indeed reversed—by the exponential suppression coming from the rise of $\bar{\sigma}$ in the factor expressing the distortion of the pion wave. Physically, this simply means that the charge exchange mechanism on a single nucleon in the nucleus would resonate, but the probability for a pion to penetrate to the site of an inner nucleon is strongly reduced by the large effective cross section for encountering some ·other nucleon on the way and undergoing some noncharge-exchange reaction, presumably $(\pi,\pi'N)$, which removes the system from the $\pi^+ A$ or $\pi^0 A'$ channels. As a consequence, the calculated cross sections, even for a light nucleus such as $^{13}C(\pi^+,\pi^0)^{13}N$, show a strong dip in the region of the 3,3 resonance while the experiment appears not to (Zai–74), an effect that is exaggerated in the eikonal estimate (Tow–75) but is present in more realistic optical potential calculations as well (War–77).

In fact, far from involving a very simple reaction mechanism $A(\pi^+, \pi^0)A'_{iar}$ embodies a feature that tends to make it quite complicated. This is the fact that, since no quantum numbers are exchanged except charge, there is no angular momentum barrier to suppress the contribution from the central nuclear region so that absorption effects play an unusually strong role and greatly reduce the contribution of the lowest-order result. Thus, central densities in Eq. 100, say, will lead to an amplitude with the structure, in the plane wave limit,

$$T_{fi} \propto \int j_\lambda(|\mathbf{k} - \mathbf{k}'|r)\rho_{fi}(r)r^2\, dr, \qquad [103]$$

with $\lambda = 0$, whereas if the transition involved transfer of orbital angular momentum ΔL, as in an excitation $0^+ \to 2^+$, say, where $\Delta L = 2$, we would have $\lambda = \Delta L$ (see Section 5.5 and, in particular, Eq. 111). Since $j_\lambda(qr) \underset{r\to 0}{\to} (qr)^\lambda/(2\lambda + 1)!!$ there is a great deal more suppression of the dense central region in the latter case and fewer effects of absorption (see also Section 5.5, where such cases are treated).

The problems of the lowest-order DWIA theory make it natural to ask about higher-order terms in the DWIA, as shown in Eq. 4.116, for instance. (Higher-order corrections to the optical potential, which distorts the pion waves, tend to have small influence on the charge exchange reaction, making it perhaps slightly *more* absorptive.) These have in fact been estimated (Eis–75a, Gal–76a, War–78b) by using closure in Eq. 4.116 to generate the correction in terms of an isovector two-nucleon correlation function. Because of the perfect overlap between initial and (isospin rotated) final state here, these corrections are quite large, of the order of a 70% addition to the cross section. While this would not appear to be enough to account for the relative absence of absorption seen in experiment, it already suffices to raise doubts about the rapidity of convergence of a generalized DWIA scheme as applied to this problem. Unfortunately, the general third-order correlation correction to the transition operator in the DWIA scheme is too difficult to be evaluated reliably. Glauber theory, which in principle could come to our rescue for this purpose, is suspect on general grounds in the 3,3 region, and, in fact, can produce quite nondiffractive results (Loh–77) for $A(\pi^+, \pi^0)A'_{iar}$ due to the intrinsic reduction of the forward peak there as opposed to the case for elastic scattering. The use of a Monte Carlo method with the full scattering series, however, yields results rather similar to those of the DWIA and thus may indicate that the first- and second-order terms in the DWIA do represent a good approximation to the full series for this case (Gib–77). Beyond the difficulties of lowest-order DWIA, there are, of course, further uncertainties for (π^+, π^0)

arising from the somewhat unknown neutron distribution in nuclei from higher-order corrections in the distorting optical potential, and so forth.

Last, we note that the comparison of $A(\pi^+, \pi^0)A'_{iar}$ with $\pi^{\pm}A$ elastic scattering for both varieties of charged pion allows the extraction of limits for the charge exchange process (Loh–77). If we ignore Coulomb effects, the charge exchange amplitude is given from isospin invariance in terms of the elastic amplitudes (for a $T = \frac{1}{2}$ target) by

$$ F(\pi^+ \to \pi^0) = \frac{1}{\sqrt{2}} \left[F(\pi^+ \to \pi^+) - F(\pi^- \to \pi^-) \right]. \qquad [104] $$

The problem in using this is, of course, that we generally do not know the phase of the elastic amplitudes from experiment; hence we must either use a model for its determination (or to constrain its range of uncertainty), exploit Coulomb-nuclear interference, or content ourselves with a triangle inequality based on Eq. 104, namely,

$$ \frac{1}{2} \left[\sqrt{\sigma(\pi^+ \to \pi^+)} - \sqrt{\sigma(\pi^- \to \pi^-)} \right]^2 \leqslant \sigma(\pi^+ \to \pi^0) $$

$$ \leqslant \frac{1}{2} \left[\sqrt{\sigma(\pi^+ \to \pi^+)} + \sqrt{\sigma(\pi^- \to \pi^-)} \right]^2. \qquad [105] $$

The double charge exchange reaction, will, of course, share many of the same physical features that we have noted for single charge exchange. It is especially interesting as a process that may allow relatively easy identification of a $\Delta T_z = 2$ double analog state—not achievable with nucleons that can only exchange one unit of charge—and as a means of producing nuclei far from the stability line, particularly proton-rich nuclei. The measurement of double charge exchange reactions is very difficult, however, due to their extremely small cross sections; on the other hand, it has some advantage over (π^+, π^0) in that it involves a charged meson in the final state. In fact these reactions only become a real possibility with the advent of high-intensity meson beams in the meson factories. The first such unambiguous and relatively precise measurement of this type then found (Mar–77b; see also Gil–64) for the double-analog transition $^{18}O(\pi^+, \pi^-)^{18}Ne$ (ground state) at 139 MeV in the forward direction $d\sigma/d\Omega|_{0°} = 1.78 \pm 0.30 \,\mu b/sr$, which is large enough for use of the reaction as a spectroscopic tool for the investigation of nuclear structure. In characteristic fashion, Glauber theory, though at best very marginally applicable at this low energy, gives a result (Liu–75) close to this value, while multiple-scattering theories tend to bracket it, within the range of uncertainties as to how best to sum the infinite series and deal with off-energy-shell effects. The issue is complicated in an interesting way by the necessity for the reaction to proceed through two charge-exchange steps involving two nucleons, thus making it sensitive to assumptions about two-nucleon correlations and their treatment (Mil–76) and holding out some hope for being able to determine at least gross properties of the correlations through a systematic study of double charge exchange. Unfortunately, the sensitivity to short-range correlation effects is not likely to be very great, since the neutral pion can travel large distances on shell between the two charge exchanging encounters; this presumably also masks mechanisms in which other channels couple in, as, for example, the conversion of the π to a ρ-meson in the intermediate state. A more quantitative estimate (Kol–69) indicates that at high energies the forward double-scattering amplitude, and therefore, $(\pi^\pm, \pi^\mp)$ at $0°$, tends to involve a nuclear average $\langle 1/r^2 \rangle$, while at low energies $\langle 1/r \rangle$ enters. In either event, at least in forward directions the sensitivity is likely to be to the nuclear size rather than to short-range features (see, for example, Lee–77b), though the latter may enter significantly at larger angles. Folded in with these sensitivities, of course, there persist uncertainties arising from the optical potential treatment of the process.

Before leaving the single and double charge exchange reactions we note that studies of transitions to states other than the analogs will eventually also be of considerable interest. These should show up with increasing strength, relative to the more coherent analog excitations, at larger scattering angles. They will measure the specifically isovector part of the transition, and then by comparing with inelastic excitation of the final state analog within the parent nucleus, separated information on the isoscalar contribution can be extracted as well. In the case of double charge exchange, there will be an especially rich collection of excitations reached, due to the exchange of two units of charge, which would otherwise be difficult to reach.

5.4b Strangeness Exchange

Instead of changing units of charge through meson scattering, one can imagine the exchange of units of strangeness, especially through the process (K^-, π^-), leading to hypernuclei in which a nucleon is replaced by a Λ hyperon. The basic, single-nucleon reaction then involves

$$K^- N \rightarrow \Lambda \pi \quad \text{or} \quad K^- N \rightarrow \Sigma \pi, \qquad [106]$$

and the production of Σ hyperons can then proceed through strong, exothermic reactions $\Sigma N \rightarrow N \Lambda$ to states with Λ hyperons only. Double strangeness exchange, with (K^-, K^0) or (K^-, K^+) is also possible and leads to double hypernuclei containing two Λ's in place of two nucleons. This process is analogous to double charge exchange in that it requires two active nucleons. The study of hypernuclei* is interesting for the information it can yield on the ΛN interaction, which is otherwise difficult to obtain (and possibly on $\Lambda\Lambda$ forces in double hypernuclei). In addition, it offers new possibilities for nuclear structure studies through the introduction of a different "contaminant"—a different particle producing new bound species. As a different particle from the nucleon, the Λ is of course not restricted by the Pauli principle so that one may produce unique nuclei, such as $^5_\Lambda$He or $^6_\Lambda$He, in which there are five baryons in the $1s_{1/2}$ shell model single-particle level, namely the Λ and four nucleons. Such an interpretation requires that the self-consistent potential obtained with a Λ present be not too different from that for nucleons only, so that a simple generalization of the shell model is possible.

A striking kinematic feature of the (K^-, π^-) process arises from its exothermic nature (Fes–66). This leads to the fact that, for $K^- + n \rightarrow \Lambda + \pi^-$ in the forward direction, the momentum transfer vanishes for an incident kaon of momentum ~ 550 MeV$/c$, and is less than 75 MeV$/c$ for kaon momenta between 300 MeV$/c$ and 1000 MeV$/c$. Thus, within that range the momentum transfer is substantially less than the Fermi momentum and the strangeness exchange reaction replaces a neutron with a Λ hyperon in the same space-spin state. Dynamically, one may expect (Lip–65) that the reaction will exhibit collective states, and in particular (Ker–71) a strangeness analog resonance that is the parallel here of the isobaric analog resonance for (π^+, π^0). The strangeness analog is generated by a single-nucleon operator that converts neutron to lambda,

$$u_+|n\rangle = |\Lambda\rangle, \quad u_-|\Lambda\rangle = |n\rangle, \qquad [107]$$

through which the full nuclear strangeness analog is constructed from the ground state by means of

$$|s\rangle = \frac{1}{\sqrt{N}} \sum_{i=1}^{A} u_+(i)|0\rangle. \qquad [108]$$

The "quality" of this state as an eigenstate of the full (hyper)nuclear Hamiltonian remains unclear thus far; it may well be inferior to the more usual isobaric analog in that respect since the hyperon-nucleon interaction is not necessarily close to the nucleon-nucleon force and this may lead to

---

*See the reviews by Gal–75a, Gal–75b, Gal–76b. Experiments on (K^-, π^-) are discussed in Pov–75 and Pov–76.

large symmetry breaking relative to the isospin case. As a consequence, the strangeness analog may suffer much splitting in a shell model sense and mixing with nonanalog states. If we imagine the transition operator for (K^-, π^-) in the plane wave impulse approximation limit, we require, in the nuclear space,

$$\langle f | \sum_{i=1}^{A} e^{i\mathbf{q}\cdot\mathbf{r}_i} u_+(i)|0\rangle \cong \langle f | \sum_{i=1}^{A} \left[1 + i\mathbf{q}\cdot\mathbf{r}_i - \tfrac{1}{2}(\mathbf{q}\cdot\mathbf{r}_i)^2 + \ldots \right] u_+(i)|0\rangle$$

$$[109]$$

where $\mathbf{q}$ is the momentum transfer, which we have assumed to be small in expanding the exponential on the grounds that we are near the "magic" region of $p_K \sim 550$ MeV/c where $q \sim 0$. Then the first term in square brackets, by comparison with Eq. 108, leads to the excitation of the strangeness analog, while the next term would contribute to the strangeness analog of a giant resonance excitation in the original nucleus, and so forth. The collectivity of the strangeness analog is especially great if one deals with the highly degenerate situation in which all neutron-hole/lambda-particle configurations $n^{-1}\Lambda$ have equal energy. Its position is then shifted considerably from its zeroth-order, single-particle location.

The analysis of the actual physical situation is complicated by the presence of a quasi-free mechanism (Dal–76a, Gal–76b) that must be removed from the (K^-, π^-) spectrum. In general, quasi-free processes arise from the ejection of a nucleon with kinematics near those for the free process, that is, essentially as if the emitted nucleon had no interaction with other nucleons. If the nucleon were truly free its "ejection" energy would simply be the recoil energy $q^2/2M$ for momentum transfer q. In the nucleus, this single, sharp energy is replaced by a large bump in the inelastic spectrum centered around $q^2/2M$ (with suitable modifications for kinematical changes) and half-width qk_F/M for a Fermi gas model with Fermi momentum k_F. These quasi-free excitations are seen very clearly in inelastic electron or proton scattering at a few hundred MeV (see for example deF–66). In the (K^-, π^-) reaction two somewhat unusual features arise: (1) There can be a quasi-free process even in the forward direction due to the exothermic nature of the process. (2) The quasi-free mechanism is possible even for $q \ll k_F$, which would normally be prohibited by the Pauli principle since the nucleon would have to recoil into an already filled state, but is here allowed since we produce a Λ hyperon that is Pauli immune.

For $qR \gtrsim 1$, where R is the nuclear radius, the incoherent, quasi-free feature tends to dominate, while for p_{K^-} near the "magic" value that makes $q \sim 0$ the quasi-free peak can be made very narrow so that coherent

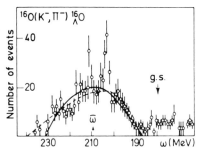 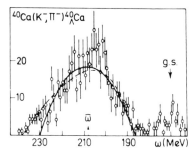

Fig. 5.14 Fits of a quasi-elastic peak to data (Pov–75, Pov–76) for $^{16}\text{O}(\text{K}^-,\pi^-)^{16}_\Lambda\text{O}$ and $^{40}\text{Ca}(\text{K}^-,\pi^-)^{40}_\Lambda\text{Ca}$. The centroid of the peak is at or near the value given in Eq. 110, depending on whether variations of the momentum transfer with the momentum of the struck nucleon are ignored (solid line) or included (dashed line). The arrow shows the position of the hypernuclear ground state (from Dal–76a).

features such as the strangeness analog resonance should stand out clearly. In intermediate situations there are contributions from the strangeness analog both as a resonance and in the quasi-free background that are difficult to disentangle. This appears to be the situation in the first data obtained, shown in Fig. 5.14 with a fitting of the quasi-free peak (Dal–76a). The energy transfer for determination of the centroid of this peak is given, for (K^-,π^-) kinematics, but ignoring the variation of q with the momentum of the struck nucleon, by

$$\omega = M_\Lambda - M_\text{N} + (U_\text{N} - U_\Lambda) - \frac{1}{4}\left(\frac{M_\Lambda - M_\text{N}}{M_\text{N}M_\Lambda}\right)k_F^2 + \frac{q^2}{2M_\Lambda} \qquad [110]$$

for neutron and lambda well depths U_N and U_Λ. Thus, from the fit in Fig. 5.14 one can extract a rough number for the difference of these effective potentials, $U_\text{N} - U_\Lambda \sim 31$ MeV. After subtraction of the quasi-free contribution there remains a narrow peak in the data (possibly with an interference shoulder) that may be the strangeness analog resonance. More refined analysis may ultimately permit the observation of other features, such as the supersymmetric state (Dal–76b) that is orthogonal to the analog and possessed of higher spatial symmetry than it, so that it should appear at considerably lower energy.

Naturally a complete analysis of (K^-,π^-) reactions will require attention to the distortion of kaon and pion waves and to the details of the transition operators. The first such efforts (Huf–74, Dal–76a) used eikonal approaches and found (Huf–74) a likelihood of sizeable corrections from various sources: Coupled-channel effects or nuclear correlations or granularity enter, perhaps quite appreciably (10–30%?). Spin terms for the $K^- + n \rightarrow \Lambda + \pi^-$ amplitude are estimated to involve only a few percent correction to the cross section. Off-shell modifications due to nucleon binding are estimated by assuming the dominance of the $\Lambda(1405)$ contribution and are found to be $\lesssim 30\%$ effects. Last, two-step processes such as $K^- + p \rightarrow \Sigma^+ + \pi^-$ followed by $\Sigma^+ + n \rightarrow \Lambda + p$ lead to possible 20% changes, so that an overall estimate of the validity of the simplest approach to DWIA for the (K^-, π^-) process suggests reliability to within about a factor of two.

5.5 Inelastic Scattering to Bound and Quasi-Bound States

The inelastic scattering of mesons on nuclei with the excitation of bound or quasi-bound states is a natural probe for the study of hadronic matter transition densities in parallel to the investigation of charge transition densities in inelastic electron scattering. As such, it is the obvious extension of the elastic scattering that investigates nuclear strongly interacting static densities. The standard theoretical tool for the analysis of such inelastic processes is the distorted wave impulse approximation (DWIA) and its generalizations, or the application of Glauber theory to the inelastic case (Chapter 4). Since these excitations generally involve the transfer of nonzero orbital angular momentum (in contrast to pion charge exchange scattering, for instance, as discussed in the previous section), they tend to have an enhanced emphasis of the nuclear surface region and a consequent reduction of complications in their description through relatively straightforward DWIA. In fact, the first inelastic pion scattering data (Bin–70) of reasonable precision were rather easily fitted* with fairly simple versions of DWIA (Edw–71, Lee–74, Hes–75), Glauber theory (Gua–71), and even eikonal approximations to DWIA (Lee–71, Rog–71).

In order to sketch the physical features of the description of (π, π') in the DWIA, we again use an eikonal model (Rog–71). The amplitude for the excitation of a nuclear state with angular momentum L, projection M, from a $J = 0$ ground state is then (cf. Section 5.4a)

$$T_{f0}(\mathbf{k} - \mathbf{k}') \equiv \langle \pi'; f(J_f = L, M_f = M) | T | \pi; 0(J_i = M_i = 0) \rangle$$

$$\cong A t_{\pi N}(0°) \int e^{i(\mathbf{k} - \mathbf{k}') \cdot \mathbf{r}} \rho_{f0}(r) Y_{LM}(\hat{\mathbf{r}})$$

$$\times e^{-iv^{-1} A t_{\pi N}(0°) \int_{-\infty}^{\infty} \rho(\mathbf{b}, \zeta) d\zeta} \, d\mathbf{b} \, dz, \qquad \textbf{[111]}$$

---

*See also the reviews in Ros–73 and Web–76.

where $\rho(\mathbf{r})$ is the nuclear density normalized so that $\int \rho \, d\mathbf{r} = 1$, $\rho_{f0}(r)$ is the radial part of the transition density, and $t_{\pi N}$ is the pion-nucleon amplitude that we have approximated everywhere as if for forward scattering. The exponential factor represents the distortion of the pion wave on its journey through the nucleus, and we have here again approximated this as if the classical pion trajectory were completely forward. This is clearly not very satisfactory for these incoherent excitations that generally vanish in the forward direction, but it emerges that even this crude approximation for the absorption factor is reasonably adequate for (π, π').

If the scattering were weak, as would be the case, for example, if we were dealing with inelastic electron scattering rather than pions, then we could of course ignore the absorption factor in Eq. 111. The result would be the form factor for the (e, e') process to the nuclear level in question. Its Fourier transform would then give us directly the transition density ρ_{f0}, provided the electromagnetic and hadronic nuclear densities do not differ. In fact, such a procedure would also automatically fold in the finite extent of the nucleon, again provided that the electromagnetic and hadronic sizes for the nucleon are similar. Simple parametrizations are then used for the measured electron scattering cross sections and for the πN amplitudes. The electron scattering form factors are taken with the structure

$$T_{\mathrm{el}}(q) = Bq^l e^{-q^2/4\gamma^2}, \quad q = |\mathbf{k}' - \mathbf{k}|, \qquad [112]$$

where detailed radial structure is suppressed in favor of surface features. The πN amplitudes are chosen to incorporate the optical theorem,

$$t_{\pi N}(q) = -\tfrac{1}{2} i v \sigma_{\mathrm{total}}(1 - i\alpha) e^{-\beta^2 q^2/2}, \qquad [113]$$

where α is the known (e.g., Hoh–64) ratio of the real to imaginary parts of the πN amplitude in the forward direction. The results of such a calculation (Rog–71) are shown in Fig. 5.15, for $^{12}\mathrm{C}(\pi, \pi')^{12}\mathrm{C}^*$ leading to the $J^\pi = 3^-$ level at 9.6 MeV, and agree quite well with the data. Moreover, it seems to be the case for many instances of hadronic scattering that the exponential distortion factor in Eq. 111 is indeed only a slowly varying function of momentum transfer, as implied there, so that a Tibell plot (Tib–69) of

$$\log(|T_{f0}|^2/q^{2l}) \cong \mathrm{constant} - q^2/2\gamma^2 \qquad [114]$$

can allow for a direct, though approximate, extraction of the size parameter γ^2 of Eq. 112 from the slope of the plot, as well as comparative information on the amplitude coefficient B from its intercept (Rog–71, Web–76).

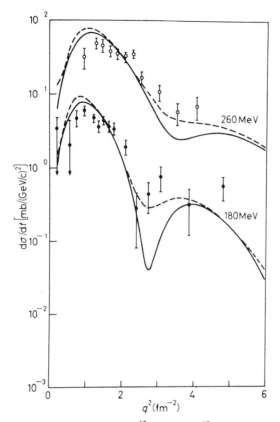

Fig. 5.15 Differential cross sections for $^{12}C(\pi^-,\pi^-)^{12}C^*(J^\pi=3^-,T=0,E_x=9.6$ MeV) at two energies in the 3,3 region in the eikonal model (Rog–71) sketched in Eqs. 111–113. The dashed curve shows the inclusion of the possible excitation of the $J^\pi=0^+$, $T=0$, level in ^{12}C at 7.6 MeV. The data are from Bin–70.

There seems to be relatively little indication for dramatic selection rules in πA inelastic scattering. A partial exception may be (Gup–76, Wal–76) the excitation of states with relatively high spin, $J^\pi=3^-,4^-$, and $T=0$. These appear to be more readily excited in (π,π') than in (e,e'), which tends to prefer $T=1$ levels for the large momentum transfers q required for large spin $L\sim qR$, or in (p,p'), where the spin-flip $T=0$ is suppressed relative to the spin-flip $T=1$, unlike the pion case where spin flips tend to enter roughly equally for both isospin cases (see forms of $t_{\pi N}$ in Chapter 2).

Another type of selection feature (Wil–74) that may be exploited in $A(\pi,\pi')A^*$ reactions depends on the relatively rapid energy variation of the πN amplitude in order to separate orbital and spin contributions in $T=1$

magnetic transitions. This is a feat that is almost impossible to accomplish in inelastic electron scattering because in the $T=1$ situation the magnetic moment coefficients in the spin or magnetization term enter with opposite sign, $K_p - K_n \sim 3.7$, and swamp the orbital or convective part. In the pion case, there is a special contribution of the same character as the orbital part that arises from kinematic averaging over the non-spin-flip piece of the πN amplitude. For incident pion momentum **k** and nucleon momentum (in the nucleus) **p**, the nonrelativistic internal energy is (cf. Eq. 22, for example)

$$
\begin{aligned}
E &= \frac{\mathbf{k}_{\mathrm{rel}}^2}{2m_{\mathrm{rel}}} = \left(2\,\frac{mM}{M+m} \right)^{-1} \left(\frac{M}{M+m}\mathbf{k} - \frac{m}{M+m}\mathbf{p} \right)^2 \\
&= \frac{M}{M+m}\left(\frac{k^2}{2m} - \frac{\mathbf{k}\cdot\mathbf{p}}{M} + \frac{m}{M}\frac{p^2}{2M} \right) \\
&\cong \frac{k^2}{2m} - \frac{1}{2M}(\mathbf{p}+\mathbf{p}')\cdot\mathbf{K},
\end{aligned}
\qquad [115]
$$

where $\mathbf{K} = \frac{1}{2}(\mathbf{k}+\mathbf{k}')$ is the mean pion momentum, and we are neglecting terms in $m/M \sim \frac{1}{7}$ and in $p^2/2M$. Then the nonspin-flip part of the πN amplitude $f(E)$ will have to be evaluated at the energy of Eq. 115, or, if we relate this to a corresponding fixed πN energy $E_\pi = k^2/2m$,

$$
f(E) \cong f\left(\frac{k^2}{2m} \right) - \frac{1}{2M}(\mathbf{p}+\mathbf{p}')\cdot\mathbf{K}\,\frac{\partial f}{\partial E}\bigg|_{k^2/2m} + \dots. \qquad [116]
$$

When this is averaged over the nucleon motion, the second term gives a contribution of exactly orbital or convective form. The spin-flip term $i\boldsymbol{\sigma}\cdot\mathbf{k}'\times\mathbf{k}$ of Eq. 2 continues to contribute, but does not swamp this convective piece in estimates for the excitation of the $J^\pi = 1^+, T=1$, level at 15.1 MeV in ^{12}C (Wil–74), thus holding out hope that both may be observed in this or some similar transition.

A somewhat similar effect can be seen in the pionic excitation of $J^\pi = 1^-, T=1$, giant dipole resonance levels, where nucleon recoil momentum terms make a significant contribution to the transition operator. For example, we consider the nonlocal form of the πN amplitudes of Eqs. 5b or 22,

$$
\begin{aligned}
\langle\mathbf{k}'|t_{\mathrm{non}}|\mathbf{k}\rangle &= b + c\mathbf{k}'\cdot\mathbf{k} \\
&\Rightarrow b + c\left(\frac{M}{M+m}\mathbf{k}'_\pi - \frac{m}{M+m}\mathbf{p}' \right) \\
&\qquad \cdot \left(\frac{M}{M+m}\mathbf{k}_\pi - \frac{m}{M+m}\mathbf{p} \right),
\end{aligned}
\qquad [117]
$$

where we have dropped the spin-flip term for clarity and are continuing to use a nonrelativistic formulation for simplicity. In Eq. 117 we have explicitly made the amplitude Galilei invariant and therefore we have introduced the nucleon momentum operators $\mathbf{p}$ and $\mathbf{p}'$ over which we must carry out the Fermi averaging of Eq. 4.118. (In the present case, however, that averaging is made in producing the DWIA amplitude for the inelastic pion excitation.) Equation 117 contains terms linear in $\mathbf{p}$ and $\mathbf{p}'$ that when taken between the nuclear ground state and a giant dipole state yield matrix elements that are just proportional to those of the giant $E1$ resonance. Therefore we may anticipate that transitions $J_0^\pi = 0^+ \to J_f^\pi = 1^-$ will be sensitive to their presence or absence, and this is indeed borne out by more detailed calculation (Kol–74). We note that in Eqs. 115 and 117 there are two different sources of velocity effects that are considered and two different consequences that are stressed: In one case velocity effects arise from the energy-dependence of the amplitude and change a magnetic type behavior. In the other, the effects come from the off-shell momentum dependence and change an electric multipole feature. In the realistic situation both effects are in fact present simultaneously. Naturally the validity of all these predictions for (π, π'), relating to the excitation of high spin states, the separation of orbital parts of magnetic transitions, and the ability to distinguish recoil effects in $E1$ transitions, depends critically on the negligibility of higher-order corrections to DWIA, which are indeed expected to be small, but which require detailed evaluation to corroborate this.

5.6 Knockout Reactions

From the very start of our discussion of pion-nucleus interactions, knockout has played a central role as a likely contender for being one of the main absorption producing mechanisms. Unfortunately, experimental information as to the fractional contributions of various absorptive channels, real and optical, and in the latter category, single-nucleon knockout, two-nucleon knockout, deuteron and helium ejection, and so forth, is as yet very fragmentary. This is particularly regrettable since the knockout processes, in addition to being interesting in their own right, could shed direct light on the absorption mechanisms. They also offer the possibility, to the extent that no major rearrangements take place as the nucleon exits, of "tagging" the nuclear region where the meson has reached. Thus, if a more deeply bound nucleon is ejected one may suppose that the meson penetrated well into the nuclear interior, where the density is higher. This region is usually masked completely by absorptive effects when we consider elastic or only slightly inelastic events, so that any insight we may obtain concerning it is especially valuable.

We shall try to develop some preliminary understanding of the quasi-free knockout mechanism by using the same kind of qualitative, heuristic model we have exploited in Section 4.1 to develop a view of the optical potential and particularly absorptive effects in it. Toward this end, we consider the contribution of a single nucleon, bound initially in a state described by the wave function $u(\mathbf{r})$, to the knockout cross section. We further make a distorted wave impulse approximation (DWIA) for the incident and outgoing pion and for the ejected nucleon.* The distorted waves are again described in the optics fashion of Section 4.1. Recalling that the index of refraction n in such a description may be complex, this gives for the quasi-free amplitude due to the single nucleon

$$T_{q-f} \cong \int e^{-i\mathbf{k}'\cdot\mathbf{r}} e^{i(n'-1)k'D'(\mathbf{r})} \chi_{\mathbf{p}}^{(-)\dagger}(\mathbf{r}) t_{\pi N} e^{i\mathbf{k}\cdot\mathbf{r}} e^{i(n-1)kD(\mathbf{r})} u(\mathbf{r}) \, d\mathbf{r},$$

$$[118]$$

where $D(\mathbf{r})$ and $D'(\mathbf{r})$ are the path lengths for the incident and outgoing pion waves, $\chi_{\mathbf{p}}^{(-)}(\mathbf{r})$ is the distorted wave for the ejected nucleon of momentum $\mathbf{p}$ (with collapsing spherical wave boundary conditions), and $t_{\pi N}$ is the πN transition amplitude. If there were no distortive effects ($n=1$, $\chi_{\mathbf{p}}^{(-)\dagger}(\mathbf{r}) \to e^{-i\mathbf{p}\cdot\mathbf{r}}$) and $t_{\pi N}$ were constant, Eq. 118 would just yield the Fourier components of the bound-state nucleon wave function for the momentum transfer $\mathbf{k} - \mathbf{k}' - \mathbf{p}$ in question. We now consider the cross section[†] in which we identify the single ejected nucleon but integrate over its momentum, measuring only the final-state pion in momentum cell $d\mathbf{k}'/(2\pi)^3$

$$d\sigma_{q-f} = \frac{1}{v} \frac{d\mathbf{k}'}{(2\pi)^3} \int |T_{q-f}|^2 \frac{d\mathbf{p}}{(2\pi)^3},$$

$$[119]$$

where v is the initial pion velocity ($=$ flux, in our units). Thus

$$d\sigma_{q-f} = \frac{1}{v} \frac{d\mathbf{k}'}{(2\pi)^3} \int \frac{d\mathbf{p}}{(2\pi)^3} \int d\mathbf{r}' e^{i\mathbf{k}'\cdot\mathbf{r}'} e^{-i(n'^*-1)k'D'(\mathbf{r}')} \chi_{\mathbf{p}}^{(-)}(\mathbf{r}')$$

$$\times t_{\pi N}^{\dagger} e^{-i\mathbf{k}\cdot\mathbf{r}'} e^{-i(n^*-1)kD(\mathbf{r}')} u^*(\mathbf{r}')$$

$$\times \int d\mathbf{r} \, e^{-i\mathbf{k}'\cdot\mathbf{r}} e^{-i(n'-1)k'D'(\mathbf{r})} \chi_{\mathbf{p}}^{(-)\dagger}(\mathbf{r}) t_{\pi N} e^{i\mathbf{k}\cdot\mathbf{r}} e^{i(n-1)kD(\mathbf{r})}$$

$$\times u(\mathbf{r}),$$

$$[120]$$

---

*Note that from the start (see the comment on the partition of the Hamiltonian in Eq. 4.44) our formalism has not been directly applicable to knockout. For a careful discussion of extensions see Tan–77.

[†] This cross section is *exclusive* in that one and only one nucleon is knocked out.

and we now commit the indiscretion of ignoring* the fact that part of our nucleon spectrum is lodged in bound states (at least one!) and treat the completeness statement for the single-nucleon space as if

$$\int \frac{d\mathbf{p}}{(2\pi)^3} \chi_{\mathbf{p}}^{(-)}(\mathbf{r}')\chi_{\mathbf{p}}^{(-)\dagger}(\mathbf{r}) \Rightarrow \int \frac{d\mathbf{p}}{(2\pi)^3} e^{+i\mathbf{p}\cdot\mathbf{r}'} e^{-i\mathbf{p}\cdot\mathbf{r}} = \delta(\mathbf{r}'-\mathbf{r}). \quad [121]$$

Then the cross section is

$$\frac{d\sigma_{q-f}}{d\Omega} \cong \frac{d\sigma_{\pi N}}{d\Omega} \int d\mathbf{r}\, e^{-2\operatorname{Im} n\cdot k' D'(\mathbf{r})} e^{-2\operatorname{Im} n\cdot k D(\mathbf{r})} |u(\mathbf{r})|^2 d\mathbf{r}. \quad [122]$$

We now use Eq. 4.6 to express $k \operatorname{Im} n$ in terms of the mean free path l and further take the case of a uniform density distribution in a spherical nucleus $[|u(\mathbf{r})|^2 = \rho_0 = 3/4\pi R^3$ as for Weisskopf-unit estimates]:

$$\frac{d\sigma_{q-f}}{d\Omega} \cong \frac{d\sigma_{\pi N}}{d\Omega} \rho_0 \int_{\text{nucl. vol.}} d\mathbf{r}\, e^{-D'(\mathbf{r})/l} e^{-D(\mathbf{r})/l}. \quad [123]$$

This expresses the single-nucleon quasi-free exclusive cross section in terms of the fundamental πN cross section, the nucleon density, and a factor expressing pion absorption along its path.

We may easily obtain an *inclusive* cross section from Eq. 123 by setting the absorption on the exit path, $D'(\mathbf{r}) \equiv 0$. This applies to a situation in which we measure the outgoing pion, and know that *at least one* nucleon has been knocked out. Then there is no loss of flux of the pion subsequent to the knockout, since all final channels are included (see Gol–64, pp. 823–824, and Kol–79). We may obtain an integrated inclusive cross section by angular integration of Eq. 123 (with $D' \equiv 0$) and multiplication by A, the number of nucleons. The factor before the integral gives $A\sigma_{\pi N}\rho_0 = l^{-1}$, the mean free path (for optical absorption). Comparison with Eq. 4.22 then gives

$$\sigma_{q-f}^{A \text{ nucleons}} \simeq \sigma_{\text{abs}}, \quad \text{(inclusive quasi-free)}; \quad [124]$$

that is, if we ignore the absorption of the outgoing pion then we find that the quasi-free inclusive cross section for the full nucleus is just the (optical) absorption cross section. However, if our measurement is exclusive

---

*If we assume there is only one bound state $u(\mathbf{r})$ and explicitly subtract it before invoking closure, and also ignore pion distortion ($n = 1$), we will have a subtractive correction involving the square of the Fourier transform of the nucleon density. Ignoring this subtraction is, of course, better justified at large angles where the elastic form factor (to which it relates) is small.

knockout, and does not allow for the ejection of more than one nucleon, we must include full final state distortion through the factor $e^{-D'(r)/l} \leqslant 1$, whence

$$\sigma_{q-f}^{A \text{ nucleons}} \leqslant \sigma_{\text{abs}}, \quad (\text{exclusive quasi-free}) \qquad [125]$$

the inequality expressing the fact that the full σ_{abs} contains *inter alia* events in which more than one nucleon is ejected, as for example in a quasi-free event followed by "absorption" along the pion trajectory through the knockout of yet another nucleon. True absorption of mesons increases the right-hand side of the inequality (Eq. 125).

The eikonal approach can be used in a more quantitative way (Huf–75c) for application to the activation experiment (Dro–75) on $^{12}C(\pi^-, \pi^- n)^{11}C$—here to bound states *only*—to arrive at an expression closely related to that of Eqs. 122 or 123, namely, for the DWIA cross section in terms of the plane-wave one,

$$\sigma_{\text{DWIA}} = \sigma_{\text{PWIA}} \int |u_{1p_{3/2}}(\mathbf{b}, z)|^2 |e^{-iv^{-1}\int_{-\infty}^{\infty} \mathcal{V}_c(\mathbf{b}, \zeta) d\zeta}|^2 d\mathbf{b} \, dz. \qquad [126]$$

Here $u_{1p_{3/2}}(\mathbf{r})$ is the wave function for a nucleon in the $1p_{3/2}$ subshell which can be ejected from ^{12}C leaving ^{11}C in a particle stable state. The optical potential $\mathcal{V}_c$ is for the pion. It is necessary (Hew–69, Ste–75) also to take account of the distortion of the ejected nucleon, incorporating several features: (1) Only for a fraction $P_T(E_\pi)$ of events does the pion transfer sufficient energy to overcome the binding of the nucleon. (2) The nucleon is ejected from the surface region and may exit immediately (probability P_E) or plow through the nucleus $(1 - P_E)$. In the latter case, (3) it has probability P_D to destroy the residual ^{11}C, and (4) a probability P_x to undergo charge exchange $^{11}C(n, p)^{11}B$. Then

$$\sigma(\pi^-, \pi^- n) = \sigma_{\text{DWIA}}(\pi^-, \pi^- n) \cdot P_T \cdot \{ P_E + (1 - P_E)[1 - P_D - P_x(n, p)] \}$$

$$+ \sigma_{\text{DWIA}}(\pi^-, \pi^- p) \cdot P_T \cdot (1 - P_E) P_x(p, n), \qquad [127]$$

a rough result that gives remarkably good agreement, as seen in Fig. 5.16, with the magnitude of the cross section for $^{12}C(\pi^-, \pi^- n)^{11}C$ and the ratio

$$R = \frac{\sigma[^{12}C(\pi^-, \pi^- n)^{11}C]}{\sigma[^{12}C(\pi^+, \pi^+ n)^{11}C] + \sigma[^{12}C(\pi^+, \pi^0 p)^{11}C]}. \qquad [128]$$

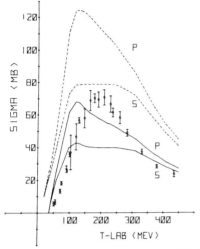
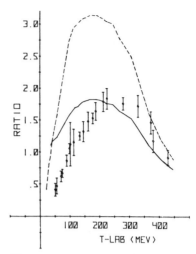

Fig. 5.16 (*a*) Cross sections for $^{12}C(\pi^-,\pi^-n)^{11}C$ as a function of pion kinetic energy (from Huf–75c). The dashed curves are eikonal results and the solid curve includes nucleon final-state interaction. "*S*" refers to a forward approximation for the πN amplitude, while "*P*" includes forward *p*-wave πN effects in a Gaussian approximation. (*b*) gives the ratio of π^- knockout to π^+-induced knockout, with a similar meaning to the curves. Data are from Dro–75.

The special importance of the final state nucleon interaction in producing reasonable results for this ratio was realized some time ago (Hew–69) and reemphasized more recently* (Ste–75). It clearly must be treated more accurately, along with the pion wave distortions, for a reliable, detailed theory of nucleon knockout. Since knockout studies are still at a fairly early stage of development, they have tended to focus primarily on inclusive features, such as the ratio of Eq. 128. One of their more promising possibilities, however, may lie in the measurement of more detailed features of the final state through the simultaneous observation of the emitted pion and nucleon. Among other aspects that may then emerge is the possibility of tagging events in which more deeply bound nucleons are involved, thus offering us a glimpse of the pion interaction in the nuclear interior, which is otherwise usually veiled by the dominance of absorptive effects there.

*See also the review by Eis–75b. A detailed study of nuclear structure effects, in particular the availability or otherwise of a particle-stable analog state in the charge exchange transition, is given in Sil–77.

References

The review articles noted at the beginning of this chapter are a good point of departure for much of the material in it: General reviews are given by Koltun (Kol–69), Hüfner (Huf–75a), and Sternheim and Silbar (Ste–74). The evolution of the subject can be seen through time-delay photography in the biennial series of international conferences on high-energy physics and nuclear structure (Cer–63, Ale–67, Dev–70, Dzh–72, Tib–74, Nag–75, Loc–77), and comprehensive summaries of pion-nucleus physics are in reviews of the topical conferences and meetings (Eri–71a, Bec–71, Gib–73, War–76, Bar–76).

6 Field Theory and
Coupled Channel Methods

The theoretical methods developed to this point have, for the most part, relied on the assumption that it is possible to describe the interacting system of a scattering meson and target nucleus by a Schroedinger equation. On the other hand, we have at several points discussed the problem that there are clearly more degrees of freedom to this system than those of the projectile and the target nucleons. The difficulties start with the Yukawa interaction that plays an important role both in meson scattering and in the nucleon-nucleon interaction, as we have seen in Sections 2.4 and 2.5. Since this interaction creates and annihilates mesons through the coupling to the meson field, the dynamical problem becomes one with (infinitely) many *meson* degrees of freedom, unlike in the Schroedinger treatment. Similarly, we have mentioned that excited states of mesons (e.g., vector mesons) or of nucleons (baryon resonances) may couple to the meson-nucleus system.

In this chapter we discuss ways to deal with these new degrees of freedom. First we survey some of the new features that arise for interactions with fields (Section 6.1). Next we develop some theoretical tools for dealing with meson-nucleus scattering with these interactions (Section 6.2). We stress an approach that produces an *effective Schroedinger equation*, since this allows us to make the most direct extension of Schroedinger methods used earlier in the book. We discuss, in this connection, the generalization of the optical potential. We then present an outline of how this method applies to multiple scattering theory, and how we can extend that theory to include new degrees of freedom, such as absorption of mesons by nuclei, or propagation of excited states of mesons or baryons in the nuclear target (Section 6.3). The detailed application to pion absorption by nuclei is postponed to Chapter 7; the application to propagation of resonances is given in Section 6.4. These methods provide the formulation that we need for Chapter 7 on meson absorption reactions.

261

6.1 Field Theory Aspects of Scattering

We now discuss some of the new features brought about by the introduction of fields or other new degrees of freedom into the interaction of a scattering meson with a nucleus. We discuss (*a*) the Yukawa interaction, (*b*) crossing symmetry, (*c*) relativity, and (*d*) other degrees of freedom.

6.1a Yukawa Interaction

The role of the Yukawa interaction in πN scattering was discussed in Section 2.4. There we took the interaction Hamiltonian to be of the form of Eq. 2.60:

$$H_I = \int d\mathbf{x} j(\mathbf{x}) \cdot \phi(\mathbf{x}) \qquad [1]$$

where ϕ is the meson field operator, and j the nucleon current density. We noted there that the interaction given by Eq. 1 induces the transitions $N + \pi \leftrightarrow N$ in which a pion is absorbed or emitted by a nucleon (see Fig. 2.7). For a single nucleon, these transitions are virtual, or off-shell, since energy is not conserved. However, for a collection of interacting nucleons, energy may be conserved in such a transition:

$$\pi + N_1 + N_2 + \cdots N_A \leftrightarrow N_1 + N_2 + \cdots + N_A.$$

These transitions can be realized in the laboratory as pion absorption by a nucleus

$$\pi + A \rightarrow A^* \rightarrow (A - x)^* + x \qquad [2a]$$

where $(A - x)^*$ is an excited (bound) state of a residual nucleus that has lost a fragment or fragments (e.g., nucleons) x. Also possible is pion production by protons

$$p + A \rightarrow (A + 1)^* + \pi. \qquad [2b]$$

(These nuclear reactions are discussed in Chapter 7.)

The first point is that the nuclear reactions given by Eqs. 2a and 2b involving pions are not included in the Watson multiple scattering theory, since that theory, as we saw, was based on a Schroedinger equation for one meson plus nucleus. The Hamiltonian (Eq. 4.44) does not contain terms like Eq. 1 that change the number of pions; only scattering of pions (by a potential) was included. It is tempting to remedy this omission by simply adding the Yukawa interaction to the Watson Hamiltonian. But this will not do, as we see shortly.

The second point about absorption and emission of mesons is that they may contribute to the scattering of a meson from a nuclear target

$$\pi + A \rightarrow A^* \rightarrow \pi + A. \qquad \qquad [3]$$

We could have an intermediate nuclear state A* either of the same energy as the initial channel, in which case A* is a reaction channel (as in Eq. 2a) coupled to the scattering channels, or A* could have a different energy, as a virtual state. In the case of πN scattering, we considered the Born term (Fig. 2.8) in which a physical nucleon with no scattering mesons constituted the intermediate state. (This is a virtual state: off-energy-shell.)

Now strictly speaking, the absorption-emission processes in Eq. 3 are also not included in the Watson multiple scattering theory, again because one and only one meson is included in the Schroedinger equation. (For the same reason the crossed Born term in Fig. 2.8 is also omitted.) One could imagine simple modifications of the Watson theory, which allowed us to include the Born terms (Fig. 2.8) in the t-matrix for πN scattering, to correct this deficiency. This would not give us all the processes of the type given in Eq. 3, however.

We illustrate the problem in Fig. 6.1. There we show several kinds of terms that could contribute to the scattering process in Eq. 3. In fig. 6.1a a

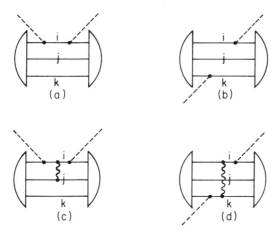

Fig. 6.1 Diagrams for contribution of pion absorption and emission to pion nucleus scattering. Dashed lines represent pions, solid lines denote the nucleons, wavy lines indicate nuclear interactions. The caps at the ends of each figure indicate the initial and final nuclear states, respectively.

pion is absorbed by the ith nucleon and is reemitted by the same nucleon. This process corresponds to the Born term (Fig. 2.8a). This term would contribute to the Watson multiple scattering series (in the first impulse term) if the Born term $T^{(0)}$ (e.g., Eq. 2.71a) is included in the $T_{\pi N}$. However, in the term represented by Fig. 6.1b, a pion is absorbed on the ith nucleon and emitted by the jth nucleon. This is not part of any term in the Watson series. Further, in c and d of Fig. 6.1, there are nuclear interactions between the absorption and emission of the pion; these terms also do not appear in the Watson series. The term c can be considered a modification of $t_{\pi N}$ in the presence of other nucleons, but it differs from the binding corrections in the Watson series, which are discussed in Chapter 4, in that the pion is not continually present. We conclude that the Yukawa interaction (Eq. 1) generates scattering processes not included in the Watson multiple scattering series.

The Yukawa interaction also generates interactions among the nucleons, independent of pion scattering. We illustrate two low-order processes in Fig. 6.2. Figure 6.2a illustrates a contribution to the "self-energy" of the ith nucleon, generated by emission and reabsorption of a pion. Fig. 6.2b illustrates the interaction of nucleons j and k through the exchange of a pion. Both terms are part of the nuclear self-energy generated by the Yukawa interaction. In the multiple-scattering theory of Chapter 4, this self-energy is given by the nuclear Hamiltonian H. One may not, therefore, simply add the Yukawa interaction to the Watson Hamiltonian, since they are not independent interactions. A specific example that illustrates this problem is shown in Fig. 6.3.

In this figure, a pion is absorbed by nucleon i, emitted by i and absorbed by k, and then reemitted by k. This is a contribution to pion scattering by the nuclear target A. We may regard this as a double-scattering term, in the Watson sense: $T_k^{(0)} G_0 T_i^{(0)}$, where the single-scattering is the Born term $T^{(0)}$, which appears once in Fig. 6.1a. However, Fig. 6.3 may also be viewed as a contribution to the term represented in Fig. 6.1d, in which the pion exchanged between nucleons i and k are part of the

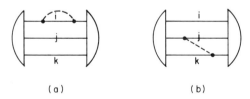

(a)   (b)

Fig. 6.2 Nuclear interactions generated by the Yukawa interaction in lowest (second) order: (a) nucleon self-energy; (b) two-nucleon interaction.

Fig. 6.3 A diagram that can be read two ways—as double scattering or as absorption-emission with an intermediate interaction.

nuclear interaction (OPE potential) between i and k. Clearly, our scattering theory must count the process represented in Fig. 6.3 only once; a multiple scattering approach in which the Born term is included in $t_{\pi N}$, and pion-exchange nucleon-nucleon potentials are included in the nuclear Hamiltonian, will overcount such terms. This illustrates the basic problem of including the Yukawa interaction in the Watson theory, which is that the Hamiltonian cannot be separated into distinct nuclear and πN interactions, such that the former describes the isolated target, and the latter, an isolated πN system.

6.1b Crossing Symmetry

The introduction of the pion field $\phi(\mathbf{x})$ into the nuclear dynamics (as in the Yukawa interaction given in Eq. 1) leads to a set of symmetry relations that are called *crossing relations* (or substitution rules), which have no analog in potential theory. We first encountered such relations (in this book) in Eq. 2.71 as a relation between the direct and crossed Born terms T_a and T_b for πN scattering. More generally, we found that the t-matrix of the Chew-Low static model (Eq. 2.92), taken fully off-energy-shell (Eq. 2.93), has terms, which can be interchanged by the replacement (substitution) $\mathbf{k} \leftrightarrow -\mathbf{k}'$, $\omega \leftrightarrow -\omega$, and $\pi^+ \leftrightarrow \pi^-$, $\pi^0 \leftrightarrow \pi^0$. The charge conjugation involved in the crossing relation of the two terms was not given explicitly in Chapter 2, but indeed indicates the field-theoretic origin of the symmetry: the field operator $\phi_v(\mathbf{x})$ that creates a meson of charge v at the point $\mathbf{x}$ may also annihilate a meson of charge $(-v)$ at the same point (see, e.g., Eqs. 2.58 and 2.59). The amplitudes for these two operations are simply related if the negative momentum $-\mathbf{k}$ of the annihilated meson is substituted for the momentum $\mathbf{k}'$ of the created meson (Eq. 2.58). The interchange of creation and annihilation interchanges the direct and crossed diagrams, for example, in Figs. 2.8–2.10. The momenta change sign as the role of incoming and outgoing mesons are interchanged. Similarly, the sign of the energies also changes (which would appear more simply had we used the field operator $\phi(\mathbf{x}, t)$ in space-time, in the Heisenberg representation).

Similar crossing relations also apply to the case of a pion scattering from a nuclear target, inducing a transition A→B in the latter (inelastic scattering):

$$\pi + A \rightarrow \pi' + B. \qquad [4]$$

If the pion has initial energy and momentum, or four-momentum $k = (\omega, \mathbf{k})$, and charge v, and final four-momentum $k' = (\omega', \mathbf{k}')$ and charge v', there will be a crossing relation between the scattering amplitudes:

$$\langle k', v', B | T | k, v, A \rangle \quad \text{and} \quad \langle -k, -v, B | T | -k', -v', A \rangle. \qquad [5]$$

The specific relation between the two amplitudes depends on the representation used to express the quantum numbers of the states A and B (spin and isospin). It is usually possible to decompose the amplitudes in a representation in which the crossing relation for the components is given simply by a sign. For forward, elastic scattering (A = B, $k' = k$) the *real* parts of the scattering amplitudes are equal, while the *imaginary* parts change sign, that is,

$$T_{\pi^+A}(\omega) = T^*_{\pi^-A}(-\omega) \quad \text{(forward: } \mathbf{k}' = \mathbf{k}). \qquad [6]$$

(For the derivation of this for πN scattering and applications to forward dispersion relations see, e.g., Gol–64 and Kal–64.)

We must remember that the crossing symmetry implied by the relation between the amplitudes of Eq. 5 is not, as it stands, a constraint on the physical scattering amplitudes, since one of the two amplitudes in Eq. 5 always involves negative energies, and therefore does not refer to a physical process.

We illustrate the operation of the crossing symmetry in Fig. 6.4, in which we compare the contribution of two processes to the scattering of π^+, from k to k', to the related process for scattering of π^- from $-k'$ to $-k$. The intermediate state C which is common to diagrams (a) and (c), may contain a π^+ plus an excitation of the target A. Similarly, D, which is common to (b) and (d), may contain a π^- plus an excitation of the target A. The amplitudes for (a) and (c) are related by crossing Eq. 5, separately from (b) and (d), which we also related by Eq. 5.

Now, in a potential theory, such as the Watson multiple scattering theory, we only encounter *uncrossed* or direct diagrams of the type shown in Fig. 6.4 *a* and *d*, where C and D include the scattering pion (π^+ or π^-) in the intermediate state. The *crossed* diagrams (Fig. 6.4 *b* and *c*) involve intermediate states with several mesons, and are not included in a Schroedinger potential theory.

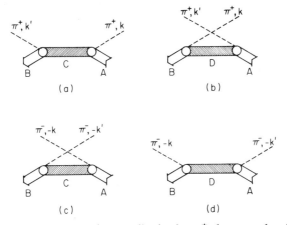

Fig. 6.4 Diagram showing scattering amplitudes for $\pi^{\pm}$ that are related by crossing relations: (a) and (c); (b) and (d).

The problem of maintaining crossing symmetry (Eq. 5) in a multiple scattering theory, actually has two parts:

1. The structure of the scattering theory such that the scattering amplitude of interest (e.g., for Eq. 4) has the desired symmetry.
2. Dealing with the *crossed* elements of a multiple scattering series.

An example of the second is illustrated in Fig. 6.5. In (a) we have a term which could be an element of a Watson series, in which a nuclear interaction acts between two π-N scatterings on a given nucleon. This was discussed as a binding correction to $t_{\pi N}$ in Chapter 4. However, the crossed process in (b) cannot be broken into successive two-body interactions in a potential theory; in a Watson series, (b) would have to be treated as an elementary scattering of π on two (interacting) nucleons. In general, the crossing of lines in scattering elements, as in (b), generates *many-body* scattering interactions.

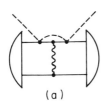
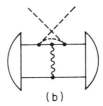

Fig. 6.5 Diagrams showing (a) lowest order binding correction to t-matrix for πN scattering in a nucleus; (b) crossed contribution related to (a).

6.1c Relativity

We come to the treatment of relativity in a multiple scattering approach to meson scattering from nuclei. We consider separately problems of relativity for the mesons, for the nucleons, and for the meson-nucleon interaction. We begin with the last.

In principle, a relativistic theory of πN scattering can be used to give $t_{\pi N}$ in a relativistic invariant form as in Section 2.1. Relativistic versions of the Chew-Low theory have been given based on the Bethe-Salpeter equation or variations of it (e.g., Aar–68), but have not been developed very far as practical models. Relativistic potential models are also possible. In any case, from the point of view of multiple scattering theory, we can take the relativistic off-shell scattering amplitude, for example, Eq. 2.27 as a given quantity, as we did for the nonrelativistic case.

The simplest question of relativity for the mesons comes from the propagation of free mesons with relativistic (kinetic) energy:

$$\omega_k = \sqrt{k^2 + m_\pi^2} \; .$$

The propagation for a meson in a Watson theory will be

$$G_0(\omega) = \frac{1}{\omega - H_0} \qquad [7a]$$

where the kinetic energy operator is given as

$$H_0 = \sqrt{k^2 + m_\pi^2} \qquad [7b]$$

in terms of the momentum operator $\mathbf{k}$. The modification in Eq. 7 can easily be adopted to the Watson approach, in spite of the analytic unpleasantness of the square-root operator. (This does cause some problems in solving the Faddeev equation for the three-body problems: $\pi - d$, $K - d$; see also Section 3.5.)

Because of the field aspects of the mesons, we also encounter a problem closely related to that of crossing which was discussed in the previous subsection. The problem is "backward propagation" of mesons, or equivalently, propagation of conjugate mesons. Consider the double-scattering diagrams of Figure 6.6. In (a) we illustrate a normal Watson double scattering, $t_j G_0 t_i$, where the π^+ propagates from i to j (forward in time) with energy ω_k (to be integrated over $\int d\mathbf{k}$). In (b) we show a related process in which a π^- propagates from j to i, or alternately, the π^+ propagates from i to j (backward in time).

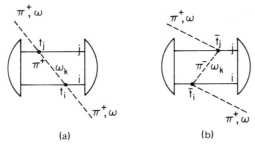

(a)   (b)

Fig. 6.6 Forward and backward propagation of pions in double scattering on a nucleus.

The amplitudes $\bar{t}_i$ and $\bar{t}_j$ are related by a kind of crossing relation (different from those in Figure 6.4b, since only one meson line is crossed) to the scattering amplitudes t_i and t_j of (a). In general, the crossing relations involve just signs ($\pm$), depending on spin and isospin. The result is that the propagators for (a) and (b) can be combined. We have

$$G_0^a(\omega) = (\omega - \omega_k)^{-1}$$

$$G_0^b(\omega) = (-\omega - \omega_k)^{-1} = (\omega - 2\omega - \omega_k)^{-1}$$

$$G_0^a + G_0^b = \frac{2\omega_k}{\omega^2 - \omega_k^2} \quad \text{or} \quad G_0^a - G_0^b = \frac{2\omega}{\omega^2 - \omega_k^2}, \qquad [8]$$

depending on the signs of the crossing relations $t \leftrightarrow \bar{t}$. (We treat the nucleons statically here.) The resulting combined propagator now has the Klein-Gordon denominator $(\omega^2 - k^2 - m_\pi^2)^{-1}$. The factors (2ω) or $(2\omega_k)$ are usually combined with the relativistic normalizing factors $(2\omega)^{-1/2}$ and $(2\omega_k)^{-1/2}$ that are included in the t-matrices. (See Eq. 2.27.) For example for a t-matrix of the simple zero-range form,

$$\langle k'|t_i(\omega)|k \rangle = t_i = \frac{a}{\sqrt{4\omega_k'\omega_k}}, \qquad [9]$$

independent of spin and isospin, the crossed amplitude $\bar{t}_i = t_i$, and the terms of Fig. 6.6 may be combined with the positive sign of Eq. 8 to obtain

$$t_j G_0^a t_i + \bar{t}_i G_0^b \bar{t}_j = \frac{a^2}{2\omega} \frac{1}{\omega^2 - \omega_k^2}. \qquad [10]$$

So, by including certain classes of crossed diagrams, we recover a form* of Watson theory with a propagator of the (modified) Klein-Gordon form (Eq. 8).

The relativity of the nucleons and the nucleus are usually handled differently from that of the mesons. Neither covariant nor relativistic potential theory has been pushed very far in nuclear theory. There are exceptions in the case of the deuteron (Gro–74a) and the Hartree-Fock theory of the nucleus (Mil–74b). In principle, it would also be possible to characterize the nucleus entirely by covariant form factors and correlation functions, in order to produce the relativistic equivalent of a Watson theory (see, e.g., Cel–75). Without a realistic relativistic theory of nuclear structure, one cannot calculate these structure functions, however.

It is also possible to leave the treatment of the nuclear dynamics to be nonrelativistic, depending on the more highly developed Schroedinger theories of the nucleus. In the rest-frame of the nucleus, nuclear motion is nonrelativistic, to a good approximation. The mesons, however, are relativistic in this frame, in the energy region of 100–1000 MeV in which we are often interested. Thus we may consider a "semirelativistic" approach in which mesons, but not nucleons, are treated relativistically.

Since in this approach one calculates in the target rest frame, one encounters a technical problem of expressing the πN t-matrix in this frame. If the t-matrix is given in terms of an invariant function, as in Eq. 2.27, the transformation is straightforward. The transformations can also be performed in the case of a relativistic potential model for $t_{\pi N}$ as was discussed in Section 5.1b (see also Hel–76). In the case of covariant t-matrices, the four-momenta squared $(p,^2 p'^2)$ of the nucleon (as well as k^2, k'^2 of the meson) are off-shell variables (see Eq. 2.28). It is not clear how these variables should be treated in a semirelativistic approach with relativistic pions and nonrelativistic nucleons. One conventional approximation is to take the nucleons on-shell: $p^2 = p'^2 = M^2$. Presumably this approximation is of the same order of accuracy as the assumption of nonrelativistic nuclear dynamics.

In the semirelativistic approximation, the amplitudes for π-nucleus scattering will not exhibit the Lorentz-transformation properties discussed in Section 2.1b. However, in the limit of low energy pions (with respect to the target rest frame) we expect Galilean invariance of $T_{\pi A}$, that is, no dependence on the velocity (or momentum) of the π + nucleus c.m. In the multiple scattering approach we require great care in maintaining consistency in the several places in which nonrelativistic approximations are made, if we are to maintain Galilean invariance in an actual calculation.

---

*This result has been obtained for the optical potential by Cam–73.

As regards the t-matrix for πN scattering, if we start with a Lorentz invariant formulation, we should obtain a Galilean-invariant t-matrix in the low energy limit, which will be independent of the c.m. velocity $v_{c.m.} = (k + p)/(m + M)$. Galilean invariance, and even translational invariance of nuclear structure functions (e.g., form factors), is often a practical problem. The difficulty is that most nuclear models, particularly those based on single-particle orbits (with or without correlations) do not treat the c.m. motion of the nucleus as a whole correctly. (The problem is well known in the case of electron-nucleus scattering, where approximate remedies have been applied.)

Another kind of problem arises in interactions for which there is a change of mass, as in the Yukawa interaction for $\pi + N \rightarrow N$. Although such an interaction may be Lorentz invariant, as in Eqs. 2.60 and 2.62, there is properly speaking, no Galilean invariant limit. This is connected to the fact that there is no well defined velocity for the c.m. in this case: although the c.m. momentum is conserved, $p + k = p'$, the c.m. velocity is not, since the mass changes in the transition, $m + M \leftrightarrow M$, and

$$v = \frac{p + k}{m + M} \neq \frac{p'}{M} = v'.$$

Thus, there are difficulties in defining a Galilean invariant interaction of the Yukawa form.

Now, the reaction in which a meson is absorbed by a nucleus, is energy conserving, and should be independent of the motion of the c.m. of the meson-nucleus system. Here there is no problem of defining Galilean invariance for the process as a whole. However, to maintain this invariance in a theory that uses a nonrelativistic Yukawa interaction, and nonrelativistic nuclear dynamics, requires a careful treatment of the approximation throughout. This problem has received considerable treatment in the literature (Eis–75c, Ho–75) and is discussed further in Section 7.2.

6.1d Other Degrees of Freedom

The Schroedinger approach to πN scattering is based on the assumption that all the physics is given in terms of interacting pions and nucleons. Including the field aspects of the pion introduces a many-pion problem, but we still deal with pions and nucleons. It is known that other particles couple strongly to the pion (or kaon) or to the nucleon, or both. For example, vector mesons may be produced in meson-nucleon collisions:

$$\begin{aligned} \pi + N &\rightarrow \rho + N & m(\rho) &= 779 \text{ MeV} \\ &\rightarrow \omega + N & m(\omega) &= 783 \text{ MeV} \\ K + N &\rightarrow K^* + N & m(K^*) &= 892 \text{ MeV}. \end{aligned} \qquad [11]$$

It is possible that these meson (excited) states may contribute in a significant way to π-nucleus (or kaonic-nucleus) reactions.

One method of incorporating the contributions of the vector mesons to the π or K scattering from nuclei is to treat them as intrinsic excitations of the pseudoscalar mesons, with appropriate interactions (or t-matrices) to describe the "off-diagonal" transitions (Eq. 11). The effect on the Schroedinger equation is simply to enlarge it to include *coupled channels*, in which the π-nucleus (K-nucleus) channel is coupled to a subspace (channel) in which a vector meson propagates in the nucleus, with no pion (or K). This enlarges the Schroedinger space, but does not essentially change the application of the Watson expansion method. Some effects of such channel coupling have been studied in simplified models (e.g., Lon–75 and Section 5.1e of this book).

One should note that the thresholds for producing the vector mesons in Eq. 11 are $T_\pi \gtrsim 1030$ MeV for the ρ, and $T_K \gtrsim 700$ MeV for the K*. Thus for lower energy scattering, the vector mesons are always virtual, that is, off their mass shells. This has the effect of limiting the range of propagation of the vector meson, which can be seen as follows. Consider a vector meson, for example, a ρ-meson, propagating between two nucleons, having been initiated by a πN collision (Eq. 11), as in Fig. 6.7a. The propagator for the ρ (Klein-Gordon) gives, for static nucleons,

$$\left(\omega_\pi^2 - \omega_\rho^2\right)^{-1} = \left(\omega_\pi^2 - m_\rho^2 - q^2\right)^{-1} = -\left(q^2 + a^2\right)^{-1}, \qquad [\,12a\,]$$

where q is the momentum carried by the ρ, and

$$a^2 = m_\rho^2 - \omega_\pi^2 \simeq m_\rho^2 \qquad [\,12b\,]$$

for $\omega_\pi \ll m_\rho$. As we saw in Eqs. 2.106 and 2.107, a propagator of the form of Eq. 12a takes the Yukawa form $r^{-1}e^{-ar}$ in position space, with a range a^{-1}. For low energy pions, this range is $a^{-1} \sim 0.26$ fm, from Eq. 12b, that is, of short range relative to the mean spacing between nucleons in a nucleus. The range is essentially the same as the vector-meson exchange to

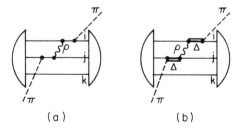

(a) (b)

Fig. 6.7 Two pion scattering processes with a ρ-meson propagating within the target: (*a*) with nucleons only; (*b*) with Δ intermediate states.

the NN interaction: $a^{-1} = m_{\rho,\omega}^{-1}$. So the contribution to π- or K-scattering is expected to involve the short distance correlation structure of nuclei. In fact there are difficulties disentangling the vector meson contribution to scattering from that of NN interactions. Consider Fig. 6.7, in which a double scattering takes place, $\pi N \to \rho N \to \pi N$, in two different ways. In (a) the intermediate state in each $\pi N \leftrightarrow \rho N$ scattering is a nucleon state: this is analogous to the Born term πN scattering. Fig. 6.7a is similar to Fig. 6.3, with the ρ replacing the intermediate π in Fig. 6.3. As in the earlier case, the ρ-exchange may be considered as part of the NN interaction between absorption and emission of the π. The problem of not counting this process more than once must be dealt with in the multiple scattering theory.

In Fig. 6.7b, the intermediate state for $\pi N \leftrightarrow \rho N$ is a nucleon excited state (isobar: N*). The term should not be considered as a contribution to the NN interaction. In a coupled channel approach, Fig. 6.7b is a standard double scattering between the channels. In a *single-channel* Watson theory, this term would have to be treated as an elementary scattering of a pion from two nucleons—distinct from $\pi N \to \pi N$ double scattering.

We now turn to another kind of excitation that occurs in πN or KN scattering, namely the excitation of baryon resonances: for example,

$$\pi + N \to N^* \ (\Delta, \text{etc.})$$
$$K + N \to Y^* \ (\Lambda^*, \Sigma^*, \text{etc.}) \qquad\qquad [13]$$

Here we may be concerned with propagation of the resonance as a real (on-shell) particle in a nuclear target. The resonance may decay back to the original $\pi + N$ (or $K + N$), or may interact, before decay, with other nucleons.

In Section 2.4 we discussed briefly the possibility of interpreting Chew-Low theory in terms of intermediate Δ states, so that the process $\pi + N \to \Delta \to \pi + N$ is included in $t_{\pi N}$. Then it might appear that many effects of Δ propagation in a nucleus could be accommodated in a Watson multiple scattering approach if carried to sufficiently high approximation. On the other hand, it may be that considering the Δ as a separate degree of freedom leads to new approximations in a more convenient approach. We consider this further in Section 6.4.

6.2 Scattering Theory for Many Degrees of Freedom

We are now ready to discuss methods to deal with the problem of meson-nucleus scattering, where there are more dynamical degrees of freedom than given in the Schroedinger equation with a Hamiltonian for one meson and A nucleons, as in Eq. 4.44. The new degrees of freedom come from the meson field, that is, from the possibility of exciting many mesons, or from the possibility of exciting different mesons (e.g., vector mesons) or different baryons (nucleon isobars) from the meson-nucleon interactions. We want to see to what extent the theoretical methods of handling the new degrees of freedom can be adopted to a multiple scattering approach. Perhaps it is not obvious why we should stick to the multiple scattering theory, which was developed with Schroedinger dynamics in mind. There are two points: first, that the partial separation of the scattering problem into an elementary physical process (πN or KN scattering) and a nuclear problem (without "real" mesons) seems to be an extremely useful approach, if it can be shown to be appropriate. Second, the basic notions of the multiple scattering theory do not seem to be tied uniquely to the Schroedinger equation, since we try to express everything in terms of scattering amplitudes and nuclear structure factors, in which the exact nature of the interactions is hidden. Therefore we expect that we are able to formulate the meson-nucleus scattering problem so that we can extend the methods introduced in Chapter 4 to the field-interaction problem.

6.2a Hamiltonian Methods

For a number of reasons, we begin with Hamiltonian methods and develop them most fully. We indicate alternatives, such as Feynman diagram methods, but do not carry then out to the same extent. First of all, this allows us to work in a language that is most similar to that used throughout the book. Second, it is in this formulation that we can most easily understand how to generalize the Watson theory of Chapter 4. And third, the Feynman methods, although most elaborated for the theory of scattering of elementary particles, are least developed for nuclear applications.

The main method we develop is a coupled-channels formulation first introduced by Feshbach, in the context of ordinary nuclear reactions, where the many degrees of freedom are those of the target nucleus; mesons were not included explicitly (Fes–58, 62). However, it seems to be a direct matter to extend the approach to include interactions with a meson field (Miz–75, 77) or with Δ-resonances produced by scattering pions (Kis–73, 76a, b), using the "doorway state" assumption of Feshbach, Kerman, and Lemmer (Fes–67).

We consider the scattering state induced by a plane wave meson (momentum **k**) impinging on the ground state of a given nuclear target.

The meson plane wave and target ground state are indicated by $|\mathbf{k},0\rangle$, and the full scattering state by $|\Psi_{\mathbf{k}}^{(+)}\rangle$, with the usual boundary conditions at large separation from the target:

$$|\Psi_{\mathbf{k}}^{(+)}\rangle \rightarrow |\mathbf{k},0\rangle + \text{outgoing spherical waves} \qquad [14]$$

as in Eq. 2.42 or Eq. 2.83 for a nucleon target. Here the outgoing waves include inelastic scattering of the meson, emission of nucleons or clusters (e.g., d or α), with or without the scattered meson, and possibly production of more mesons, baryon resonances, and so forth if the initial energy is sufficient to produce these in the asymptotic final state. The kinds of outgoing waves at large distance are classified as *channels*. There are scattering channels (one meson with any excitation of the target), absorption channels (no mesons emitted, but any nuclear excitation), meson production channels, and so on. These are *open* channels, that is, they represent physically possible reaction products that can be measured asymptotically. *Closed* channels refer to all the virtual processes that may occur during a reaction but are not allowed in the asymptotic waves by energy conservation: virtual multiple meson production or N* production below the threshold for emitting these particles are examples.

The starting point of the coupled channels method is the separation of the state-space of the interacting system (in which $\Psi_{\mathbf{k}}^{(+)}$ is defined) into two parts, corresponding, in a sense we discuss in a moment, to the open and closed channels. We do this by defining projection operators P and Q, such that, in the entire space, $P + Q = 1$, $P^2 = P$, $Q^2 = Q$, and $PQ = QP = 0$ (Fes–62; see also Aus–70). The main purpose of P is to single out those degrees of freedom with which we deal explicitly in the scattering theory. For example, we should like to have a formulation that is similar to that of the Schroedinger theory, in which only the scattering meson, and the target nucleons appear explicitly. For this we would choose P such that it defined a state-space (vector-space) corresponding to the degrees of freedom of one meson and A nucleons, for example, through functions of their coordinates. Then although $|\Psi_{\mathbf{k}}^{(+)}\rangle$ is a very complicated state vector, with an enormous number of degrees of freedom, the projected state vector $P|\Psi_{\mathbf{k}}^{(+)}\rangle$ has the same complexity as a Schroedinger wave function for one meson and A nucleons. In fact, we can define the $(A + 1)$-body wave function in the P-projected space, for any state-vector $|\Psi\rangle$ of the system by

$$\psi(\mathbf{r}_1,\ldots,\mathbf{r}_A;\mathbf{r}) = \langle \mathbf{r}_1,\ldots,\mathbf{r}_A;\mathbf{r}|P|\Psi\rangle. \qquad [15]$$

If we apply P to the scattering state vector (Eq. 14) in the asymptotic region, it gives

$$P|\Psi_{\mathbf{k}}^{(+)}\rangle \rightarrow |\mathbf{k},0\rangle + \text{outgoing spherical waves of one meson} \quad [16]$$
with any number of nucleons (or nuclear clusters);

that is, it projects onto the scattering channel at large distance. At short distance, within the interaction region, the channels cannot be separated, but we may use the same definition (viz. Eq. 15) to define the projected state everywhere.

There is considerable freedom in choosing the projection operator P; the choice in general reflects the channels that are to be treated explicitly. For example, if meson absorption is a possible reaction (open channel), as it is for pions of any energy, then we might include that channel with no pion coordinate in the P-space. If we denote the scattering projection (Eq. 15) as P_1 and the projection onto A nucleons with no meson (absorption) as P_x, then the P-space is given by the sum $P = P_1 + P_x$, and $P_1^2 = P_1$, $P_x^2 = P_x$, and $P_1 P_x = P_x P_1 = 0$. For energies below the threshold for production of more mesons, these two spaces completely contain the open channels, since only scattering and absorption of the meson are then possible, and in the asymptotic region

$$P|\Psi_{\mathbf{k}}^{(+)}\rangle \rightarrow |\Psi_{\mathbf{k}}^{(+)}\rangle. \qquad [17]$$

In this case the complementary projection Q involves only the closed channels.

The system is assumed to obey a dynamical equation

$$(E_k - H)\Psi_{\mathbf{k}}^{(+)} = 0 \qquad [18]$$

where

$$E_k = \omega_k + E_0$$

with E_0 the target ground state energy. The Hamiltonian in Eq. 18 involves all degrees of freedom of the meson field interacting with the nucleus. For example, it may contain, in addition to the kinetic energies of all the particles, a Yukawa interaction (Eq. 1) that provides both an interaction among the nucleons, as well as a mechanism for the scattering of the mesons from the nucleus. Additional coupling of the mesons and nucleons to vector mesons could also be present in the Hamiltonian of Eq. 18 (e.g., all the interactions that we discussed in Sections 2.4 and 2.5).

Having partitioned the state-space into two parts by projectors P and Q, we now decompose the Hamiltonian itself:

$$
\begin{aligned}
H_{PP} &\equiv PHP & H_{PQ} &\equiv PHQ \\
H_{QP} &\equiv QHP & H_{QQ} &\equiv QHQ.
\end{aligned}
\qquad [19]
$$

The scattering equation (Eq. 18) may be written as a pair of coupled equations:

$$
(E_k - H_{PP})|P\Psi_{\mathbf{k}}^{(+)}\rangle = H_{PQ}|Q\Psi_{\mathbf{k}}^{(+)}\rangle
$$

$$
(E_k - H_{QQ})|Q\Psi_{\mathbf{k}}^{(+)}\rangle = H_{QP}|P\Psi_{\mathbf{k}}^{(+)}\rangle,
\qquad [20]
$$

which may be solved formally for $|P\Psi_{\mathbf{k}}^{(+)}\rangle$

$$
\left[E_k - \mathcal{K}(E_k) \right]|P\Psi_{\mathbf{k}}^{(+)}\rangle = 0.
\qquad [21a]
$$

This is an *effective* Schroedinger equation for the projected scattering state $|P\Psi_{\mathbf{k}}^{(+)}\rangle$ with an *effective* interaction

$$
\mathcal{K}(E_k) = H_{PP} + H_{PQ}(E_k + i\eta - H_{QQ})^{-1}H_{QP}.
\qquad [21b]
$$

The equation and the interaction in Eq. 21 depend on the scattering energy E_k and on the choice of P that gives the projected space in which the equation is defined. Equation 21 formally reduces the original interaction problem, Eq. 18, to one with the degrees of freedom with which we deal—the meson and nucleons. The practical problem is not only the solution of the simpler dynamical equation (Eq. 21a), but the specification of the effective interaction (Eq. 21b), which now contains all the complexities of the coupling of the many degrees of freedom. We return to the question of the practical use of Eq. 21 below.

We need an expression for the transition amplitudes into any of the open channels included in $|P\Psi_{\mathbf{k}}^{(+)}\rangle$ of Eqs. 16 or 17. It is convenient to use a result of formal scattering theory for scattering with many channels. We assume the final channel has a plane wave meson of momentum $\mathbf{k}'$, with the nuclear target in a state labeled by n—inelastic scattering. We write the t-matrix for the transition (Gol–64)

$$
\langle \mathbf{k}',n|T(E_k)|\mathbf{k},0\rangle = \lim \langle \mathbf{k}',n|H - E_k|\Psi_{\mathbf{k}}^{(+)}\rangle
\qquad [22]
$$

where the limit is on the $\eta \to 0^+$ in the definition of the scattering solution (Eq. 14). The expression given as Eq. 22 is the analog of Eq. 2.43 in potential theory, for one channel, with $(H - E_k)$ replacing V in the present case.* Now we may combine the fact that $\langle \mathbf{k}', n | P = \langle \mathbf{k}', n |$, since the final channel is in the P-space, with the identity $P + Q = 1$, to write the right-hand side of Eq. 22 as

$$\lim \langle \mathbf{k}', n | P (H - E_k)(P + Q) | \Psi_{\mathbf{k}}^{(+)} \rangle. \qquad [\mathbf{23a}]$$

If we use the formal solution for $| Q \Psi_{\mathbf{k}}^{(+)} \rangle$ from Eq. 20, and Eqs. 19 and 21b in Eq. 23a, we can express Eq. 22 in the form

$$\langle \mathbf{k}', n | T(E_k) | \mathbf{k}, 0 \rangle = \lim \langle \mathbf{k}', n | \mathcal{K}(E_k) - E_k | P \Psi_{\mathbf{k}}^{(+)} \rangle. \qquad [\mathbf{23b}]$$

Thus we can calculate the scattering amplitude completely within the P-space, if we know the effective interaction and its solution (Eqs. 21a and 21b) in that space.

The effective Schroedinger equation method just discussed starts with physical states of the meson-nuclear system, but projected into a simpler space. This, as we see below, allows us to follow the general structure of the Watson expansion in developing a multiple scattering theory. Another Hamiltonian method, which also deals directly with physical quantities, is the Chew-Low theory, which we discussed for the theory of πN scattering in Chapter 2. The theory for π-nucleus scattering may be developed in close parallel to that of Section 2.4c with the Hamiltonian (Eqs. 2.80 and 2.82) extended to many nucleons; nonpionic interactions among the nucleons could be included in H_0. With ψ_0 taken as the target ground state, Eq. 2.83 gives the scattering state given in Eq. 14. The derivation of Section 2.4c then leads to an expression analogous to that of Eq. 2.92 for the t-matrix for inelastic scattering, as in Eqs. 22 and 23:

$$\langle \mathbf{k}', n | T(E_k) | \mathbf{k}, 0 \rangle = (2\pi)^3 N_{k'} N_k \{ \langle \psi_n | j^\dagger(\mathbf{k}')(\omega_k + E_0 - H + i\eta)^{-1} j(\mathbf{k}) | \psi_0 \rangle$$
$$- \langle \psi_n | j(\mathbf{k})(\omega_k - E_0 + H)^{-1} j^\dagger(\mathbf{k}') | \psi_0 \rangle \}, \qquad [\mathbf{24}]$$

where ψ_n is the final state of the nuclear target. This becomes an equation for amplitudes, if we insert a complete set of states $\{\psi_m\}$ of the Hamiltonian H:

$$\langle \mathbf{k}', n | T(E_k) | \mathbf{k}, 0 \rangle = (2\pi)^3 N_k N_{k'} \left\{ \frac{\langle \psi_n | j^\dagger(\mathbf{k}') | \psi_m \rangle \langle \psi_m | j(\mathbf{k}) | \psi_0 \rangle}{\omega_k + E_0 - E_m + i\eta} \right.$$
$$\left. - \frac{\langle \psi_n | j(\mathbf{k}) | \psi_m \rangle \langle \psi_m | j^\dagger(\mathbf{k}') | \psi_0 \rangle}{\omega_k - E_0 + E_m} \right\}. \qquad [\mathbf{25}]$$

---

*Because of the limiting procedure, Eq. 18 does not apply in this expression; see Gol–64, pp. 177–186 for discussion.

The set $\{\psi_m\}$ of states is large indeed: in place of the single nucleon states that give the Born terms for πN (Eq. 2.94), there is a whole spectrum of nuclear states with no asymptotic pions; these are the nuclear absorption states. In place of the one-meson/one-nucleon states that lead to Eq. 2.97, one has states in which a meson scatters from *any* initial target state. States with many asymptotic mesons (meson production) *and* nuclear excitation will also occur in Eq. 25. As in the πN case, the scattering t-matrix (Eq. 25) can also can be written in terms of a current amplitude (Eq. 2.98):

$$\langle \mathbf{k}',n|T(E_k)|\mathbf{k},0\rangle = (2\pi)^3 N_k \langle \Psi_{\mathbf{k}',n}^{(-)}|j(\mathbf{k})|\psi_0\rangle. \qquad [26]$$

Just as for the effective Schroedinger equation method, the Low equation given as Eq. 24 or Eq. 25 deals with physical amplitudes only, in the sense of renormalization. We find this approach substantially less amenable to a systematic multiple scattering development because of the nonlinear form of the equations.

6.2b Optical Potentials and the Green Functions

The idea of using an optical potential to characterize the propagation of the elastic wave of a scattering meson in a nucleus has been introduced and discussed in Section 4.1. In Section 4.2 we developed the multiple scattering expansion for the optical potential as a way of summing formally the Watson series for elastic scattering. The usefulness of these methods leads us to generalize the formulation of elastic scattering so that we may make use of optical potential ideas, in terms of the theories of scattering for many degrees of freedom that we have discussed in the previous section. We also discuss the closely related idea of the Green function, which describes the propagation of elastic waves in the target and which may also be used to define the optical potential.

Since we are interested in elastic scattering, it is useful to remember some results of potential theory, which were introduced in Section 2.3. We gave a linear equation (Lippmann-Schwinger) for the transition operator (Eq. 2.49)

$$T(E) = V + V G_0(E) T(E) \qquad [27]$$

in terms of the potential V and the free propagator or Green operator $G_0(E) = (E + i\eta - H_0)^{-1}$ (Eq. 2.45b). The propagation of the meson in the presence of the potential is given by the Green operator for the full Hamiltonian $H = H_0 + V$:

$$G(E) = (E + i\eta - H)^{-1}. \qquad [28]$$

The two Green operators may be related through the potential by

$$G_0^{-1}(E) - G^{-1}(E) = V. \qquad [29]$$

These operators may also be related directly to $T(E)$, given in Eq. 27, in the form*

$$G(E) = G_0(E) + G_0(E)T(E)G_0(E). \qquad [30]$$

This result may be used as follows. We define the Green functions as the momentum matrix elements of $G_0(E)$ and $G(E)$:

$$\langle \mathbf{k}'|G_0(E)|\mathbf{k}\rangle = (2\pi)^3 \delta(\mathbf{k}' - \mathbf{k}) G_0(E, k)$$
$$\langle \mathbf{k}'|G(E)|\mathbf{k}\rangle = G(E; \mathbf{k}', \mathbf{k}), \qquad [31]$$

where $G_0(E, k)$ is the diagonal matrix element for the free propagator: $G_0(E, k) = (E + i\eta - E(k))^{-1}$. If we take plane wave matrix elements of Eq. 30 and invert the resulting equation for the t-matrix, we find the following expression in terms of the Green functions (Eq. 31):

$$\langle \mathbf{k}'|T(E)|\mathbf{k}\rangle = (2\pi)^3 \delta(\mathbf{k}' - \mathbf{k}) G_0^{-1}(E, k) + G_0^{-1}(E, k') G(E; \mathbf{k}', \mathbf{k}) G_0^{-1}(E, k).$$
$$[32]$$

This last equation allows us to calculate the amplitude for elastic scattering for theories that are formulated in terms of Green functions, rather than Hamiltonians. This helps us later to make contact with the methods of quantum field theory, which has such a form. Equation 32 is the nonrelativistic potential theory analogy of what is called the *reduction formula* for calculating scattering amplitudes from propagator functions in field theory. (See, e.g., Bjo–65, Section 6.7; Gol–64, Chapter 10.)

Now let us return to the problem of elastic scattering of a meson from the complex nuclear target. The method of the optical potential is to restrict attention to a single channel (elastic), accounting for the flux lost to other channels through the optical potential. We may adopt the projective method used earlier in this section, only now we restrict P to project onto the space of one pion moving with respect to the nuclear ground state. With this definition, the asymptotic projected state (Eq. 16) has only elastically scattered waves. The effective Schroedinger equation (Eq. 21) is now a one-channel wave equation for the elastic scattered wave function

---

*Starting with $T = V + TG_0 V$ and $G = G_0 + G_0 VG$, we invert $T = V(1 - G_0 V)^{-1}$, $G = (1 - G_0 V)^{-1} G_0$, to obtain Eq. 30.

$\psi_{\mathbf{k}}^{(+)} = |P\Psi_{\mathbf{k}}^{(+)}\rangle$. It is useful to define a kinetic energy operator H_0 for the elastic channel, which has as eigenstates the plane wave states of the meson with the target in its ground state

$$H_0|\mathbf{k},0\rangle = E_k|\mathbf{k},0\rangle, \quad \text{with } E_k = \omega_k + E_0. \qquad [33]$$

(The operator and states in Eq. 33 are in the P-space.) The effective Schroedinger equation may now be written

$$(E_k - H_0 - \mathcal{V}(E_k))|\psi_{\mathbf{k}}^{(+)}\rangle = 0 \qquad [34]$$

where we have defined the *optical potential* for elastic scattering from the effective interaction (Eq. 21b)

$$\mathcal{V}(E) = \mathcal{K}(E) - H_0. \qquad [35]$$

The elastic scattering amplitude is given in terms of functions defined in the elastic channel, using Eq. 23b with $n = 0$:

$$\langle \mathbf{k}',0|T(E_k)|\mathbf{k},0\rangle = \langle \mathbf{k}',0|\mathcal{V}(E_k)|\psi_{\mathbf{k}}^{(+)}\rangle. \qquad [36]$$

The t-matrix can be obtained, as usual, from a linear equation in $\mathcal{V}(E_k)$, of the form given in Eq. 27:

$$T(E_k) = \mathcal{V}(E_k) + \mathcal{V}(E_k)G_0(E)T(E_k), \qquad [37]$$

where the potential is now *complex*, reflecting the loss of flux to other open channels (which are now in the Q-space), as well as dependent on the total energy E_k of the reaction. All the dynamical complexity of the scattering has been put into the potential $\mathcal{V}(E)$, which couples the P and Q spaces (Fes–58).

Now it is instructive to obtain the optical potential from the Green functions. First we must define the full Green function for a pion propagating in the nuclear ground state:

$$G(E;\mathbf{k}',\mathbf{k}) = \langle \mathbf{k}',0|(E + i\eta - H)^{-1}|\mathbf{k},0\rangle, \qquad [38]$$

an obvious generalization of Eqs. 28 and 31. Since the plane wave states are in P-space, that is, $P|\mathbf{k};0\rangle = |\mathbf{k},0\rangle$, we may replace the operator in Eq. 38 by $P(E + i\eta - H)^{-1}P$. This may be evaluated by inverting the 2×2 matrix (Eq. 19), and we find

$$P(E + i\eta - H)^{-1}P = [E + i\eta - \mathcal{K}(E)]^{-1} \equiv G(E), \qquad [39a]$$

$$G(E;\mathbf{k}',\mathbf{k}) = \langle \mathbf{k}',0|G(E)|\mathbf{k},0\rangle, \qquad [39b]$$

where we have defined a Green operator $G(E)$ which operates entirely within the P-space. Similarly, the free Green operator $G_0(E) = (E + i\eta - H_0)^{-1}$ also operates in the P-space. We may *define* the optical potential to be the P-space operator that obeys the relation in Eq. 29,

$$\mathcal{V}(E) = G_0^{-1}(E) - G^{-1}(E). \qquad [40]$$

This definition clearly agrees with that given earlier in Eq. 35. The main point here is to show that the Green functions for elastic scattering in a many-channel case behave exactly as in the one-channel potential scattering case discussed earlier, with the optical potential $\mathcal{V}(E)$ replacing V. We have not yet discussed how to calculate $\mathcal{V}(E)$.

The development of the connection of elastic scattering with the Green functions in Eqs. 31 and 39 allows us now to make contact with the methods of quantum field theory. In its modern form, this approach does not discuss effective wave equations in a Hamiltonian form. However, the Green functions do play a natural role in the dynamical description, and we make use of that fact.

The Green functions are defined in terms of the field operators (see, e.g., Bjo–65). For the case of elastic scattering of a meson, we are interested in the *causal* Green function with respect to the nuclear ground state, which we define in position space:

$$G(x,y) = -i\langle 0| T[\phi(x)\phi(y)]|0\rangle, \qquad [41]$$

where $\phi(x)$ is the Heisenberg field operator for the meson, and the nuclear ground state is denoted by $|0\rangle$. (We do not describe the nucleons by fields here.) The T in Eq. 41 denotes a time-ordered product of the field operators. This Green function describes the full dynamics of a meson propagating from y to x in the nuclear target; the Heisenberg operators carry the dynamics in their time (x_0) dependence. To make connection with the previously defined Green functions (Eqs. 31 and 39) we Fourier transform to four-momentum space, with $k = (E, \mathbf{k})$:

$$G(E; \mathbf{k}', \mathbf{k})(2\pi)\delta(E - E') = \int d^4x \, d^4y \, e^{i(k' \cdot x - k \cdot y)} G(x,y). \qquad [42]$$

The $\delta(E - E')$ reflects the time-translation invariance of Eq. 41.

To define a free propagator or Green function for the meson, one needs a noninteracting field, but one which gives the proper asymptotic properties to the meson and nuclear target so that they carry their proper physical masses when separated at large distance. In field theory this is done by defining *in* (or *out*) fields $\phi_{in}(x)$ (see Bjo–65, Chapter 16). The free Green function is then given by

$$G_0(x,y) = -i\langle 0|T[\phi_{in}(x)\phi_{in}(y)]|0\rangle$$

$$= \Delta_F(x-y,m^2).$$ [43]

In the second line we have given the standard notation for the free Green function as the Feynman propagator for a boson of mass m. The momentum-transform of Eq. 43 has the well-known Klein-Gordon form

$$G_0(E,k) = G(m;k^2) = (k^2 - m^2 + i\eta)^{-1}$$

$$= (E^2 - k^2 - m^2 + i\eta)^{-1}.$$ [44]

We do not go into the techniques of field theory, such as Feynman diagrams, which are used for calculation: We assume that it is possible to obtain the Green function (Eq. 41). It is then possible to obtain the t-matrix for elastic scattering from this Green function by a form of the *reduction formula*, which in our case reads, for the on-energy-shell t-matrix:

$$\langle k',0|T(E_k)|k,0\rangle = \frac{1}{2E_k} G_0^{-1}(E,k')G(E;k',k)G_0(E,k),$$ [45]

for $k' \neq k$. This, with the relativistic Green functions (Eq. 41 and 44) is analogous to the nonrelativistic result previously given in Eq. 32. Thus we find that it is not necessary to have an optical potential or an effective wave equation to obtain the elastic scattering: the same information is contained in the Green functions.

To complete the story, we may now construct an optical potential for meson-nucleus elastic scattering, from the field-theoretic Green functions given in Eqs. 41 and 44. We use the relation given in Eq. 40 to define

$$\langle k'|\Pi(E)|k\rangle = G_0^{-1}(E,k)(2\pi)^3\delta(k'-k) - G^{-1}(E;k',k).$$ [46]

The notation Π corresponds to the field-theoretic quantity, which is also called the proper self-energy function of the propagating meson. Although Π corresponds to the optical potential (40) of the nonrelativistic theory, the dimensions of Π are (energy)^2 compared to $\mathcal{V} \propto$ (energy); this follows from the difference between the Schroedinger (Eq. 31) and Klein-Gordon (Eq. 44) forms for $G_0(E,k)$. For the case of a uniform, infinite nuclear target (nuclear matter), all the Green functions are diagonal in k so that Eq. 46

simplifies to

$$G(E,k) = (E^2 - \mathbf{k}^2 - m^2 + i\eta - \Pi(E,k))^{-1}. \qquad [47]$$

In the nonrelativistic limit, we see that

$$\Pi(E,k) \simeq 2m\mathcal{V}(E,k), \qquad [48]$$

where $\mathcal{V}$ is given by Eq. 40.

6.3 Multiple Scattering

We now return to the question of practical approximate methods for dealing with the various forms of scattering theory we have just discussed. We make contact with the multiple scattering expansion methods discussed in Chapter 4, and see how they may be applied in the present context. The method of coupled channels, which leads to an effective Schroedinger equation (Eq. 21), is perhaps the simplest and most direct way of making contact with the Watson expansion, discussed previously, and extending it, where needed.

Let us proceed in two steps. First we consider a case for which only scattering channels are open, for example, K^+ scattering by a nuclear target below the threshold for producing other mesons (π, K^*, etc.). The target cannot absorb the K^+, since that would require the production of a strangeness $S = 1$ baryonic state which, so far as is known, does not exist. Later we consider the extension to nonscattering open channels: meson absorption (for π or K^-) and meson production (for higher energies).

We begin with the projection technique of the previous section; we take the P-projector onto a Schroedinger-like space, as in Eq. 15. Since only scattering channels are open, these channels all lie entirely in the P-space; there is no difference between Eqs. 16 and 17. The effective Schroedinger equation in this space was given in Eq. 21. To develop a multiple-scattering expansion in the Watson form, we have to separate the effective interaction (Eq. 21b) into three parts:

$$\mathcal{K}(E) = H_0 + \mathcal{V}_N(E) + \mathcal{V}_{KN}(E), \qquad [49]$$

where H_0 is the kinetic energies of the meson plus A nucleons, $\mathcal{V}_N(E)$ is the interaction of the nucleons with each other, and $\mathcal{V}_{KN}(E)$ is the interaction of the meson with the nuclear system. Once the kinetic energy is defined, the second and third terms may be separated uniquely: the part of $[\mathcal{K}(E) - H_0]$ that does not depend on the coordinates of the meson is $\mathcal{V}_N(E)$. Now we may write an equation of the Lippmann-Schwinger form for a t-operator:

$$T(E) = \mathcal{V}_{KN}(E) + \mathcal{V}_{KN}(E) G_N(E) T(E), \qquad [50]$$

whose matrix elements give the K-nucleus scattering amplitudes given in Eqs. 22 and 23b. The nuclear Green operator $G_N(E)$ is given in the P-space by

$$G_N(E) = [E + i\eta - H_0 - \mathcal{V}_N(E)]^{-1}. \qquad [51]$$

Equations 50 and 51 are now almost of the form in which we began the development of the Watson expansion, (Eqs. 4.45 and 4.46) with two exceptions: First, the various interactions are *energy-dependent*, because of the suppression of the extra degrees of freedom, for example, represented by coupling of the meson field to the nucleons (e.g., $N \leftrightarrow K + \Lambda$ or Σ). Second, the meson-nuclear interaction $\mathcal{V}_{KN}(E)$ is not simply a sum of two-body KN interactions, as in our potential theory (Eq. 4.44); there are also terms involving two or more nucleons (three-body, etc. interactions):

$$\mathcal{V}_{KN}(E) = \sum_{i=1}^{A} \mathcal{V}_i(E) + \sum_{i \neq j} \mathcal{V}_{ij}(E) + \cdots \qquad [52]$$

An example of the kind of process that can contribute to the meson-two-nucleon term $\mathcal{V}_{ij}$ is illustrated in Fig. 6.5(*b*), where in the present case, the crossed meson lines could be K^+, the propagating meson K^-, and some (exchange) interaction between two nucleons in the middle. More complicated processes contribute other many-body interactions to Eq. 52. The general feature of all such processes is that the intermediate states involved are in the Q-projected space, for example, because they involve several mesons. As a consequence of Eq. 52, if we want to eliminate the interaction $\mathcal{V}_{KN}(E)$ from Eq. 50 by substituting elementary t-matrices, as in Eq. 4.52, we must define a whole sequence of such t-matrices:

$$t_i(E) = \mathcal{V}_i(E) + \mathcal{V}_i(E) G_N(E) t_i(E)$$

$$t_{ij}(E) = \mathcal{V}_{ij}(E) + \mathcal{V}_{ij}(E) G_N(E) t_{ij}(E). \qquad [53]$$

One can now generate a multiple scattering series expansion of Eq. 50, which takes the form

$$T(E) = \sum_i t_i(E) + \sum_{i \neq j} t_i(E) G_N(E) t_j(E) + \cdots$$

$$+ \sum_{i \neq j} t_{ij}(E) + \sum_{i \neq j,k} t_i(E) G_N(E) t_{jk} + \cdots$$

$$+ \sum_{(ij) \neq (kl)} t_{ij}(E) G_N(E) t_{kl}(E) + \cdots. \qquad [54]$$

The first line leads to the ordinary Watson expansion in the KN t-matrix $t_i(E)$, as in Eq. 4.55. The new terms are many-body corrections to the Watson series, and might be expected to contribute at the same level as many-body correlations. For example, the coherent approximation, which leads to the lowest order optical potential (Eq. 4.67) in the Watson approach, would now include linear terms in all the t-matrices (Eq. 53):

$$\mathcal{V}_c^{(1)} \simeq \langle 0 | \sum_i^A t_i + \sum_{i \neq j} t_{ij} + \cdots | 0 \rangle. \qquad [55]$$

However, the second term of Eq. 55 depends on the close proximity of two nucleons, and would be expected to be of the same order of magnitude as the two-body correlation (second) term of $\mathcal{V}_c^{(2)}$ (Eq. 4.71). The importance of these terms has not generally been investigated. (An exception is a calculation of the contribution of the process of Fig. 6.7b for π scattering in which t_{ij} involves the exchange of a ρ-meson (Lev–77); the estimate is $\sim 10\%$ of the single-scattering term, for d or ^{4}He targets.) Further, although the single-scattering term in Eq. 55 could be obtained from the free KN t-matrix, through the impulse approximation, the t_{ij} and other terms are not directly so obtainable.

So for the case of low-energy K^+-nucleus scattering, where the only open channels are already those given by the Schroedinger equation, the multiple scattering theory is not greatly altered by the presence of other degrees of freedom. The two new features of the dependence of interactions on the total reaction energy (which is a fixed quantity for all operators as in Eq. 54 and the presence of many-body interactions in $\mathcal{V}_{KN}(E)$ (Eq. 52) can be accommodated. (We might note that $\mathcal{V}_N(E)$, the nuclear interaction, also contains many-body terms, but this fact plays no special role in the Watson theory, particularly when approximations, such as closure, are involved.)

Now let us return to the more general case in which there are open channels available to the meson-nucleus system that are not included in the Schroedinger wave functions. The most obvious case is the meson absorption channel, which is always energetically allowed for nuclear targets, for the cases of π or $\overline{K}$ scattering.

We return to the notation of Section 6.2a, in which P_1 was the projector onto the scattering space (Eq. 15) only, while P_x was the projector onto the absorption space. (We may also use P_x for any other specific channel of interest, not in the P_1 space.) We have $P = P_1 + P_x$, $P + Q = 1$, as before. We formally remove the Q-space from the scattering equations as in Eqs. 19–21. Now we use the separation of channels P_1 and P_x to split the P-space equation (Eq. 21a) into coupled equations

$$[E - \mathcal{K}_1(E)]\phi_1 = \mathcal{V}_{1x}(E)\phi_x$$

$$[E - \mathcal{K}_x(E)]\phi_x = \mathcal{V}_{x1}(E)\phi_1 \qquad [56]$$

with

$$\phi_1 = P_1\Psi_k^{(+)} = P_1 P\Psi_k^{(+)}, \qquad \phi_x = P_x\Psi_k^{(+)} = P_x P\Psi_k^{(+)},$$

$$\mathcal{K}_1(E) = P_1\mathcal{K}(E)P_1 \qquad \mathcal{V}_{1x}(E) = P_1\mathcal{K}(E)P_x$$

$$\mathcal{V}_{x1}(E) = P_x\mathcal{K}(E)P_1 \qquad \mathcal{K}_x(E) = P_x\mathcal{K}(E)P_x. \qquad [57]$$

Here $\phi_1 = \phi_1(r_1, \ldots, r_A; r)$ in coordinate space (Eq. 15) has the variables of the meson plus nucleons, while $\phi_x = \phi_x(r_1, \ldots, r_A)$ has only the nucleons.

Now $\mathcal{K}_1(E)$ is the effective interaction in the scattering space of the meson and nucleons; we may decompose it just as in Eq. 49 into parts representing the kinetic energy, the nuclear interaction, and the meson-nuclear interaction,

$$\mathcal{K}_1(E) = H_0 + \mathcal{V}_N(E) + \mathcal{V}_{\pi N}(E), \qquad [58]$$

where we here consider π-meson scattering. We may consider the homogeneous equation for the scattering channel,

$$[E - H_0 - \mathcal{V}_N(E) - \mathcal{V}_{\pi N}(E)]\psi_1^{(+)} = 0. \qquad [59]$$

This is an effective Schroedinger equation for π-nucleus scattering, but does not give the complete scattering in this space, since the coupling to the absorption channel (P_x) has been removed. We denote the scattering amplitude for this uncoupled solution by $T_1(E)$:

$$\langle k', n | T_1(E_k) | k, 0 \rangle = \langle k', n | \mathcal{V}_{\pi N}(E) | \psi_1^{(+)} \rangle. \qquad [60]$$

We may now follow the development of an extended Watson multiple scattering expansion for $T_1(E)$, as developed in Eqs. 50–55. All the new features found in those equations also apply here: the dependence of operators on E_k, and the presence of many-body terms in $\mathcal{V}_{\pi N}(E)$. There is one new feature in the present case, which affects the use of the impulse approximation for the single scattering term $t_i(E)$ in the expansion given in Eq. 54 of $T_1(E)$. The problem is that for this expansion, based on Eq. 59, we have decoupled the absorption space P_x from the scattering space P_1. That means that we do not consider any process in π-scattering in which the meson is absorbed and then later emitted, as in Eq. 3. Some contributions of this sort are illustrated in Fig. 6.1. Of these, the term represented in Fig. 6.1a is the (direct) Born contribution to πN scattering, which is clearly a one-nucleon term. According to the present method, this contribution must be *excluded* from $t_i(E)$ in Eq. 54, since it involves coupling of the P_1 to the P_x space. The result is that we may only use the impulse approximation for $t_i(E)$ if we subtract the contribution of the direct Born term

$$t_i(E) \simeq t_{free}(\pi N) - t(\text{direct Born}). \qquad [61]$$

Similarly the other contributions of Fig. 6.1 should be excluded from the t_{ij} terms of Eq. 54.

With these changes, we may use the generalized Watson expansion (Eq. 54) for $T_1(E)$. The formal solution for the scattering function $\psi_1^{(+)}$ of Eq. 59 may be written

$$|\psi_1^{(+)}\rangle = [1 + G_N(E_k)T_1(E_k)]|\mathbf{k}, 0\rangle. \qquad [62]$$

This is a solution to the uncoupled channel problem only. We may, however, solve the coupled equations (Eq. 56) in the following form, using the homogeneous solution $\psi_1^{(+)}$:

$$\phi_1 = \psi_1^{(+)} + [E + i\eta - \mathcal{K}_1(E)]^{-1}\mathcal{V}_{1x}(E)\phi_x,$$

$$\phi_x = [E + i\eta - \mathcal{K}_x(E)]^{-1}\mathcal{V}_{x1}(E)\phi_1. \qquad [63]$$

We note that there are no homogeneous terms in ϕ_x, since we assumed in defining $\Psi_k^{(+)}$ that there were incoming waves only in the elastic channel. We may further solve for ϕ_x in terms $\psi_1^{(+)}$ by writing

$$[E - \mathcal{K}_x(E) - W_x(E)]\phi_x = \mathcal{V}_{x1}(E)\psi_1^{(+)}$$

$$W_x(E) = \mathcal{V}_{x1}(E)[E + i\eta - \mathcal{K}_1(E)]^{-1}\mathcal{V}_{1x}(E). \qquad [64]$$

The first line of Eq. 64 can be inverted to find ϕ_x, which may then be substituted in the first line of Eq. 63 to obtain

$$\phi_1 = \psi_1^{(+)} + \left[E + i\eta - \mathcal{K}_1(E) \right]^{-1} K_1(E)\psi_1^{(+)}$$

$$K_1(E) = \mathcal{V}_{1x}(E)\left[E + i\eta - \mathcal{K}_x(E) - W_x(E) \right]^{-1}\mathcal{V}_{x1}(E). \quad [65]$$

The operator $K_1(E)$ contains all the coupling between the scattering and absorption channels, and is clearly very complicated. Among many other contributions, $K_1(E)$ will include those shown in Fig. 6.1.

Finally, we may use Eqs. 63–65 to find expressions for the full amplitude for scattering and for absorption transitions. For the first we may use Eq. 23b to write

$$\langle \mathbf{k}', n | T(E_k) | \mathbf{k}, 0 \rangle = \lim \langle \mathbf{k}', n | (\mathcal{K}(E_k) - E_k) P | \Psi_{\mathbf{k}}^{(+)} \rangle$$

$$= \langle \mathbf{k}', n | \mathcal{V}_{\pi N}(E_k) | \phi_1 \rangle + \langle \mathbf{k}', n | \mathcal{V}_{1x}(E_k) | \phi_x \rangle$$

$$= \langle \mathbf{k}', n | T_1(E_k) | \mathbf{k}, 0 \rangle + \langle \psi_{1,n}^{(-)} | K_1(E_k) | \psi_1^{(+)} \rangle.$$
$$[66]$$

In the second line we have used $P = P_1 + P_x$ and Eq. 57. In the last line we have used Eqs. 63 and 65 and regrouped terms using Eq. 60. We have also introduced a final state scattered wave (with ingoing boundary conditions) in analogy to Eq. 62:

$$\langle \psi_{1,n}^{(-)} | = \langle \mathbf{k}', n | \left\{ 1 + \mathcal{V}_{\pi N}(E_k)\left[E_k + i\eta - \mathcal{K}_1(E_k) \right]^{-1} \right\}$$

$$= \langle \mathbf{k}', n | \left[1 + T_1(E_k) G_N(E_k) \right]. \quad [67]$$

Equation 66 is our principal result for elastic or inelastic scattering of a meson from a nuclear target, where absorption is a possible channel. The first term (T_1) gives the scattering with the coupling to the absorption channel excluded; this term may be treated by the extended Watson multiple scattering expansion given in Eqs. 54 and 55. The coupling of the absorption channel to the elastic scattering is completely in the second term (K_1) of Eq. 66, for which an explicit calculation of absorption and emission of the meson, as well as propagation of the nuclear system in the intermediate (x) channel (Eq. 65) is required. Equation 66 is similar to one we encounter again in Section 6.4, in connection with the "doorway-state approximation," for which it was derived in Fes–67.

The transition amplitude to the P_x-space, that is, for absorption of the incoming meson by the nuclear target, may also be obtained from Eq. 23b, but with channel states $\langle f_x |$, which have plane waves for the particles that are asymptotically free in the final channels:

$$\langle f_x | T(E_k) | \mathbf{k}, 0 \rangle = \lim \langle f_x | (\mathcal{H}(E_k) - E_k) P | \Psi_{\mathbf{k}}^{(+)} \rangle$$
$$= \langle f_x | \mathcal{V}_{x1}(E_k) | \phi_1 \rangle + \langle f_x | \mathcal{V}_x(E_k) | \phi_x \rangle, \qquad [68]$$

where we have introduced an effective interaction for the final channel

$$\langle f_x | \mathcal{V}_x(E) = \langle f_x | [\mathcal{H}(E_k) - E] P_x. \qquad [69]$$

Again, substituting Eqs. 63 and 65 and regrouping terms, we find that the absorption amplitude may be written

$$\langle f_x | T(E_k) | \mathbf{k}, 0 \rangle = \langle \psi_f^{(-)} | \mathcal{V}_{x1}(E_k) | \psi_1^{(+)} \rangle \qquad [70]$$

where we have introduced a scattering state in the absorption channel, by

$$\langle \psi_f^{(-)} | = \langle f_x | \{ [\mathcal{V}_x(E_k) + W_x(E_k)] [E_k + i\eta - \mathcal{H}_x(E) - W_x(E)]^{-1} + 1 \}.$$
$$[71]$$

This final state is defined to include all the effects of coupling of the absorption and scattering channels, unlike the initial state $\psi_1^{(+)}$ which was defined for the uncoupled scattering channel given in Eqs. 59 and 62.

To give a theory of meson absorption in this formulation, we must provide a theory of the three different components: the initial and final states $\psi_1^{(+)}$ and $\psi_{f,n}^{(-)}$, and the effective absorption operator $\mathcal{V}_{x1}$. The initial state may be expressed in terms of the Watson expansion in the scattering channels in Eqs. 54 and 62. The other components represent different problems. We discuss them further in Chapter 7, where we take up the subject of meson absorption again, but in more detail.

The method of coupled channels is not the only theoretical approach available for dealing with field degrees of freedom in meson-nucleus scattering. Some authors (Ing–74, Cam–75, Mil–77) have taken as a starting point the Low equation (Eqs. 24 and 25) for the meson-nuclear system. If we specialize to elastic scattering, we may follow a similar development to that of the Chew-Low equations for πN elastic scattering, as given in Section 2.4. We consider Eq. 25 with $n = 0$; from the complete set of states ψ_m we single out those $\Psi_{1,0}^{(-)}$ with outgoing waves only in the elastic channel: mesons of energy ω_l and momentum $\mathbf{l}$, and the nuclear target in the ground state (0). Since these are eigenstates of the complete Hamiltonian H, there are incoming waves in all channels, including the absorption channel. We write an equation of the Chew-Low form:

$$\langle \mathbf{k}', 0 | T(E_k) | \mathbf{k}, 0 \rangle = \langle \mathbf{k}' | D(E_k) | \mathbf{k} \rangle$$

$$+ \int d\mathbf{l} \left\{ \frac{\langle \mathbf{k}', 0 | T(E_k) | \mathbf{l}, 0 \rangle * \langle \mathbf{l}, 0 | T(E_k) | \mathbf{k}, 0 \rangle}{\omega_k - \omega_l + i\eta} \right.$$

$$\left. - \frac{\langle \mathbf{k}, 0 | T(E_k) | \mathbf{l}, 0 \rangle * \langle \mathbf{l}, 0 | T(E_k) | \mathbf{k}', 0 \rangle}{\omega_k + \omega_l} \right\}$$

$$[72]$$

where the inhomogeneous term $\langle \mathbf{k}' | D(E_k) | \mathbf{k} \rangle$ represents the sum over all states $\{\psi_m\}$ in the complete set, other than the elastic channel state $\{\Psi_{\mathbf{l},0}^{(-)}\}$. We have used the relation in Eq. 26 for $n = 0$ (and Eq. 2.98) to obtain the t-matrices on the right-hand side of Eq. 72. We now have a nonlinear integral equation for the elastic scattering amplitude, of the Chew-Low form, in terms of the "driving" term $D(E_k)$, which contains information about all other channels of the system including the open channels: inelastic scattering and meson absorption. In the πN case, there were no other open channels, and the driving term was taken to be the Born term $T^{(0)}$ in Eq. 2.95. In the present case, this term is part of the meson absorption contribution to $D(E_k)$. A complete specification of these states $\{\psi_m\}$ is very awkward: remember that they contain parts in all channels, and must be orthogonal to the $\{\Psi_{\mathbf{l},0}^{(-)}\}$. However, one may treat the driving term from a multiple scattering point of view and develop appropriate approximations.

First we may note that the nucleon current $j(k)$ that appears in Eqs. 24 and 25 is a sum of one-nucleon terms:

$$j(k) = \sum_{i=1}^{A} j_i(k). \qquad [73]$$

Then we may separate the terms in $D(E)$ which involve the same particle twice $(j_i j_i)$ or different particles $(j_i j_j, i \neq j)$. We now make a drastic assumption: that there are no correlations (not even antisymmetry!) in the states ψ_0, ψ_m. In this case we can drop the $i \neq j$ terms in $D(E)$, since no uncorrelated state ψ_m can be connected to ψ_0 by both j_i and j_j in that case.

We further assume closure on the intermediate states $\{\psi_m\}$, ignoring excitation of nucleons (or mesons), and also assume that, for these states, there is little overlap with the elastic channel states $\{\Psi_{\mathbf{l},0}^{(-)}\}$. The result of all of these assumptions is an impulse approximation,

$$\langle \mathbf{k}' | D(E_k) | \mathbf{k} \rangle \simeq \sum_{i=1}^{A} \langle \mathbf{k}' | D_i(E_k) | \mathbf{k} \rangle$$

$$\langle \mathbf{k}' | D_i(E_k) | \mathbf{k} \rangle = \langle \mathbf{k}' | t_i(\omega_k) | \mathbf{k} \rangle, \qquad\qquad [74]$$

where the last expression is the Chew-Low t-matrix for πN elastic scattering in the static approximation (from the closure assumption). This approximation to the driving term is analogous to the lowest order approximation (Eq. 4.67) to the optical potential, $\mathcal{V}_c^{(1)} \simeq \langle 0 | \Sigma_i t_i | 0 \rangle$. In the present form, D determines the elastic t-matrix through the nonlinear Eq. 72, while $\mathcal{V}^{(1)}$ is used in the Lippmann-Schwinger (or Schroedinger) equation. Equation 72 has the formal advantage of exhibiting the crossing symmetry expected for the scattering amplitude (see Section 6.1b), since that symmetry is built into the Low equation given in Eqs. 24 and 25 from the start. (This symmetry is not explicit in Eq. 66, for which it also holds in principle.) The disadvantage of the nonlinear form of Eq. 72 is that of solving the equations. (Some problems of multiplicity of solutions for the πN case are discussed in Cas–56.) Further problems with the method develop with any attempt to improve the approximations systematically. (See, however, Cam–75 and Mil–77.)

It is interesting to compare the scattering formula in Eq. 65, obtained from the coupled channel method, with the Chew-Low form given in Eq. 72, from the point of view of the interplay of the scattering and absorption mechanisms. In the former, the scattering *without* absorption is treated as a separate problem, for which a scattering amplitude (Eq. 60) and scattering functions (Eqs. 62 and 67) are calculated by multiple scattering methods. The contribution of the absorption channel is calculated from these quantities in Eq. 66. The process is reversed in the Chew-Low theory, since the absorption channels appear in the driving term $D(E_k)$ in Eq. 72, which must be calculated first in order to obtain the elastic scattering amplitude from the integral equation. The total amplitude does not separate as in Eq. 66.

It is also possible to develop a multiple scattering method in terms of Feynman diagrams with effective vertices. These techniques are borrowed from quantum field theory, which lies beyond the scope of this book. We refer the reader to some of the methods proposed in the literature: Tay–65,66, Dov–71,73a,b, Cel–74, and Bro–75. We sketch briefly a technique based on the Green function approach of J. G. Taylor (Tay–65,66), whose structure most closely parallels that of the method of coupled channels we have outlined (see Miz–75).

The method of effective vertices limits the degrees of freedom of the interacting system, much as the projection method limits the degrees of freedom in the effective Schroedinger equation. Each vertex stands for a complicated set of processes that involve more degrees of freedom than allowed—for example, many mesons. We illustrate the method for the case of πN scattering, in which the degrees of freedom are to be pion and nucleon, and nucleon with no pion. The nucleons are dressed—physical particles. We write the t-matrix for πN scattering in the form

$$t_{\pi N} = \gamma^\dagger g_N \gamma + t_2 \qquad [75]$$

in which the first term represents $\pi + N \rightarrow N \rightarrow \pi + N$, where the system may propagate as a nucleon part of the time. The second term represents processes in which the πN system never contains fewer particles than one pion and one nucleon. The decomposition is illustrated in Fig. 6.8a, where the circular vertices in the first figure on the right-hand side stand for effective Yukawa vertex functions $\gamma^\dagger, \gamma$ of Eq. 75, and the single line stands for the dressed propagation of a single nucleon, with propagator g_N. The simplest contribution to this term is simply the Born term (direct only) of Chew-Low theory (Section 2.4), which gives the pole of the nucleon propagator g_N. The second term (t_2) is represented by the square vertex in Fig. 6.8a. This in turn may be decomposed into a part in which the pion and nucleon propagate, and one in which there are always more particles propagating than just $\pi + N$ (e.g. $2\pi + N$). This decomposition leads to an integral equation for t_2:

$$t_2 = V_2 + V_2 g_\pi g_N t_2, \qquad [76]$$

where V_2 is the term with always more than two particles ($\pi + N$) propagating. The equation is illustrated in Fig. 6.8b where the solid square vertex

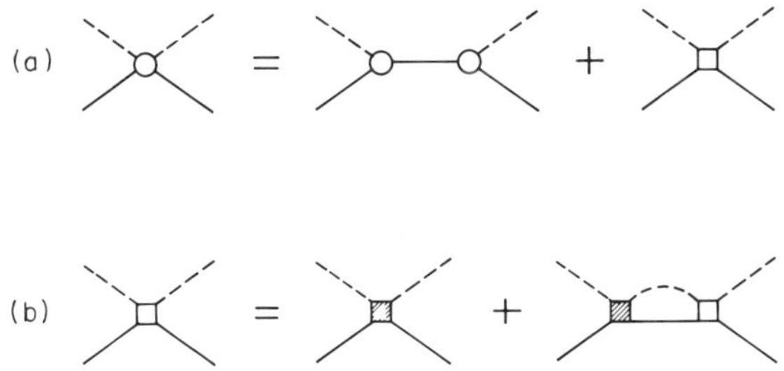

Fig. 6.8 Feynman diagrams for πN scattering with effective vertices: (a) $t_{\pi N}$ as given by Eq. 75; open square vertex represents t_2; (b) represents Eq. 76 for t_2.

294 Field Theory and Coupled Channel Methods

represents V_2, which acts as an effective interaction for the πN system. Equation 76 is then a form of Bethe-Salpeter equation for the scattering of two particles. As opposed to the Lippmann-Schwinger equation, the propagator in Eq. 76 is the product of the separate propagators for a free pion (g_π) and nucleon (g_N). This difference is typical of Feynman diagram methods. It is easier in this case to accommodate relativistic propagation of the particles than in the Lippmann-Schwinger case, where explicit "backward" diagrams must be added to get Klein-Gordon propagation, as in Eqs. 8 and 10.

Let us pass directly to the application to pion scattering from a nuclear target; we consider elastic scattering. In analogy to the $\pi - N$ case (Eq. 75) we write the elastic scattering amplitude in the form

$$\langle \mathbf{k}',0|T|\mathbf{k},0\rangle = \langle \mathbf{k}',0|T_1|\mathbf{k},0\rangle + \langle \mathbf{k}',0|\Gamma^\dagger G\Gamma|\mathbf{k},0\rangle, \qquad [77]$$

which is illustrated in Fig. 6.9a. The first term (T_1, square vertex) involves propagation of A nucleons plus any number (greater than zero) of mesons, while the second term represents the possibility of having only A nucleons propagating at some time. This decomposition is the Feynman analog of the result in Eq. 66 for the coupled channel method: T_1 is the scattering amplitude decoupled from absorption, while the second term in Eq. 77 is like the $K_1(E)$ term of Eq. 66 in which the absorption channel enters. Here the effective (many-body) vertices for absorption (Γ) and emission ($\Gamma^\dagger$) of the pion include the multiple scattering which is given by $\psi_1^{(\pm)}$ in Eq. 66. This is illustrated in Fig. 6.9c, where the solid vertex is equivalent to the absorption operator $\mathcal{V}_{x1}$ in Eq. 65. The scattering amplitude T_1 in Eq. 77

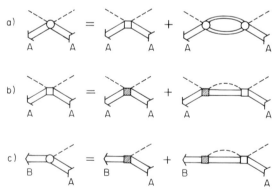

Fig. 6.9 Representation of scattering and absorption of a meson by a nuclear target: (a) decomposition of scattering as in Eq. 77; (b) integral equation for T_1; (c) amplitude for meson absorption.

may be given by an integral equation (Bethe-Salpeter) as shown in Fig. 6.9b: this is the analog of an optical potential Lippmann-Schwinger equation for elastic scattering.

This brief sketch simply shows that we can develop scattering equations of the relativistic integral equation form, which closely parallel those of the coupled channel theory. We do not attempt to develop the multiple scattering expansion method further here. What is required is a reexpression of the nuclear amplitudes in Eq. 77 in terms of elementary πN amplitudes such as those of Eqs. 75 and 76.

6.4   Propagation of Resonances

We now discuss methods in which meson-nucleon resonances, or isobars, are treated as separate degrees of freedom. We discuss specifically the $P_{3/2,3/2}$ resonance in πN scattering, $\Delta(1232)$, which dominates the lower energy πN interactions. The methods may also be applied to other resonances, such as Y^* resonances of the $\overline{K}N$ system. The approaches under discussion begin with a simple model for resonant πN scattering,

$$\pi + N \rightarrow \Delta \rightarrow \pi + N, \qquad \qquad [78a]$$

and then formulate π-nucleus scattering on terms of coupling to a Δ-nucleus system

$$\pi + A \rightarrow \Delta + (A - 1) \rightarrow \pi + A. \qquad \qquad [78b]$$

The πN model assumed for Eq. 78a generates the coupling of the systems in Eq. 78b. The main physical problem in these approaches is an adequate description of the Δ-nuclear interacting system.* This may be thought of as the problem of describing the propagation of a Δ through a nucleus, once the Δ has been created by πN scattering. There are many parallels in these methods with the conventional πA multiple-scattering theory, but we are here more interested in stressing the different features introduced.

A simple example may be used to illustrate the approach. Let the resonant $\pi - N$ amplitude be represented by a separable t-matrix,

$$\langle \mathbf{k}'|t(\varepsilon)|\mathbf{k}\rangle = g(k')\tau_\Delta(\varepsilon)g(k). \qquad \qquad [79]$$

(We ignore angular momentum coupling.) The function $\tau_\Delta(\varepsilon)$ gives the energy dependence of the πN resonant state; it may be considered to be

---

*This problem is even more acute for the Y^*-nuclear system.

the propagator for the Δ in free space. Now consider the lowest order Watson approximation to the optical potential (Eq. 4.67)

$$\mathcal{V}(E) = \langle 0| \sum_i t_i |0\rangle \qquad [\mathbf{80}]$$

where t_i is the πN amplitude in the nuclear target. If we assume that t_i differs from t_{free} (Eq. 79) only in the modification of the Δ-propagator $\tau_\Delta(\varepsilon)$, we have

$$\langle k'| \mathcal{V}(E)|k\rangle = g(k')\langle 0| \sum_i \delta(\mathbf{k}' + \mathbf{p}'_i - \mathbf{k} - \mathbf{p}_i)\tau_i^\Delta(\varepsilon)|0\rangle g(k) \quad [\mathbf{81}]$$

where $\tau_i^\Delta(\varepsilon)$ is the Δ-propagator in the nuclear medium. (In the Watson expansion, such a modification of the Δ-propagator would appear as a sum of interactions of either the ith nucleon or the pion with other target nucleons, between the initial and final interaction of the pion with the ith nucleon. These contributions would be classified as "binding effects" and "local field correction," respectively.) In what follows, we see how these effects, and also corrections due to nuclear correlations, can be incorporated into the modified propagator for the Δ in the target.

6.4a Doorway States

The basic idea here is that the Δ-nucleus states form a set of states (doorway states) that intervene between the elastic channel for π-nucleus scattering, and all other states, such as inelastic states, continuum states, and so on of the system. The formulation is most easily developed using the projected space-coupled channel theory given earlier in Sections 6.2 and 6.3. We follow Kisslinger and Wang, who have developed this method for πA (Kis–76a). [This formulation was first applied to doorway states in low energy nuclear reactions, in the original work of Feshbach, Kerman, and Lemmer (Fes–67).] We break up the space by projection into three parts, as in Section 6.3: $P_1 + P_x + Q = 1$. For the present application, we choose the spaces as follows:

P_1 the elastic πA channel: contains one meson and target ground state;

$P_x \equiv D$ doorway states: set of excited states of the $\Delta + (A-1)$ system;

Q other states, including πA inelastic channels.

The Hamiltonian equation may be separated as in Eqs. 20, 56, and 57. The homogeneous scattering solutions in the P_1 space, $\psi_1^{(\pm)}$, are given by $[E - \mathcal{H}_1(E)]\psi_1^{\pm} = 0$, as in Eq. 59, and the corresponding elastic scattering amplitude $\langle \mathbf{k}', 0 | T_1(E_k) | \mathbf{k}, 0 \rangle$ is given by Eq. 60. The doorway states are to be chosen (see below) so that there is no resonant scattering in the P_1 space.

The amplitude for elastic scattering may be obtained from Eqs. 65 and 66 (with $x \equiv D$)

$$\langle \mathbf{k}', 0 | T(E_k) | \mathbf{k}, 0 \rangle = \langle \mathbf{k}', 0 | T_1(E_k) | \mathbf{k}, 0 \rangle$$
$$+ \langle \psi_1^{(-)} | \mathcal{V}_{1D}(E) [E_k - \mathcal{H}_D(E_k) - W_D(E_k)]^{-1} \mathcal{V}_{D1}(E) | \psi_1^{(+)} \rangle$$

$$[\mathbf{82}]$$

with

$$W_D(E) = \mathcal{V}_{D1}(E) [E + i\eta - \mathcal{H}_1(E)]^{-1} \mathcal{V}_{1D}(E).$$

An alternative way of calculating the t-matrix is through the optical potential, which in this case may be written as a P_1-space operator:

$$\mathcal{V}_{\text{opt}}(E) = \mathcal{H}_1(E) + \mathcal{V}_{1D}(E) [E - \mathcal{H}_D(E)]^{-1} \mathcal{V}_{D1}(E). \quad [\mathbf{83}]$$

Both Eqs. 82 and 83 are exact results; so far there has been no approximation, nor have we made any specific use of the doorway states.

The main assumption of the model is that, for a specific set of ("doorway") states (projected by D), there is no direct coupling between the elastic space P_1 and $Q = 1 - P_1 - D$; that is,

$$P_1 H Q = 0 = Q H P_1. \quad [\mathbf{84}]$$

It follows from this that

$$\mathcal{V}_1(E) = P_1 H P_1 \equiv V_1,$$

$$\mathcal{V}_{1D} = P_1 H D \equiv V_{1D},$$

$$\mathcal{V}_{D1} = D H P_1 \equiv V_{D1}; \quad [\mathbf{85}]$$

that is, the coupling interaction between the elastic and doorway spaces (and the interaction in the elastic space) are independent of the Q-space (and energy). In the present model, we assume that (V_{1D}, V_{D1}) is the interaction that couples the free πN system to the Δ; that is, we can write the πN t-matrix in the form

$$t(E) = t_{\text{N.R.}} + t_{\text{res}}(E)$$

$$\langle \mathbf{k}', \mathbf{p}' | t_{\text{res}}(E) | \mathbf{k}, \mathbf{p} \rangle = \delta(\mathbf{k}' + \mathbf{p}' - \mathbf{k} - \mathbf{p}) \frac{\langle \mathbf{k}', \mathbf{p}' | V_{1D} | \Delta \rangle \langle \Delta | V_{D1} | \mathbf{k}, \mathbf{p} \rangle}{E - M_\Delta + \frac{i}{2}\Gamma_\Delta(E)},$$

$$[86]$$

where M_Δ and $\Gamma_\Delta(E)$ are the mass and width of the Δ-resonance, and $t_{\text{N.R.}}$ is the nonresonant part of the t-matrix.

Now let us consider a specific set of doorway states, $\{\phi_j\}$, which diagonalize $(E - \mathcal{K}_D(E))$ at fixed E:

$$\left(\mathcal{K}_D(E) - E_j(E) + \frac{i}{2}\Gamma_j(E)\right)\phi_j = 0,$$

$$[87]$$

where E_j and Γ_j are real, and depend on the coupling of the D and Q spaces. Using this set in Eq. 83 for the optical potential with Eq. 85 we find

$$\mathcal{V}_{\text{opt}}(E) = V_1 + \sum_j \frac{V_{1D}|\phi_j\rangle\langle\phi_j|V_{D1}}{E - E_j(E) + \frac{i}{2}\Gamma_j(E)}.$$

$$[88]$$

The first term V_1 is a nonresonant contribution to the potential, involving P_1-space interactions only. The second term reflects the coupling to the doorway states, each of which appears with a resonancelike energy denominator, or propagator, in the sum.

The potential matrix for the target ground state is obtained from Eq. 88, using plane wave states $|\mathbf{k}, 0\rangle$:

$$\langle \mathbf{k}' | \mathcal{V}_{\text{opt}}(E) | \mathbf{k} \rangle = \langle \mathbf{k}', 0 | V_1 | \mathbf{k}, 0 \rangle + \sum_j \frac{\langle \mathbf{k}', 0 | V_{1D} | \phi_j \rangle \langle \phi_j | V_{D1} | \mathbf{k}, 0 \rangle}{E - E_j(E) + \frac{i}{2}\Gamma_j(E)}.$$

$$[89]$$

We would like to evaluate the numerator of the resonant term in Eq. 89 by relating it to the numerator of the resonant πN amplitude of Eq. 86. The doorway states ϕ_j are supposed to be states of Δ coupled to the residual target $(A - 1)$. The simplest approximation (Kis-76a) is to assume that the numerators of Eqs. 86 and 89 are simply proportional. Then the numerator of Eq. 89 becomes

$$\langle \mathbf{k}', 0 | V_{1D} | \phi_j \rangle \langle \phi_j | V_{D1} | \mathbf{k}, 0 \rangle$$

$$= A \int d\mathbf{p} \langle \mathbf{k}', \mathbf{p}' | V_{1D} | \Delta \rangle \langle \Delta | V_{D1} | \mathbf{k}, \mathbf{p} \rangle \rho_N(\mathbf{p}', \mathbf{p})\rho_\Delta^{(j)}(\mathbf{p} + \mathbf{k})$$

$$\simeq \langle \mathbf{k}' | t_{\text{res}}(E) | \mathbf{k} \rangle \left[E - M_\Delta + \frac{i}{2}\Gamma_\Delta(E) \right] A F_j(\mathbf{k}', \mathbf{k}) \quad [90]$$

with $p' = p + k - k'$, where in the last line, we remove the t-matrix (Eq. 86) from the integral, ignoring the nucleon momenta p and p', as in the usual impulse approximation (e.g., Eq. 4.124). The density matrix for nucleons is $\rho_N(p', p)$, and the diagonal momentum density for the Δ in the state ϕ_j is $\rho_\Delta^{(j)}(p + k)$; both are normalized to unity. The form factor F_j is defined

$$F_j(k', k) = \int dp\, \rho_N(p + k - k', p)\rho_\Delta^{(j)}(p + k'). \qquad [91a]$$

This may be contrasted with the normal form factor for elastic scattering (Eq. 4.125),

$$\rho_0(k' - k) = \int dp\, \rho_N(p + k - k', p), \qquad [91b]$$

which is a function of $q = k' - k$ only. This latter form leads to a local optical potential in the forward scattering approximation, as we have seen in Eq. 4.131. The extra term $\rho_\Delta^{(j)}$ in Eq. 91a gives that form factor more complicated dependence on k' and k, leading to a *nonlocal* optical potential when Eq. 90 is transformed to position space.*

In this simplest approximation, the "isobar doorway model" gives an optical potential of the form

$$\langle k' | \mathcal{V}_{opt}(E) | k \rangle = \langle k', 0 | V_1 | k, 0 \rangle + \sum_j \langle k' | t_{res}(E) | k \rangle$$

$$\times \frac{E - M_\Delta + \frac{i}{2}\Gamma_\Delta(E)}{E - E_j(E) + \frac{i}{2}\Gamma_j(E)} F_j(k', k). \qquad [92]$$

The first, nonresonant term is to be evaluated, presumably, from the $t_{N.R.}$ of Eq. 86, by standard (Watson) approximations. The second term is given in terms of the resonant t-matrix (off-shell), the energy denominators (propagators) for the Δ in the states (ϕ_j) and the form factors F_j. This equation can serve as a basis for a phenomenological treatment of the optical potential, in which E_j, Γ_j, and F_j are given simple functional forms adjusted to fit scattering data. Some applications of this have been made to elastic and charge exchange pion scattering, as well as $\bar{K}$-nucleus scattering (Kis–76a, b, Nag–74, Aue–77).

The strength of this approach is its simplicity. It presumes to have put into the propagators and form factors all the complications of correlations and binding corrections of the multiple scattering expansion. Whether very simple parametric forms will prove adequate is to be seen. Without the

---

*See discussion in Chapter 5, Eqs. 29–38, on the origin of this nonlocality (Len–75).

simplest assumptions, even Eq. 92 is not so easy to evaluate. Further systematic corrections to the approximations that led to Eq. 92 would presumably be as complicated as in any other theory. In particular, Friedman (Fri–75) has investigated the relation between the doorway approach and the conventional Watson expansion. Another approach would be to give a dynamical model for the doorway states, Eq. 87, and use the results to calculate the required quantities for the optical potential (Eq. 92). One class of such models is discussed in the following subsection.

6.4b Δ-Hole Model

Let us consider a simple model for the doorway states (ϕ_j) introduced in Eq. 87: these are to be particle-hole states (Δh) obtained by exciting a nucleon to a Δ in a single particle state, coupled to the remaining single hole in the target nucleus. We now write the optical potential in Eq. 83 or Eq. 85 in the form

$$\mathcal{V}_{\text{opt}}(E) = V_1 + \mathcal{V}_{\Delta h}(E)$$

$$\mathcal{V}_{\Delta h}(E) = V_{1D} \Pi_{\Delta h}(E) V_{D1} \qquad [93a]$$

where $\Pi_{\Delta h}$ is the Green (or polarization) function for the Δ-hole propagation, given in terms of the diagonal (Δh) states by

$$\Pi_{\Delta h}(E) = \sum_{\Delta h} \frac{|\Delta h\rangle\langle\Delta h|}{E - E_{\Delta h} + i\eta} . \qquad [94]$$

The Green function for noninteracting particle and hole may be written

$$\Pi^0_{\Delta h}(E) = \sum_{\alpha\beta} \frac{|\alpha\beta^{-1}\rangle\langle\alpha\beta^{-1}|}{E - \varepsilon_\alpha + \varepsilon_\beta + i\eta} = \left(E - H^0_\Delta - H^0_N + i\eta\right)^{-1} \qquad [95]$$

where α labels the state of the Δ, and β, the nucleon hole state, with $H^0_\Delta|\alpha\rangle = \varepsilon_\alpha|\alpha\rangle, H^0_N|\beta^{-1}\rangle = -\varepsilon_\beta|\beta^{-1}\rangle$. (Note that $\varepsilon_\alpha = \varepsilon_\alpha(E)$ can be energy-dependent and complex.) The contribution of Π^0 to the optical potential (Eq. 93a) is illustrated in Fig. 6.10a. We may also include Δh terms in which the π-lines are crossed, as in Fig. 6.10b. For general $\Pi(E)$, we obtain the crossed terms by adding to Eq. 93a

$$\mathcal{V}_{\Delta h}(E) = V_{1D} \Pi_{\Delta h}(E) V_{D1} + V_{D1} \Pi_{\Delta h}(-E) V_{1D}. \qquad [93b]$$

This effectively includes some of the Q-space (2 mesons-Δh) along with the

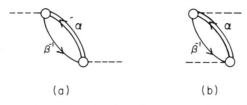

(a)   (b)

Fig. 6.10 Diagram for Green function for pion propagation in Δ-hole model: (a) direct, as in Eq. 95; (b) crossed diagram.

Δh D-space. If we use the noninteracting Green function in Eq. 95 in Eqs. 93a or 93b, we obtain a slightly modified version of the Watson first-order optical potential. Without interaction, the Δ would be assigned its free value of energy $\varepsilon_\alpha(E)$, as given, for example in Eq. 86.

Now we consider the inclusion of interactions of the Δ and hole, which we illustrate using Feynman-Goldstone diagrams in Fig. 6.11. First we consider the interaction of the Δ with the nucleus as a whole, independent of the hole. In Fig. 6.11 we show a first-order potential interaction of the Δ with each nucleon in the target nucleus; this gives a Hartree or average field interaction. This interaction is not known, but may be calculated for a model Δ-N interaction, illustrated in Fig. 6.11b, which could for example involve exchange of π or vector mesons. Higher order contributions of the Δ-N interaction involve excitations beyond the Δ-h, which are beyond the present model. They could be included by modifying the hole-state energy ε_β, for example, by a shift and width, as in the phenomenological doorway model. We also illustrate in Fig. 6.11 a $\Delta N \leftrightarrow N\Delta$ exchange interaction, here the exchanged particle is a π. This does not contribute to the Hartree potential, but does play a role, as we see below.

A further interaction of the Δ with the nucleus involves the Δ self-energy. In Fig. 6.11d we illustrate a contribution to the self-energy due to coupling to the πN state. This interaction is present for the free Δ; in the nucleus, it may be considered to be modified by the Pauli exclusion principle, which prevents dissociation into πN if the nucleon momentum is occupied by target nucleons.* In the Feynman diagram method, this Pauli correction comes from the exchange interaction of (c); for $\mathbf{p}' = \mathbf{p}$, diagram (c) exactly cancels diagram (d). Further interactions of the π or N in (d) with the nucleus, also change the Δ energy $\varepsilon_\alpha(E)$.

Let us now consider the interactions of the Δ and hole. In Fig. 6.11(e) we illustrate the interaction $V'(E)$ which takes a Δh state to another Δh

*See also Fig. 5.7 and discussion at the end of Section 5.1d.

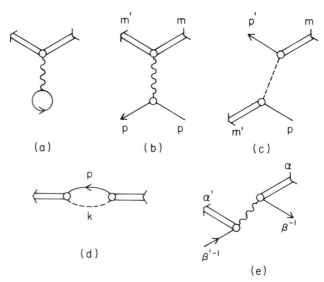

Fig. 6.11 Various terms involving the interaction of a propagating Δ with the target nucleus.

state:

$$\langle \alpha', \beta'^{-1} | V'(E) | \alpha, \beta^{-1} \rangle. \qquad [96]$$

This interaction also involves the $\Delta N \leftrightarrow N\Delta$ exchange interaction. However, the exchange of a pion, as in (c) is not allowed to contribute to the optical potential through diagram (e). This is because the one-pion state in (e) is in the elastic scattering space P_1, given by a single pion plus target ground state, and not in the doorway D-space, or Q-space in which the Green function Eq. 94 and the interaction $\mathcal{H}_D$ or Eq. 83 are calculated. Therefore, diagram (c) is to be excluded from V', although exchanges of other mesons (e.g., ρ-mesons) are not excluded. Because (c) must be excluded from the optical potential, the Pauli exclusion violating part of (d) is not automatically canceled, and now must be removed by calculation. There is an alternative approach to this separation of one-pion exchange diagrams (c): that is to calculate the elastic scattering amplitude directly, starting with Eq. 82 rather than with Eq. 83 for $\mathcal{V}_{\text{opt}}$. Now all diagrams (e) are included (technically through the addition of $W_D(E)$ in Eq. 82). In general the calculation of T is more complicated than for $\mathcal{V}$, but in the case of the Δ-h model, it is about as simple to proceed directly to T (Hir–77 a, b, Wei–77).

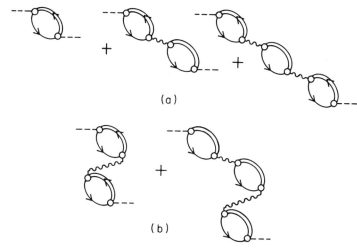

$$(a)$$

$$(b)$$

Fig. 6.12 (*a*) Sum of Δ-hole contributions given in Eqs. 97 and 98; (*b*) includes multiple Δ-hole excitations.

Returning to the theory of the optical potential, we can include the Δh interaction $V'(E)$ in the Δh propagator: this is the TDA (Tamm-Dancoff approximation) in which only one Δh pair propagates. We can write $\Pi_{\Delta h}$ as a series in $\Pi_{\Delta h}^0$ and V', which can be summed (Fig. 6.12*a*) to give

$$\Pi_{\Delta h}(E) = \Pi_{\Delta h}^0 + \Pi_{\Delta h}^0 V' \Pi_{\Delta h}^0 + \Pi_{\Delta h}^0 V' \Pi_{\Delta h}^0 V' \Pi_{\Delta h}^0 + \cdots$$

$$= \left[\left(\Pi_{\Delta h}^0 \right)^{-1} - V' \right]^{-1}$$

$$= \left[E - H_\Delta^0 - H_N^0 - V'(E) + i\eta \right]^{-1}. \qquad [97]$$

The optical potential (Eqs. 93a or 93b) can be calculated directly from this expression.

This can be put into a simpler form for the case of nuclear matter (for the target) if we use a few further assumptions. (We here follow Bay–75, Bro–75, and Wei–77.) In this case, the $\alpha\beta^{-1}$ representation becomes diagonal by momentum conservation. If we neglect the nucleon momentum throughout (static approximation) the quantum numbers $\alpha\beta^{-1}$ are replaced by $\mathbf{k}$, the pion momentum, and we may write Eq. 97 in a $\mathbf{k}$-diagonal representation:

$$\Pi_{\Delta h}(\mathbf{k}, E) = \frac{\Pi_{\Delta h}^0(\mathbf{k}, E)}{1 - V'(\mathbf{k}, E)\Pi_{\Delta h}^0(\mathbf{k}, E)}. \qquad [98]$$

Using Eq. 93a, the optical potential becomes

$$\mathcal{V}_{\Delta h}(\mathbf{k}, E) = \frac{|V_{1D}(k)|^2 \Pi_{\Delta h}^0(\mathbf{k}, E)}{1 - V'(k, E) \Pi_{\Delta h}^0(\mathbf{k}, E)}.$$ [99]

The numerator in Eq. 99 is simply the first order (Watson) result for noninteracting Δh. The denominator reflects the Δh interaction V' (Fig. 6.11c). Baym and Brown (Bay–75) have shown that in a zero range model with $|V_{1D}(k)|^2 \propto k^2$, and a short-range repulsive correlation between the Δ and hole-nucleon, $V'(k, E) \to \frac{1}{3}$ (approximately). This recovers the Lorentz-Lorenz correction to $\mathcal{V}_{opt}$, which was discussed earlier in Section 5.1e.

The summation of Eq. 97 can also be extended to include multi-Δh pairs, as in Fig. 6.12b. This leads to a treatment analogous to the RPA (random phase approximation) theory of particle-hole ($p-h$) excitations in nuclei (see, e.g., Bro–71).

Another variation of this model is given by including the ordinary nucleon $p-h$ states as doorways (in addition to the $\Delta-h$ states). These states are reached through the absorption reaction $\pi + N \leftrightarrow N$ in addition to the $\pi + N \leftrightarrow \Delta$ interaction already introduced, for example in Eq. 86. The doorways now include all the intermediate states of the Chew-Low theory: one meson and no meson (including crossed terms). (This is the model considered in Bro–75.)

The doorway states reached by the $\pi N \leftrightarrow N$ interaction involve direct π-absorption, without resonant scattering of the incoming pion wave. To include the possibility of absorption after resonant (Δ) scattering, one must introduce coupling of the Δh states to states in the Q-space (e.g., $2p-2h$ states). In the approach of Lenz (Len–75) this is included parametrically through an absorptive complex part of the potential W_D in the Δh space. Practical applications of this approach to the propagation of Δ in nuclei have been made recently (Hir–77 a,b).

References

Most of the questions discussed in this chapter are subjects of current research; we have indicated a sample of the modern literature for the reader. We can recommend a few works for general background for this chapter: for methods of quantum field theory, Bjo–65; for coupled channel methods in nuclear reaction theory, Fes–58, 62 and Aus–70; for doorway-state theory, Fes–67. Several approaches to Δ-propagation in nuclei are presented in Len–75, Bro–75, and Kis–76a,b.

7 Absorption and Production of Mesons

7.1 General Description

Absorption reactions refer to a large class of nuclear reactions in which a meson (π or K) enters but no meson leaves the target. Sometimes we refer to these reactions as "true absorption" to distinguish them from "optical absorption," which denotes the loss of meson flux from the elastic channel by inelastic scattering. This loss is an apparent absorption, from the point of view of the optical model, but the inelastically scattered mesons do leave the nuclear target and can be detected in the laboratory. We devote a separate chapter to the theory of the "true" absorptions (and the related reaction—meson production) for several reasons: We have already noted in Chapters 3, 4, and 6 that the formal treatment of absorption leads us beyond the realm of multiple-scattering theory in a Schroedinger framework. We outlined an approach to the theory in Chapter 6, which we now apply. Second, we find that the kinematics of absorption reactions are in many ways different from those of the scattering reactions (e.g., those treated in Chapter 5) so that we have to consider different approximate methods. And third, there is a rich variety of absorption reactions available to experimental study, some of which have high cross sections or branching ratios, relative to scattering reactions. For kinematical or dynamical reasons, these may provide probes for particular nuclear features. These reactions deserve individual attention.

The point of view of the present chapter is somewhat different from that of Chapters 4 and 5, which dealt with scattering reactions. This is by necessity, since the theory of absorption and production of mesons is not nearly as well developed or understood as that of, say, elastic scattering. A number of pictures or descriptive models have emerged over the years, each of which may be of some use in interpreting certain aspects of the subject. Theoretical work on particular reactions has often begun in the middle, with a number of *ad hoc* assumptions, from which a specific

305

calculational model then emerges. The accepted test of the validity of such work has generally been the ability of the model to match some interesting feature of the experimental data.

We do not intend to survey all possible approaches. There are review articles, which include Kol–69 and Huf–75a, as well as periodic reports at conferences (e.g., Kol–71, Eis–75b, Nob–76, Wal–77). We first survey some qualitative features of the nuclear reactions induced and discuss them in terms of simple pictures of the processes involved. We then present a general formulation of the microscopic theory of absorption and production, based on the coupled-channels method discussed in the previous chapter (Sections 6.2 and 6.3), which provides us with a unified formulation in which various models can be discussed and which is a natural extension of the multiple scattering theory we have employed for scattering reactions. The presentation is an outline of a possible theory; it is by no means the only method that may work. We do not go into very much technical detail here: for the most part we limit ourselves to discussing the kinds of approximation that are usually made. In the third section we discuss the application of the microscopic theory to several interesting absorption reactions, and see to what extent this helps us understand the general features seen experimentally.

7.1a Absorption and Production Reactions with π or K Mesons

We may characterize nuclear absorption of π-mesons by

$$\pi^{\pm} + A \to A^* \to B^* + X. \qquad [1]$$

Normally only charged mesons are studied since these are available as particle beams. Negative mesons may also be absorbed at rest, from orbits in π^--atoms, into which they have been previously captured by some combination of atomic scattering and radiative processes. We have denoted the state of the target A after absorption as A^*; this is a nuclear state of high excitation, since the mass plus kinetic energy of the meson is now entirely given to the nucleus (minus a small kinetic energy of recoil of A^*). Since that excitation will be at least $m_\pi = 140$ MeV, the state A^* is always unstable to the emission of nuclear particles. (We here exclude *radiative* absorption, in which a photon may carry away excess energy; see Eq. 3 below.) Therefore, the nuclear reaction must continue to some final state, $B^* + X$, where B^* denotes a particle-stable nuclear daughter state and X some combination of (stable) nuclear emissions: $p, n, d, t, {}^3He, {}^4He,$ and so on, which carry away the excitation energy as kinetic energy. In principle, the final state may have many fragments (fission): $B_1^* + B_2^* + X$.

Finally, the particle-stable final states B* may decay by EM or β-transitions to nuclear ground states.

For K$^-$ mesons, we have

$$K^- + A \rightarrow {}_Y A^* \rightarrow \begin{cases} {}_Y B^* + X \\ B^* + X_Y. \end{cases} \qquad [2]$$

Here the excited nucleus ${}_Y A^*$ is a *hypernucleus*, containing one hyperon—for example, Λ or Σ—to conserve the strangeness ($S = -1$) quantum number brought in by the K-meson. The particle-stable final nucleus may either by a hypernucleus ${}_Y B^*$ or a normal nucleus B*, depending on whether it has kept the hyperon, emitting only normal nuclear particles in X, or has emitted the hyperon along with nuclear particles in X_Y. Again, the emitted particles in X or X_Y carry away the excess excitation energy. The remaining hypernucleus ${}_Y B^*$ or emitted hyperon will eventually decay to a nonstrange state by weak-interaction or electromagnetic transitions. (Recall the elementary decays: $\Lambda \rightarrow N\pi$, $\Sigma^\pm \rightarrow N\pi$ (weak), and $\Sigma^0 \rightarrow \Lambda\gamma$ (EM).)

The K$^+$ meson cannot be absorbed by nuclear targets through strong interactions, which conserve strangeness, since no final nuclear state of strangeness $S = 1$ is possible. Such a reaction could proceed through a weak interaction, but it is not considered here.

There is a class of reactions that may be considered to be absorption but are somewhat closer to inelastic scattering. One example is radiative absorption; for example,

$$\pi^\pm + A \rightarrow A^* + \gamma$$
$$K^- + A \rightarrow {}_Y A^* + \gamma. \qquad [3]$$

In this case, the emitted γ may carry away all the excess excitation energy, so that no particles are emitted from A*(${}_Y A^*$), or a number of nuclear particles may be emitted to leave a stable final nucleus, as in Eqs. 1 or 2. In the first case the extra nonnuclear particle (γ) changes the kinematics of the reaction very much from that of nonradiative absorption. (Because of the electromagnetic coupling, the rates or cross sections for Eq. 3 are smaller than the total rates for the nonradiative processes in Eqs. 1 or 2.) A very similar case is the strangeness exchange reaction

$$K^- + A \rightarrow {}_Y A^* + \pi^-, \qquad [4]$$

which has been discussed earlier (in Section 5.4) as a scattering reaction, to which it may be very similar in kinematics. However, this reaction has some aspects of an absorption, as we discuss again later.

Last, we should mention the reactions in which a meson is produced by an incident nuclear particle. Commonly, production is induced by protons, for example, to make a source of mesons for a meson beam:

$$p + A \rightarrow \pi + (A + 1)^* \qquad\qquad [5a]$$

$$p + A \rightarrow K^+ + {}_Y(A + 1)^*. \qquad\qquad [5b]$$

(Other nuclear beams can in principle also produce mesons.) The reactions in Eq. 5 are endothermic; the final nuclear targets $(A + 1)^*$ or ${}_Y(A + 1)^*$ may or may not emit further particles depending on the energies of the incident particle and produced meson. The reaction in Eq. 5a is an inverse of the absorption reaction in Eq. 1, in which only a proton is emitted; but it leads to an excited nuclear target $(A + 1)$ (in general). The reaction in Eq. 5b is not the inverse of Eq. 2, however, since it is ${}_Y(A + 1)$ and not A that carries the strangeness; Eq. 5b represents "associated production" of K^+ $(S = 1)$ and Y $(S = -1)$.

7.1b Qualitative Aspects of Absorption

Some insight into the features particular to the absorption reactions in Eqs. 1 and 2 is obtained from consideration of the conservation of energy and momentum for these cases. In the lab frame, the target A is at rest initially; the meson brings in linear momentum **q** and energy $\omega_q = (q^2 + m^2)^{\frac{1}{2}}$, which are both absorbed by the target. The resulting target state A^* (or ${}_Y A^*$) recoils with momentum **q** and kinetic energy $E_R = q^2/(2MA)$. For moderate momenta ($q \lesssim$ few 100 MeV/c) and all but the lightest nuclear targets, $E_R \sim$ few MeV. For the K^- case, energy is required to transform a target nucleon to a hypernucleon: $\Delta E_Y = M(Y) - M(N)$ ($\sim$175 MeV for Λ, $\sim$250 MeV for Σ). The remaining energy goes into target excitation,

$$\Delta E^* = \omega_q - E_R - \Delta E_Y \quad \text{(for } K^-\text{).} \qquad\qquad [6a]$$

The angular momentum brought in is limited by the momentum and nuclear radius

$$J(A^*) \lesssim qR, \qquad\qquad [6b]$$

which could be as high as $10(\hbar)$ for $q \sim 300$ MeV/c on Pb, but will be less for many reactions.

The energy and angular momentum may be taken off by one or a few energetic nuclear particles, or by a large number of less energetic particles. The maximum number of nucleons emitted may be estimated, using the fact that it takes $\gtrsim 8$ MeV to remove a nucleon on the average. For mesons at rest, the maximum numbers are of order

$$\frac{\Delta E^*}{8 \text{ MeV}} \sim \frac{m_\pi}{8 \text{ MeV}} \sim 16\text{-}17 \quad (\text{for } \pi^-)$$

$$\sim \frac{m_K - \Delta E_Y}{8 \text{ MeV}} \sim 37\text{-}38 \quad (\text{for } K^-) \qquad [7]$$

Of course these limits could be reached only if all nucleons were emitted with about zero kinetic energy and if the excitation were quickly "thermalized" so as to be shared among all the nucleons, which is highly improbable.

How does the energy and momentum of the meson actually get distributed among the particles of the target nucleus? For example, can the meson be absorbed by the first nucleon it encounters, depositing all its energy and momentum on that nucleon, which then escapes the target? Clearly, if the nucleon were free, outside the nucleus, it could not absorb the meson, since it cannot absorb the initial energy. (This is clear in the c.m. frame: initial energy is $(\omega_q + M + q^2/2M)$ while the final energy is just the nucleon mass M.) Similarly, no absorption is possible by a Fermi gas nucleus (with no interactions), since the same plane wave kinematics applies here. What is required for absorption is the possibility of exciting the nucleon by ω_q by a transfer of momentum $\mathbf{q}$: for free nucleons, the energy transfer is too large. For nucleons bound in a finite nucleus, the states are no longer plane waves, and it becomes possible to match the excitation energy with the momentum transfer, as we can see in a simple example. Suppose a nucleon is initially bound in a potential, with energy ε_α. After absorption, it escapes in a state of momentum $\mathbf{p}$, and with energy given by

$$\mathscr{E}(p) = p^2/2M = \omega_q + \varepsilon_\alpha. \qquad [8]$$

Since the meson brings in momentum $\mathbf{q}$, the nuclear potential has to supply a momentum of $|\mathbf{p} - \mathbf{q}|$ to the nucleon. For a zero-energy pion, this amounts to $|\mathbf{p} - \mathbf{q}| = |\mathbf{p}| \cong 480$ MeV/c; for higher-energy pions, $|\mathbf{p} - \mathbf{q}|$ increases, although not rapidly for forward absorption: $\mathbf{p} \| \mathbf{q}$. This momentum transfer required is large for a single-particle potential of nuclear range, and is therefore not very probable. (In the limit of infinite nuclei, the probability goes to zero, as we saw above for the plane wave case.) However, there are other mechanisms (than a single nucleon potential) that can affect the matching of momenta and energy and therefore play a role in the dynamics of absorption.

One simple process is elastic scattering of the incoming meson by the nuclear target before absorption. In order to affect the matching conditions for absorption, the meson must scatter off its momentum shell, so that $\omega \neq \sqrt{q^2 + m^2}$, which can happen within the target. For example, if $q \to q' > q$ in such a scattering, so that $\omega_q < \sqrt{q'^2 + m^2}$, the momentum supplied to the nucleon becomes $|\mathbf{p} - \mathbf{q}'|$, which may be smaller and therefore more probable than in the no-scattering case.

Another simple process is *inelastic* scattering of the incoming meson. This transfers some energy to the target before absorption, leaving less than ω_q to be transferred in absorption, say, by a single nucleon. However, for on-shell scattering, the momentum is reduced along with the energy ($\omega_q' = \sqrt{q'^2 + m^2}$). This has a small effect on reducing $|\mathbf{p} - \mathbf{q}|$, but as in the elastic case, the matching is improved by scattering off-momentum-shell, $\omega_q' < \sqrt{q'^2 + m^2}$.

More complicated processes may also change the matching conditions for energy and momentum transfer: for example, interactions among the nucleons. For the initial nuclear target, these would correspond to ground state correlations, which can supply additional momentum to nucleons, thereby increasing the probability of absorption. Interactions of nucleons after absorption similarly may provide the momentum transfer required for absorption.

In Section 7.2 we show how to formulate these general notions. To summarize our qualitative view of the absorption process: Absorption of a free meson by a single bound nucleon, which then carries off all the energy and momentum, is not a very probable process. Any or all of a number of other interactions, elastic or inelastic scattering of the meson or of nucleons in the target, may take part in the absorption process. In general, these processes lead to several nuclear particles sharing the energy and momentum of the incoming meson; some or all of these particles may be emitted during the reaction.

Rather similar arguments apply to the sharing of energy and momentum for the inverse reactions in Eq. 5 of meson production by single nucleons on nuclei. The case of radiative absorption, given in Eq. 3, or the strangeness exchange reaction (K^-, π^-) in Eq. 4, are rather different from the kind of absorption just discussed, because of the extra particle (γ or π^-) released in the reaction, which can carry off both momentum and energy. In fact, both

$$\pi + N \to N + \gamma$$

and

$$K^- + N \to Y + \pi^- \qquad [9]$$

can take place for free nucleons.

7.1c Simple Pictures of Absorption

We here consider some highly oversimplified models of the absorption process, which focus on one or another of the dynamical aspect we have just discussed. A more quantitative treatment begins in Section 7.2.

The first picture is that of direct absorption by a cluster of nucleons in a nucleus. The cluster, denoted by C, might be any correlated subgroup of nucleons, which dynamically share the energy and momentum of the incoming meson (see Fig. 7.1). We then understand a reaction like $\pi + A \to B^* + X$ (Eq. 1) to be given by

$$A \to B^* + C, \qquad\qquad [10a]$$

$$\pi + C \to X. \qquad\qquad [10b]$$

The final target B^* is a "spectator" in the reaction, supplying only recoil momentum and binding energy to C; the dynamics of the absorption is given by Eq. 10b. If the cluster C has only a few particles, these will be emitted energetically in X. Further particles could be emitted by the spectator B^*, but these would be expected to be less energetic. If the cluster is identified with a real subnucleus—for example, d, t, ^{3}He, ^{4}He—then the direct picture (Eq. 10) leads to a kind of impulse model for absorption, in which the measured cross sections for $\pi + C \to X$ are used to obtain cross sections for absorption on A. For example, the absorption cross section integrated over kinematics in this approximation would be of the form

$$\sigma(\pi A \to B^* X) = \sigma(\pi C \to X) N_A(C + B^*), \qquad [11]$$

where $N_A(C + B^*)$ is the probability of decomposition, or a spectroscopic strength, corresponding to Eq. 10a.

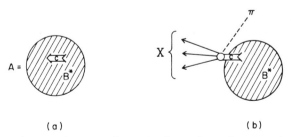

(a)   (b)

Fig. 7.1 Representation of meson absorption by a cluster in a target nucleus, as in Eq. 10: (*a*) cluster separation of target; (*b*) absorption.

This kind of model is appealing for its simplicity. The dynamics of absorption is not explicitly given, since the sharing of meson energy and momentum is essentially the same for the target as for the cluster C, with some kinematical correction for the motion of the cluster in the initial target. Both kinematical and dynamical effects of scattering of the meson by the nuclear target, or of the emitted particles in X by the final nucleus are ignored. These might be added to the picture in some way, using a very simple scattering model, as we discuss below. The extreme simplicity of this picture has kept it in use since the first studies of meson absorption in approximately 1948. There is experimental evidence that supports the picture in part, in that a large fraction of the energy is shared by a small number of nucleons, at least in the case of heavier nuclear targets. However, it is hard to make the picture work as a quantitative model. Aside from the multiple scattering problem already mentioned, there are several problems connected with the clusters. First, it is not clear that any one cluster should apply to all nuclei (or even to all states B* corresponding to one target A). There is some evidence that for stopped π^- absorbed by ^{12}C, the absorption is by ^{4}He clusters (e.g., Vai–64, 70, Cas–71) while for ^{14}N, slow π^+ are absorbed by ^{2}H clusters (Afa–65). For ^{7}Li, one has evidence of absorption both by ^{4}He and ^{2}H. For other nuclei—for example, odd A—^{3}H or ^{3}He may play a role. Second, clusters other than those corresponding to actual bound subsystems may contribute (e.g., neutron (or proton) pairs in a $^1S_0$ relative state). Third, the clusters may not behave in nuclei as they do in free space, invalidating an approximation like Eq. 11: np pairs in a nucleus need not look very much like ^{2}H (even when JT-quantum numbers correspond), and further, for example, the free deuteron is more diffuse than an np pair in a nucleus, with a mean separation of ~ 1.2 fm. The kinematics of absorption by a bound cluster differ in some respects from a free one, as we see further in Section 7.3.

The second picture of the absorption reaction is that of a statistical process. The target nucleus is usually treated as a Fermi gas of nucleons in a box or in a single-particle potential. The distribution of energy and momentum is accomplished by a statistical series of two-body collisions (cascade) treated as on-shell (i.e., elastic). This method allows the reaction to emit any number of particles energetically allowed. The problem is that such a picture does not allow for absorption *per se*, since this cannot take place in a Fermi gas as discussed earlier. Therefore some separate mechanism is invoked to imitate the cascade. For example, the first step might be absorption by a cluster, as in Eq. 10:

$$\pi + C \rightarrow X$$
$$X + B^* \rightarrow \text{cascade.} \qquad [12]$$

The collision of the X particles with the remaining Fermi-gas nucleons in B* proceeds until all energetic particles escape, leaving a bound residual target. The last steps could be "thermal" emission of particles from a nucleus at equilibrium.

An alternative scattering mechanism that allows for absorption in this picture, is excitation of resonances in meson-nucleon collisions, for example,

$$\pi + N \rightarrow \Delta, \qquad [13a]$$

which can proceed with free kinematics, conserving energy and momentum. The resonance can then scatter elastically or de-excite by collisions with target nucleons:

$$\Delta + N \rightarrow \Delta + N \qquad [13b]$$

or

$$\Delta + N \rightarrow N + N. \qquad [13c]$$

The last step completes the absorption of the π; the cascade then proceeds as in the previous case (see Har–73 and Gin–78).

The main problem with this picture is that because it is statistical, it loses most details of the absorption process and cannot give much idea of the cross sections to particular nuclear excited states (e.g., before the thermal emissions take over). However, the cascade process may well be a major part of the whole chain leading to stable final states, and the statistical picture may provide a method of understanding this part of the picture.

7.2 Transition Amplitudes: Microscopic Theory

We now put the descriptive ideas of the previous section into a more concrete theoretical form and examine in some detail the components of such a theoretical analysis. We use the theoretical framework of the coupled channels approach of Chapter 6, which was developed for the application to absorption and emission in Section 6.3.

We discuss, for the most part, the absorption reactions in Eqs. 1 and 2, which can be characterized by

$$m(\mathbf{q}) + A_i \rightarrow B_f + X_f \qquad [14]$$

for a meson of initial momentum q, absorbed by target nucleus A in state i, leaving a particle stable nucleus (or hypernucleus) B in state f, and one or many particles (or nuclear fragments) emitted in a wave state X_f. The transition amplitude

$$\langle B_f, X_f | T | q, A_i \rangle = (2\pi)^3 T_{fi}(q) \delta \big[K(B_f) + K(X_f) - q - K(A_i) \big] \quad [15]$$

leads to an expression for the partial cross section for absorption

$$d\sigma = \frac{2\pi}{v(q)} \Sigma \int |T_{fi}(q)|^2 \delta \big[E(B_f) + E(X_f) - E(A_i) - \omega_q \big] dn_f \quad [16]$$

where $v(q)$ is the c.m. velocity of the meson, and dn_f is the element of multidimensional phase space for the emitted particles in X_f—namely, $d\mathbf{k}/(2\pi)^3$ for each emitted particle* and a factor conserving momentum: $\delta[\Sigma p_{\text{final}} - q - K(A_i)]$. The sum and integral are over quantities not measured in a given partial cross section: spin projections, energies, and solid angles of emitted particles not detected.

The transition amplitude in Eq. 15 for absorption may be expressed as the matrix element of an effective absorption interaction $\mathcal{C}(\omega_1)$, using the coupled channels theory as developed in Section 6.3. We may rewrite Eq. 6.70 for the amplitude in Eq. 15 as

$$\langle B, X | T | q, A \rangle = \langle \psi_{B,X}^{(-)} | \mathcal{C}(\omega_q) | \psi_{q,A}^{(+)} \rangle. \quad [17]$$

The state $\psi_{q,A}^{(+)}$ represents the solution to the scattering problem for $m(q) + A \rightarrow m(q') + A^*$, which includes elastic and inelastic scattering of the incoming meson on the initial nuclear target, *without absorption* (as discussed in Chapter 6, see Eq. 6.62). Note that although Eq. 17 resembles the DWIA approximation in form, Eq. 17 is exact; the wave functions are, of course, much more complex than the optically distorted waves of the DWIA.

The final state $\psi_{B,X}^{(-)}$ is the solution to a scattering equation (Eq. 6.71) for the final nuclear products $X + B \rightarrow$ other *nuclear* states. There are no mesons (explicitly) in this channel.

The effective absorption operator $\mathcal{C}(\omega)$ is identical with $\mathcal{V}_{x1}(E)$ of Eq. 6.70 (with $\omega = E$). This is the effective interaction in the projected space (P) that connects the channels with one meson (scattering channels) to those with no mesons (absorption channels), as defined in Eqs 6.21 and 6.57:

$$\mathcal{C}(\omega) = \mathcal{V}_{x1}(E) = P_x H(E) P_1$$

$$= P_x H P_1 + P_x H Q (E^+ - H_{QQ})^{-1} Q H P_1. \quad [18]$$

*For plane waves normalized $\langle \mathbf{k}' | \mathbf{k} \rangle = (2\pi)^3 \delta(\mathbf{k}' - \mathbf{k})$.

A similar formulation applies to the meson production amplitude; for reactions like Eq. 5, which we write as

$$p + A_i \rightarrow m(\mathbf{q}) + B_f + X_f, \qquad [19]$$

where A_i, B_f, and X_f have the same meaning as in Eq. 14. The transition amplitude takes the form

$$\langle \mathbf{q}, B, X | T | \mathbf{p}, A \rangle = \langle \psi_{\mathbf{q},B,X}^{(-)} | \mathcal{Q}^\dagger(\omega_q) | \psi_{\mathbf{p},A}^{(+)} \rangle \qquad [20]$$

where $\psi^{(+)}$ now represents the no-meson channel, with an incoming proton of momentum $\mathbf{p}$ in the present case. The final state $\psi^{(-)}$ contains one meson, of asymptotic momentum $\mathbf{q}$. The effective production operator $\mathcal{Q}^\dagger(\omega)$ is simply the hermitian conjugate of $\mathcal{Q}(\omega)$ defined in Eq. 18. For the special case that the final *nuclear* state $(B + X)$ in Eq. 20 is the ground state of the nucleus $(A + 1)$, the amplitude in Eq. 20 is just the conjugate of that for the absorption reaction $m(\mathbf{q}) + (A + 1) \rightarrow A + p$.

We now discuss in detail how one may handle the several components $\psi^{(+)}$, $\psi^{(-)}$, and $\mathcal{Q}(\omega)$ that are involved in the transition amplitude. We note here that one could choose to formulate absorption in terms of Feynman amplitudes, or other covariant descriptions,* as discussed briefly in Section 6.3. Although there are some advantages to such an approach, particularly in dealing with Lorentz invariance, it does not seem suitable for dealing with the nuclear aspects of the problem, for which a nonrelativistic Schroedinger method seems more appropriate.

7.2a   Models for the Absorption Operator

We do not know a great deal about the absorption operator for mesons in a nuclear system; therefore, we begin with the simplest possibilities, consistent with what we do know. We generally assume that at the elementary level, there is an interaction that involves one nucleon (or baryon) and that induces transitions:

$$\pi + N \leftrightarrow N$$
$$K^- + N \leftrightarrow Y$$
$$K^+ + Y \leftrightarrow N. \qquad [21]$$

We first encountered this kind of interaction in connection with meson-nucleon scattering in Section 2.4. What little we know about this interaction comes partly from the interpretation of the Born contribution of this one-nucleon absorption (and emission) interaction.

*See also Sha–67.

316 Absorption and Production of Mesons

In Section 2.4 we discussed interactions of the Yukawa type, which can be described in terms of a Hamiltonian linear in the meson field (see Eq. 2.60)

$$H' = \int dx\, j(x)\phi(x) \qquad [22]$$

where $j(x)$ is a current operator for nucleons, a pseudoscalar function, as given, for example, for pions in Eqs. 2.61 and 2.62. We are not here directly concerned with H' but rather with the effective operator $\mathcal{C}(\omega_q)$, which depends on H' in a complicated way, as indicated in Eq. 18. This effective operator reflects the full interaction* of the meson with each nucleon, as well as the simultaneous interaction of the nucleons with each other.

A natural first approximation for $\mathcal{C}(\omega_q)$ is to neglect the many-body aspect, and consider only the renormalized Yukawa interaction, such as is appropriate for single nucleons (see Fig. 7.2). This assumption is analogous to making the impulse approximation for meson-nucleon scattering in the presence of other nucleons in a nucleus. In both cases, we expect there to be corrections due to the nuclear system; we come to these later. There is, of course, a major difference between the present approximation of $\mathcal{C}(\omega_q)$ by the single-nucleon term, and the impulse approximation for scattering; namely, for the latter we have the free scattering process on which to base our approximation, while absorption cannot occur for free nucleons. Thus even the single-nucleon term is accessible to us only indirectly.

We write the "impulse" approximation to the effective operator in the form (see Eq. 2.82)

$$\mathcal{C}(\omega_q) = \sum_i \int d\mathbf{k}\, N_k j^{(i)}(\mathbf{k}) a(\mathbf{k}) + \text{corrections} \qquad [23]$$

where $j^{(i)}(\mathbf{k})$ is an operator on the ith nucleon, corresponding to the absorption of a meson of momentum $\mathbf{k}$. (The operator for meson emission is given by the hermitian conjugate of Eq. 23.) The operator $\mathcal{C}(\omega_q)$ may be specified by giving the plane wave matrix elements of $j^{(i)}$ for a single nucleon, using Eqs. 2.63 and 2.82.

The simplest nonrelativistic form for $j^{(i)}$ was given in Eq. 2.65 for the approximation of static nucleons:

$$j^{(i)}(\mathbf{k}) = -(4\pi)^{\frac{1}{2}} \frac{f}{m} \boldsymbol{\sigma}^{(i)} \cdot \mathbf{k} \tau_\mu^{(i)} F(\mathbf{k}). \qquad [24]$$

*The coupled channel method of Chapter 6 is based on physical channels, that is, dressed, in the field theoretic sense. See the discussion following Eqs. 6.23 and 6.25.

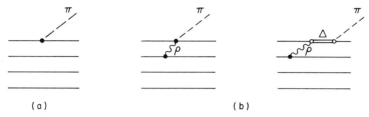

Fig. 7.2 Contributions to absorption operator $\mathcal{C}$: (a) impulse or one-nucleon term; (b) two-nucleon terms with ρ-meson exchange.

For the static approximation, the spin-isospin variables of the nucleon, but not the momenta, enter. This implies, of course, a preferred frame fixed in the nucleus in which the motion of the nucleons may be neglected. Such neglect may be a very poor approximation, particularly for small values of the meson momentum k, which may be smaller than the actual momenta of the nucleons in the nucleus.

The dependence of the static form in Eq. 24 on an absolute velocity (of the meson) would seem to be a theoretical shortcoming: we would expect only dependence on the relative (meson-nucleon) velocity, from considerations of Galilean relativity. In fact, we cannot apply this principle of relativity unambiguously to the case of absorption, because velocities are not properly defined for all the particles. This is because of the change of mass in absorption, which, as we have seen earlier, requires that at least one plane wave particle involved in the interaction be off-energy shell. We note that it is only the overall amplitude (Eq. 17), not any single element, such as Eq. 24, which need be invariant. In attempts, not completely justified, to recover some sort of Galilean invariance, some authors have treated the momentum **k** in Eq. 24 as the c.m. momentum for the initial meson + nucleon, taken as on-shell. Others, interpreting $\mathbf{v}_m = \mathbf{k}/M$ to be the meson velocity, calculate the nucleon plane wave matrix elements of Eq. 24, such that

$$\frac{1}{m}\langle \mathbf{p}'|\mathbf{k}|\mathbf{p}\rangle \rightarrow \frac{1}{m_{\text{red}}}\left[\mathbf{k} - \frac{m}{2M}(\mathbf{p}'+\mathbf{p})\right], \qquad [25]$$

which behaves like an "average" relative velocity of the meson and nucleon. Relative velocities are, of course, Galilei invariant, as discussed in Chapter 5, Eq. 5.23. Presumably the same approximations should be applied to the argument of the form factor $F(\mathbf{k})$.

The problems of relativistic invariance could certainly be more directly treated by starting with a Lorentz-invariant interaction. For example, in

Eq 2.62a we considered the pseudoscalar current of the form

$$j^{(i)}(\mathbf{x}) = ig_0 \tau_\mu^{(i)} \gamma_5. \qquad [26a]$$

In terms of plane wave matrix elements, we may write, following Eq. 2.67 and suppressing isospin,

$$\langle p' | j^{(i)}(0) | p \rangle = ig \langle p' | \gamma_5 | p \rangle = g\Gamma(p',p) \qquad [26b]$$

here $\Gamma(p',p)$ is a *vertex function* that plays a role in the relativistic theory analogous to the form factor $F(\mathbf{k})$ in the static theory (with $k = p' - p$). The difficulty with the covariant formulation is that it requires that the rest of the theory—multiple scattering and nuclear structure—be similarly formulated. Without a relativistic theory of the nucleus, we cannot immediately apply it to absorption.

One conventional method of recovering the nucleon velocity dependence of the interaction, for a nonrelativistic nuclear theory, starting from a Lorentz-invariant formulation, is the Foldy-Wouthuysen transformation, carried out to low orders in M^{-1}. There is a small literature on the application of this method to pion absorption and related questions of Galilean invariance* (e.g., Fri–74,77). Without going into detail, it is sufficient to remark that even to lowest order in the nucleon velocities, the form of the interaction is not unique, and further, that the form may depend on *nuclear* interactions.

To summarize: for the impulse approximation in Eq. 23 to the absorption operator, we do not really have enough detailed information about forms to specify the operator. The static form in Eq. 24 is often used, not only because it is simple, but because it does also appear in the Chew-Low theory of πN scattering, which has some successes of its own. For cases where nucleon motion cannot be neglected (as for mesic atoms, where the meson is almost at rest in the nucleus) some approximate dependence on nucleon velocities, as in Eq. 25, has been introduced. Relativistic forms have not generally been used, except for the two-nucleon case—for example, $\pi^+ d \leftrightarrow pp$ (Del–77) and for $A(p,\pi)A'$ (Mil–78).

The inadequacies or ambiguities of the impulse approximation are not really the central problem in determining the absorption operator, however. The "corrections" to Eq. 23 include all the effects of the nuclear medium during absorption, which may well be as important as velocity effects. These corrections play a role in absorption analogous to that of "exchange currents" in the theory of electromagnetic interaction with

*See also the discussion at the end of Section 6.1c.

nuclei.* The physical origins of both involve many-nucleon interactions, through mesonic exchange. Little of this has been worked out for the case of meson absorption, so we do not discuss it further. Without a more complete theory of the absorption operator, including its many-body components, one may resort to the simpler one-body operator forms in Eqs. 24 and 25 as *phenomenological*, perhaps with strengths and form factors that depend on the "average" nuclear physics. This would be akin to the use of "effective charges" for nuclear EM transitions.

7.2b Incoming Wave

We return now to the absorption amplitude of Eq. 17 and discuss the state $\psi_{\mathbf{q},A}^{(+)}$, which describes the scattering of the incoming meson, in an equation that excludes the possibility of absorption. This scattering problem is, in principle, the same as we have discussed in Chapter 4, namely, one meson scattering from a complex nuclear target with Schroedinger dynamics. (We use an *effective* Schroedinger equation for this channel, as in Eq. 6.59.) Therefore, we have at our disposal the formal apparatus of multiple scattering theory (with possible modification introduced in Section 6.3.). However, we do not find it necessarily useful to follow the order of approximations followed in Chapter 4, particularly the coherent approximation that heavily stresses the role of the nuclear ground state in meson scattering and that figures heavily in the Watson development of the optical potential. The development of Chapter 4 is particularly suited for the treatment of elastic scattering and for a number of direct nuclear reactions in which the ground state plays a special role (discussed in Chapter 5).

For the case of absorption, two distinct differences should be noted. First, as we have discussed in Section 7.1, the kinematics of absorption requires transfer of momentum and energy in such a way that *inelastic* scattering of the incoming meson may contribute at least as significantly as *elastic* scattering. That is, the components of $\psi_{\mathbf{q},A}^{(+)}$ that contribute to the amplitude in Eq. 17 may well include considerable excitation of the nuclear target. Second, if this is the case, the optical potential is probably not appropriate for the treatment of the incoming wave. The optical potential method keeps explicitly only the elastic component of the meson-nuclear scattering wave; all inelastic components are removed and accounted for as lost probability flux. But this "lost flux" may well be important in absorption, and should therefore not be suppressed from the beginning.

*See Chapter 8.

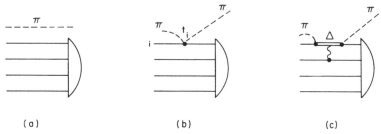

Fig. 7.3 Diagrams representing the initial state in the absorption process, as in Eq. 27: (*a*) plane wave meson; (*b*) single scattering; (*c*) Δ-propagation.

With this in mind, let us return to a formal expression for the incoming wave (see Eq. 6.62):

$$|\psi_{\mathbf{q},A}^{(+)}\rangle = (1 + G_N(E)T_1(E))|\mathbf{q},0\rangle, \qquad [27]$$

where $|\mathbf{q},0\rangle$ is the initial asymptotic state with a plane-wave meson and nuclear ground state, $T_1(E)$ is the meson-nucleus scattering amplitude (without meson absorption), and $G_N(E)$ is the Green function for a free meson and the nuclear target in Eq. 6.51.

Now, the simplest approximation to Eq. 27 is to keep only the plane wave (see Fig. 7.3*a*):

$$|\psi_{\mathbf{q},A}^{(+)}\rangle \simeq |\mathbf{q},0\rangle. \qquad [28]$$

Although we do not expect this to be a good approximation, for reasons already discussed, the plane-wave term alone can contribute significantly to the absorption amplitude in Eq. 17. First, the nuclear target ground state may have correlations among the nucleons. In terms of the qualitative discussion of Section 7.1, the correlations supply high-momentum nucleons, which are more suitable for absorbing mesons, in terms of the matching conditions. Second, some of the required momentum and energy transfer may be supplied by interactions in the final state.

The next stage of development is to expand $T_1(E)$ in Eq. 27 in a multiple scattering series, as in Eq. 4.55. Keeping just the first (single scattering) term, $T_1 = \sum_i t_i$, allows one scattering before absorption (see Fig. 7.3 *b*):

$$|\psi_{\mathbf{q},A}^{(+)}\rangle \simeq \left[1 + G_N(E)\sum_{i=1}^{A} t_i(E)\right]|\mathbf{q},0\rangle. \qquad [29]$$

This is about as far as the multiple scattering has been taken (to date) for detailed calculations of nuclear absorption (except for the coherent or optical approximation, which we discuss below). The single-scattering term was first introduced in the study of the $\pi^+ d \to pp$, where it was found to be important over a large range of π energy. The scattering provides a momentum transfer to the struck nucleon and to the incoming pion, which then carries that momentum to a second nucleon, by which the meson is absorbed, for example through the absorption operator in Eq. 23. This provides a way to share the transfer of the initial pion energy between the two nucleons and has been thought to be important in the reaction $\pi + A \to B + N_1 + N_2$ in which two fast nucleons are emitted, carrying away most of the initial energy (see Section 7.3).

[A technical matter: In the process just described, the scattering takes place on one nucleon and is absorbed on another. In the calculation of the scattering amplitude in Eq. 17, if one uses Eqs. 23 and 29, the two sums over nucleon number should be restricted, that is, $i \neq j$. This is consistent with the treatment of the current operator $j^{(i)}$ of Eq. 23 as fully renormalized: scattering before absorption on a *single* nucleon is already included in $j^{(i)}$. In general, if T_1 is expanded in a multiple scattering series and grouped in terms of the nucleon last struck, $T_1 = \Sigma_j T^{(j)}$, then the absorption amplitude is calculated (using Eqs. 23 and 29) as

$$\mathcal{C}(\omega_q)|\psi_{q,A}^{(+)}\rangle \simeq \sum_{i \neq j} \int d\mathbf{k}\, j^{(i)}(\mathbf{k}) a(\mathbf{k}) \left[1 + G_0(E) T_1^{(j)}(E) \right] |\mathbf{q},0\rangle.] \quad [30]$$

For very low energy pions, for which the single πN scattering is relatively weak, the single-scattering approximation in Eq. 29 may be valid, and it has, in fact, been applied to absorption of stopped π^- in light nuclei (Kol–68, Kop–71, Par–71). However, at higher energies, for which t_i is not a small amplitude, Eq. 29 may not be adequate. For a deuteron target, the full multiple scattering series for $T^{(j)}$ (Eq. 30) may be summed by solution of the (Faddeev) scattering problem for $\pi + d$, as discussed in Chapter 3 (see Eq. 3.103). For complex nucleus, the complete series cannot be given exactly, so an approximate method of summation is desirable.

It might seem useful to approximate the incoming wave by the solution for the optical potential—the potential obtained by methods discussed in Chapter 4 (e.g., coherent approximation). This raises two objections, already mentioned earlier. The first is the optical potential wave has flux diminishing inside the target (see Section 4.1) to account for inelastic processes. This removal of flux is inappropriate here, since the inelastically scattered meson may still be absorbed. Therefore, the plane wave approximation (Eq. 28) may be better than the optical wave approximation.

The second is that the inelastically scattered wave may be absorbed *more strongly* than the elastic, as mentioned in Section 7.1 because of the improved momentum-energy matching conditions. For this reason (as well as that of not removing flux) the inelastic components should also be kept in the incoming wave. In this sense Eq. 29 could well be a better approximation than is the optical wave.

The wave function of the fixed-scatterer approximation (Sections 4.1 and 4.3d) does not have these objections, since the full flux is kept including inelastic waves (but only in the closure approximation—all target states degenerate). This wave function has to be expressed in the position space of all the nucleons, as well as of the meson (Eqs. 4.42 and 4.181):

$$\langle \mathbf{r}_1, \ldots, \mathbf{r}_A; \mathbf{r} | \psi_{\mathbf{q},A}^{(+)} \rangle = \Phi_0(\mathbf{r}_1, \ldots, \mathbf{r}_A) \psi_\mathbf{q}^{(+)}(\mathbf{r}), \qquad [31]$$

where the meson wave $\psi_\mathbf{q}^{(+)}(\mathbf{r})$ depends implicitly on the nucleon positions. The absorption amplitude would then have to be calculated as an integral over all coordinates: the final state wave functions must also be given in terms of nucleon positions. Since the fixed scatterer approximation has been applied to elastic scattering from light nuclei ($A \lesssim 16$), it should be possible in principle to perform such a calculation for absorption (e.g., Gib-76).

Another way of handling inelastic scattering before absorption is in a statistical model; we discuss this further in Section 7.2d.

7.2c Final Nuclear States

The final states $\psi_{B,X}^{(-)}$ for absorption provide many more possibilities than does the initial state $\psi_{\mathbf{q},A}^{(+)}$. First of all, there are many different final channels possible, as we have discussed in Section 7.1, each corresponding to a different partition of the final nuclear system into a daughter nucleus B* in a particular state and emitted particles or fragments X. Second, we may be interested in less than complete information about the final channels, and therefore some details may not be important. For example, if we want to calculate the total absorption cross section, we sum over all final channels; presumably much simpler approximations will obtain than for the rate leading to any specific channel. Similarly, if we are interested in, say, the specific reaction $A(\pi^+, 2p)B^*$, we may not be interested in the possibility that the daughter state B^* be particle unstable with further (slow) emissions. So we would treat the final state as a three-body system (asymptotically): $p_1 + p_2 + B^*$, even though further division into more complicated subchannels might be possible.

The simplest approximation for any final state is a product of plane waves for each of the outgoing particles in X and a final bound state B*. The plane waves depend, of course, on the channel; if any emitted particles are complex $(d, t, \alpha, \text{etc.})$, an internal wave function for that fragment is also included. Then we write, for example,

$$\langle \psi_{B,X}^{(-)}| \simeq \langle \mathbf{p}_1, B^*| \qquad \text{for } A(\pi, N)B^*$$
$$\simeq \langle \mathbf{p}_1, \mathbf{p}_2, B^*| \qquad \text{for } A(\pi, 2N)B^*$$
$$\simeq \langle \mathbf{p}_1, \mathbf{p}_2, \phi_t, B^*| \quad \text{for } A(\pi^-, nt)B^*, \qquad \qquad [32]$$

and so forth, where ϕ_t is the triton ground state function.

The plane wave approximation (Eq. 32) omits any correlation among the outgoing particles, or between any of these and nucleons in the final target. This omission may neglect dynamical effects of some importance, since these correlations reflect interactions that supply energy and momentum transfer, which makes absorption possible. The missing final state interactions can be included in a number of approximations, which rapidly become quite complicated. For example, in the case of $A(\pi, 2N)B^*$, the correlations of the two emitted nucleons may be included relatively easily in the form

$$\langle \psi_{B,2N}^{(-)}| \simeq \langle \psi_p^{(-)}P, B^*|, \qquad \qquad [33]$$

where $\psi_p^{(-)}$ is the relative wave function for two scattering nucleons with ingoing scattering boundary conditions, and asymptotic relative momentum $\mathbf{p} = \mathbf{p}_1 - \mathbf{p}_2$. The c.m. of the pair moves as a plane wave of momentum $\mathbf{P} = \frac{1}{2}(\mathbf{p}_1 + \mathbf{p}_2)$. In this case, there is no account taken of correlations with the target nucleus. To add these means to pass to a three-body scattering description of the final state. This is a formidable problem to treat fully, even with simplified (e.g., separable) interactions, since many partial waves will in general be involved. (See, however, Mor–73 and Gar–74.)

In some cases there may be strong coupling among the final channels, which would be difficult to handle completely, since each channel may involve many particles. Some very simplified models of the final state interactions, as given, for example, by a statistical calculation, are discussed in the next section. It may well make sense in such a treatment to introduce an intermediate state (or states) that has sufficient structure for the absorption reaction, for example, in terms of short range correlations. Then any particular final channel f can be reached from this intermediate state $\psi_n^{(-)}$, by further scattering transitions $\langle f|T'|\psi_n^{(-)}\rangle$, and

$$\langle \psi_f^{(-)}| \simeq \langle f|T'|\psi_n^{(-)}\rangle \frac{1}{E^+ - E_n + i\eta} \langle \psi_n^{(-)}| \qquad [34]$$

(with a possible sum over n). The $n\rightarrow f$ transitions could be treated, for example, statistically.

7.2d Other Forms—Simple Approximations

The preceding discussion in Sections 7.2a, b, and c has stressed the separation of the transition amplitude into three parts, following the formulation introduced in Section 6.3. This division is motivated by the desire to treat separately the multiple-scattering of the incoming meson, the absorption process itself, and the interactions in the final state. It may be more useful, for some purposes, to treat the absorption amplitude with a different separation of parts, or under a different kind of approximation. This might be suitable for particular final states, or for a simplified model of meson scattering, or for a less detailed calculation of the absorption process.

1. For the particular reaction $\pi + A \rightarrow B^* + 2N$, the following simplifying assumptions may be considered. A pair of nucleons is emitted: It is reasonable to assume that the incoming pion was absorbed, after scattering, on one of those two nucleons. Further, one might expect that if momentum transfer is supplied by the nuclear dynamics, it comes largely from interactions between the same nucleon pair—from short-range (or tensor) correlations in the initial target state, or from interaction of the pair in the final state, as in Eq. 33. This leads one to consider a pair-absorption approximation, in which the two nucleon interactions, *and* the absorption interaction, *and* any scattering of the incoming meson between the pair, are collected into one term, a two-nucleon modified absorption operator a_{ij}, such that the amplitude in Eq. 17 may be written as

$$\langle B, k_1 k_2|T|q, A\rangle \simeq \sum_{i \neq j} \langle B, k_i, k_j|a_{ij}|q, \tilde{A}\rangle, \qquad [35]$$

where the initial state is a plane wave pion, with a simplified target ground state $\tilde{A}$, with no short-range correlations included (e.g., a shell-model state). Similarly, the final state has two plane wave nucleons and the residual target nucleus B, for example, in a two-hole shell-model state. The effective absorption operator a_{ij} might be given in a simple parametric

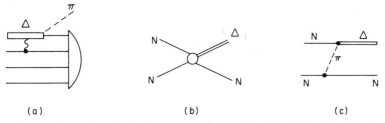

(a)   (b)   (c)

Fig. 7.4 Pion absorption in the Δ-propagation model: (a) initial state with Δ + hole in target; (b) ΔN→NN vertex; (c) one-pion exchange contributions to (b).

form, for example,

$$a_{ij} \sim a(q)\,\delta(\mathbf{r}_i - \mathbf{r}_j) \times f(\sigma_i, \sigma_j, \tau_i, \tau_j) \qquad [36]$$

or might be calculated separately. Such a method could be useful for comparing the same reaction on different targets. A similar treatment could also be developed, for example, for quartets of nucleons ("α-clusters")*.

2. For π-nucleus scattering in the 3,3 resonant energy region, we may wish to use some version of the model of a Δ propagating in the nucleus, as discussed in Section 6.4. This suggests a different separation of the absorption amplitude (Eq. 17), since now the initial state may have a Δ present (see Fig. 7.4a). The absorption process might then be described as an inelastic collision: Δ + N→N + N (Fig. 7.4b). In fact this transition is often treated in a one-pion-exchange (OPE) model (Fig. 7.4c); in this case the propagation-Δ formulation does not differ greatly from the treatment of Sections 7.2a, b, and c, with the Δ included in the πN scattering.

3. *Statistical Calculation.* This approach offers a shortcut to the rather involved and detailed handling of scattering and correlations in the previous sections. Such calculations are generally too crude for treating specific reaction modes but may be useful for gross estimates, as for integrated or total cross sections. The formulation differs from that given above, first in that the quantities dealt with are probabilities, not amplitudes; it is $|T_{fi}|^2$ (Eq. 16) that is calculated. Second, some very simplified nuclear dynamics is introduced: a Fermi gas model or schematic particle-hole ("exciton") model. Third, the scattering of mesons, or particles excited by the mesons, can include inelastic (nuclear) scattering, but will always be on-shell, that

---

*The "pole-approximation" methods of Shapiro (Sha–65,67, Kol–66) are closely related to this approach.

is, energy conserving. So effects that depend on off-shell propagation must be neglected.

On the other hand, it is possible to keep many channels simultaneously both before and after absorption, in which, for example, any number of nucleons is allowed to escape. In fact, the usual methods apply best, in some sense, to the treatment of final channels. There is no way of including the one-nucleon absorption vertex in the method, since only energy-conserving interactions are allowed. Therefore, it is necessary to introduce at least a two-body absorption process, which could be a pair-operator like Eq. 36, or could involve $\Delta + N \rightarrow N + N$, as in part 2 above, if Δs are produced (statistically) in earlier stages of the process by $\pi + N \rightarrow \Delta$ (on-shell).

It seems reasonable to apply a statistical method to calculate only the final channel processes in the sense of Eq. 34. We write

$$|T_{fi}|^2 = \sum_n |\langle f|T'|n\rangle|^2 G_n^2(E) |\langle n|T|\mathbf{q},A\rangle|^2 \qquad [37]$$

where the probability $|T|^2$ for absorption $A \rightarrow n$ may be treated by the methods of Section 7.2a, b, and c, above. The intermediate states n might be all two-nucleon plus two-hole (in target) states. The propagation $G_n^2(E)$ and final scattering $|T'|^2$ are then treated statistically. This represents a *cascade* initiated by a specific, perhaps direct, reaction.

New methods have been introduced recently for treating such statistical processes in the formulation of *transport* equations. Applications to high energy nuclear reactions have been developed by Aga–75c, 77, Rem–75, and Huf–78, and may well be useful for pion-induced reactions as well.

7.3 Applications to Reactions

We now discuss specific aspects of a few particular absorption reactions with nuclear targets. We stress types of reactions for which experiments have been or can easily be done. We concentrate more on those cases for which the microscopic theory, as developed in Section 7.2, may apply. We also see in the course of the discussion what interest there may be in these reactions with respect to nuclear structure information, such as hole states, correlations, or clusters in nuclei. With these interests in mind, the cases we discuss are *direct reactions* with *few* particle emission, where the emitted particles are directly measured. We do not deal therefore with the statistical aspects of absorption reactions.

7.3a One-Nucleon Reactions: (π, N) and (N, π)

This kind of reaction is relatively easy to study experimentally, since there is one particle in, one particle out. Normally these are charged particles, as in (π^+, p) or (p, π^+). It has often been easier to study the production reactions (p, π^+) than the absorption (π^+, p) because of the greater intensity of proton beams. However, there is considerable similarity between the two reactions, when the final nuclear states are particle stable. (For the case of ground state to ground state transitions, the (p, π^+) and (π^+, p) reactions are exactly related by "detailed balance," or time reversal.) The differential cross sections for the bound-state reactions are quite small, typically 1–100 nb/sr for (p, π^+). For final nuclear states that can emit nuclear particles, the cross sections are higher, but the reaction kinematics may be quite different than in the bound state case. The reaction $(p, \pi^+ X)$ to unbound states is the conventional reaction for producing pions in an accelerator. The reaction (π, pX) is not necessarily a close relative of (π, p), unless the former reaction goes through a well-defined (resonant) intermediate state that subsequently emits the particle(s) X.

There are by now experimental measurements of differential cross sections for (p, π^+) to bound states for a number of nuclear targets, for $T_p \cong 185$ MeV. An example is illustrated in Fig. 7.5 (Dah–73).

The one-nucleon reactions with charge exchange (π^-, p) and (p, π^-) are also interesting, although less common (lower cross section), because they must involve more than just the simple absorption interaction $\pi + N \leftrightarrow N$, which can only accomplish one unit of charge exchange: $\pi^- + p \leftrightarrow n$ or $\pi^+ + n \leftrightarrow p$.

The simplest approach to a microscopic description of the one-nucleon reactions is in terms of the Distorted Wave Born Approximation (DWBA). (The Plane Wave Born Approximation is even simpler, and is discussed below.) The transition amplitude in Eq. 17 takes the form [for (π^+, p)]

$$\langle B, \mathbf{p} | T | \mathbf{q}, A \rangle = \langle \psi_{\mathbf{p}}^{(-)}, B | \mathcal{C}(\omega_q) | \psi_{\mathbf{q}}^{(+)}, A \rangle, \qquad [38]$$

where B denotes a bound state of the final nucleus [with $(A-1)$ nucleons] and $\mathbf{p}$ is the asymptotic value of the momentum of the outgoing proton. The approximation in Eq. 38 states that in the initial scattering state, the target remains in its ground state (A) and therefore the incoming pion wave is the *coherent* wave, given by the optical potential for elastic scattering of the pion on the target (see Chapter 4). In the final scattering state, the nucleus remains in its final state, while the proton leaves in a coherent wave $\psi_{\mathbf{p}}^{(-)}$, given by an optical potential appropriate to "elastic" scattering of a proton of the same energy on the (excited) final nucleus B.

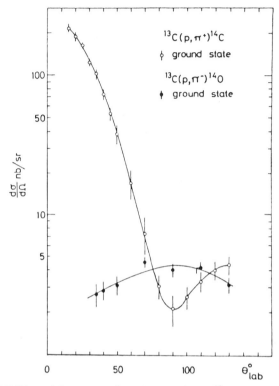

Fig. 7.5 Differential cross sections for $(p, \pi^{\pm})$ on ^{13}C (from Dah-73).

Again taking the simplest approach, it is consistent to use the impulse approximation for the absorption operator $\mathcal{Q}(\omega_q)$ as given in Eq. 23, with, for example, one of the specific forms for the nucleon operator $j^{(i)}(\mathbf{k})$ given in Eqs. 24–26. The absorption operator takes a neutron of momentum $\mathbf{l}$ in a target A into a proton of momentum $\mathbf{l}+\mathbf{k}$, where $\mathbf{k}$ is the momentum of the (scattered) pion at absorption. Then Eq. 38 may be written

$$\langle B, \mathbf{p} | T | \mathbf{q}, A \rangle \simeq \int \frac{d\mathbf{l}\, d\mathbf{k}}{(2\pi)^6} \psi_{\mathbf{p}}^{*(-)}(\mathbf{l}+\mathbf{k}) \langle \mathbf{l}+\mathbf{k} | j(\mathbf{k}) | \mathbf{l} \rangle \langle B, \mathbf{l} | A \rangle \psi_{\mathbf{q}}^{(+)}(\mathbf{k})$$

$$[39]$$

where $\psi^{(-)}$ and $\psi^{(+)}$ are the scattering (distorted) wave functions for the proton and pion, respectively, $\langle \mathbf{l}+\mathbf{k} | j(\mathbf{k}) | \mathbf{l} \rangle$ is the plane wave matrix element of the current for one nucleon, and $\langle B, \mathbf{l} | A \rangle$ is the overlap (integral) of the initial target state with the final target plus a plane wave proton (of momentum **l**). If the nucleons are described in terms of single particle wave functions $u_\alpha(\mathbf{l})$ (orbitals—in momentum), the overlap may be written in the form

$$\langle B, \mathbf{l} | A \rangle = \sum_\alpha \mathcal{S}_\alpha(B, A) u_\alpha(\mathbf{l}), \qquad [40]$$

where $\mathcal{S}_\alpha$ is the *spectroscopic factor* for the (hole) orbital α.

In the DWB approximation, the (π^+, p) process is similar to a number of one-nucleon transfer reactions of nuclear physics, for example, the deuteron "pickup" reaction, $A(n, d)B$, with the pion in the former playing the role of the neutron in the latter, and the (γ, p) reaction. The overlap integral in Eq. 40 appears similarly in all one-nucleon transfer reactions (in DWBA) and contains the nuclear structure information; the other factors in the amplitude in Eq. 39 depend on the specific properties of the projectile. We illustrate the DWBA picture of (π^+, p) in Fig. 7.6a.

If we pass to the plane wave limit, Eq. 39 may be further simplified, since now $\psi_\mathbf{q}^{(+)}(\mathbf{k}) \rightarrow (2\pi)^3 \delta(\mathbf{k} - \mathbf{q})$, $\psi_\mathbf{p}^{(-)}(\mathbf{l} + \mathbf{k}) \rightarrow (2\pi)^3 \delta(\mathbf{l} + \mathbf{k} - \mathbf{p})$, and

$$\langle B, \mathbf{p} | T | \mathbf{q}, A \rangle \rightarrow \langle \mathbf{p} | j(\mathbf{q}) | \mathbf{p} - \mathbf{q} \rangle \langle B, \mathbf{p} - \mathbf{q} | A \rangle, \qquad [41]$$

up to a constant factor (see Fig. 7.6b). In this limit, the momentum transfer $(\mathbf{p} - \mathbf{q})$ must come from the initial nucleon orbital $u_\alpha(\mathbf{l})$ with $\mathbf{l} = \mathbf{p} - \mathbf{q}$. (This approximation neglects the orthogonality of the final proton wave to the bound orbital state.) Now **p** and **q** are fixed by kinematic considerations of conservation of energy, as discussed in Section 7.1 above in connection with Eq. 8. The transfer $|\mathbf{p} - \mathbf{q}|$ is never less than about 450 MeV/c, which

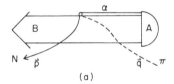

(a)

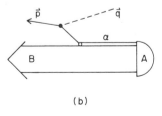

(b)

Fig. 7.6 Representation of $A(\pi^+, p)B$ reaction: (*a*) distorted wave approximation; (*b*) plane wave approximation.

is larger than typical target nucleon momenta, and it can be considerably higher for $|\mathbf{q}| > 0$ at backward angles. (For nuclear matter, $k_F \simeq 1.4$ fm$^{-1} \simeq$ 280 MeV/c). For these high momenta, the orbitals $u_\alpha(\mathbf{l})$ are expected to be monotonically decreasing functions of $|\mathbf{l}|$ with amplitudes much smaller than for low momenta, for example, for $|\mathbf{l}| \lesssim 150$ MeV/c. It is not surprising, from this consideration above, that the cross sections for (π^+, p) to bound states are rather small (microbarns). However, the plane-wave limit here appears to be a very poor approximation to the DWBA, just because of the very high momentum transfer required. The inclusion of distortion of the π^+ and p waves can make considerable difference in the full amplitude in Eq. 39.

The optical distortions have two major effects on the π and p waves: momentum transfer and (optical) absorption (loss of flux). The first might be expected to *increase* the reaction amplitude by improving, on the average, the momentum matching in Eq. 39. The overlap in Eq. 40 will be increased by decreasing $|\mathbf{l}|$, which may be accomplished by lowering $|\mathbf{l} + \mathbf{k}|$ relative to $|\mathbf{p}|$, or by increasing $|\mathbf{k}|$ relative to $|\mathbf{q}|$. However, if orthogonality of final and initial nucleon states is maintained, the effect may not be so simple (Nob–78, Eis–79). The optical distortions will also affect the angular distribution of the (π, p) amplitude.

The effect of absorption in the optical distortion is complicated: one would expect a lowering rate due to this effect, but absorption also contributes to momentum transfer, which may work in the opposite direction. One may also question whether optical absorption should be included in the distorted waves: May not the excluded flux actually contribute to the (π, p) reaction through inelastic scattering? (See the discussion in Section 7.2 following Eq. 28 and the related discussion in Section 5.4) This question may well have different answers for different combinations of initial and final nuclear target states (A, B). Transitions with small spectroscopic factors $S_\alpha(B, A)$, such as in Eq. 40, may well be dominated by multistep processes, as in other one-nucleon transfer reactions.

There have been a number of calculations of the (π^+, p) or (p, π^+) reactions in the DWBA approximation (see, e.g. Nob–76 for references). These show that the theoretical results are indeed sensitive to assumptions about the form of the orbital $u_\alpha(\mathbf{l})$, as well as of the optical potentials— particularly of the pion potential, which may have large effects from momentum dependence (see Chapter 5, the Kisslinger form, p. 196). There is also sensitivity to the assumed form of the absorption operator $j(\mathbf{k})$. This last is more of interest for low pion energy, since then the dependence of j on the nucleon motion becomes important, as we discussed in Section 7.2a. For example, for the nonrelativistic form in Eq. 24 we discussed the use of "relative velocity" (Eq. 25) to give "Galilean invariance" to the operator. For low pion energies, $T_\pi \sim 30$–40 MeV (or proton energies $T_p \sim 180$ MeV), it turns out that at $0°$, the "velocities" of the pion and proton are similar, that is, $q/m \sim p/M$ (for plane-wave momenta). The relative velocity is almost zero therefore, and the PWBA amplitude is anomalously small. However, for DWBA the momenta are considerably spread out, and the cancellation is much reduced. Similarly, using relativistic forms for the current operator, one may find considerable sensitivity to assumptions about the (relativistic) momentum dependence of the nucleon wave functions (Mil–78, Nob–78). However, it seems likely that such effects, although present, are obscured by the optical distortion, as well as by further corrections to be discussed, to an extent that they cannot be directly looked for in the (π^+, p) reaction.

Let us return to the question of adequacy of the DWBA in Eqs. 38 and 39 for the (π^+, p) reaction. In Fig. 7.7 we illustrate some of the multiple scattering effects, to the next order of approximation, that could compete with the DWBA mechanism. In Fig. 7.7*a* the incoming pion collides with the bound nucleon, ejecting it from the target; the pion is absorbed by the residual target, leaving it in the final state B. This process could be treated by modifying the absorption operator $\mathcal{Q}(\omega_q)$ in Eq. 38, replacing the single-nucleon current $j(\mathbf{k})$ by a two-nucleon operator, which represents the scattering on one nucleon and absorption by a second (with the pion propagating off-energy shell in between). If one restricts the intermediate $(A-1)$-nucleon state C, to be the final state C = B, then one can still follow the DWBA method in Eq. 38 with the new absorption operator replacing $j(\mathbf{k})$. If one does not restrict intermediate states (C ≠ B), the contribution of Fig. 7.7*a* takes a more complicated form.

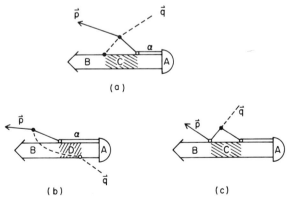

Fig. 7.7 Multiple scattering effects in $A(\pi^+, p)B$ reaction.

More multiple scattering terms are illustrated in Figs. 7.7*b* and *c*, in which the initial or final particles scatter from the initial or final target nuclei. For Fig. 7.7*b*, if the intermediate nuclear state D = A (initial target ground state), this effect is included in the pion optical potential scattering in the DWBA in Eqs. 38 and 39. Similarly, for Fig. 7.7*c*, if the intermediate (A − 1)-nucleon state C = B, the effect is included in the outgoing proton distorted wave in Eqs. 38 and 39. However, with D≠A, or C≠B, new effects are introduced. Given that the single-nucleon transfer process in DWBA is limited by the large momentum transfers required, there is no reason to ignore the contribution of multinucleon, multistep processes, like those of Fig. 7.7.

The most popular method of treating the two-step process of Fig. 7.7 has been in a two-nucleon model, in which it is assumed that for all three processes the scattering involves only one nucleon (as in the impulse approximation) and the absorption one more nucleon, so that the combined effect is given by a two-nucleon effective absorption operator, as shown in Fig. 7.8*a*. The contributions to this would include pion scattering (Fig. 7.8*b*), in which the scattering nucleon (1) is ejected, as in Fig. 7.7*a*, or the absorbing nucleon (2) is ejected, as in Fig. 7.7*b*. The interaction of the outgoing nucleon in Fig. 7.7*c* is given by a diagram like Fig. 7.8*c*, with a nucleon-nucleon interaction after absorption. Such two-nucleon models have been introduced for (π^+,p) either treating the effective interaction of Fig. 7.8*a* phenomenologically (Rei–72), or calculating the individual terms in Fig. 7.8*b* and *c* in a microscopic model (Gro–74b). For the former case, the effective absorption operator is taken directly from the on-shell amplitudes for $\pi + NN \leftrightarrow NN$, where they are available from $pp \rightarrow pp\pi^0$ and $pn\pi^+$ —a kind of two-nucleon "impulse" approximation.

On the one hand, because the two-nucleon models have more sharing of momentum built into the mechanism through the scatterings of Fig. 7.8*b*

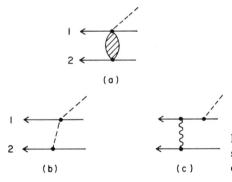

(a)

(b)   (c)

Fig. 7.8 (*a*) Two-nucleon effective absorption operator, with two possible contributions: (*b*) and (*c*).

and c, there is less sensitivity to details of the momentum distribution of the nucleus through the orbital wave functions or the current operator $j(\mathbf{k})$, such as are important in the one-nucleon DWBA model. However, there is more to be calculated in the two-nucleon models, and they have not been pushed as far. It is interesting to note that the two-nucleon models (e.g., Gro–74b) can give the order-of-magnitude, and to some extent the shape of the angular distribution, of some of the experimental (π^+,p) or (p,π^+) data. In this, they are not superior to the one-nucleon transfer DWBA calculations; but they do appear to demonstrate the role of the multistep processes. To date, there has been no calculation of (π^+,p) that has tried to incorporate both one-nucleon and two-nucleon effects. This is not a trivial extension, since there is some overlap of effects of the optical distortion with the two-body scattering, as noted above, with regard to Fig. 7.7b and c. It is quite likely that the relative role of the processes differs from case to case, depending on the initial and final target states A and B.

Finally, it is interesting to consider the charge exchange absorption reaction (π^-,p). This reaction has been seen, as has its inverse, (p,π^-); the cross section for (p,π^-), $\sigma(\theta)$ for small θ, or integrated σ, is smaller than that for (p,π^+) by about $\frac{1}{50}$ [for ^{9}Be, $T_p \sim 185$ MeV (Nob–76)]. See Fig. 7.5 for ^{13}C$(p,\pi^-)^{14}$O. The single-nucleon model in PWIA, or in DWIA with only elastic distortion, gives no possibility for this reaction. However, (π^-,n) followed by *charge-exchange scattering* (n,p) of the outgoing neutron, or preceded by charge exchange (π^-,π^0) of the incoming pion, could easily give a contribution within the DWBA approach, suitably modified to include isobaric analog channels. Within this approach, the (π^-,p) amplitudes will generally be smaller than the (π^+,p). This might seem to support the notion that all two-(or more) step processes must be small, but that is not necessarily so. In fact, the two-nucleon models also predict amplitudes for (π^-,p) to be much smaller than (π^+,p): in this case, the result follows from either (a) the known ratios for $\pi^-pp{\to}pn/\pi^+pn{\to}pp$, or ($b$) from the microscopic amplitudes for πN and NN charge exchange. Data on the $(\vec{p},\pi)$ reaction with polarized protons also exist and may well provide new information (Au1–78).

7.3b Two-nucleon Reactions: $(\pi, 2N)$

Absorption reactions in which two fast nucleons are emitted have been of interest for some years. The varieties generally studied are

$$\pi^- + A \to n + n + B_1^*$$
$$\to n + p + B_2^* \qquad\qquad [42a]$$

and

$$\pi^+ + A \rightarrow p + p + B_3^*$$

$$\rightarrow p + n + B_4^*. \qquad \qquad [42b]$$

The π^- modes can be studied with stopped π^-, which are absorbed from π-atomic orbits; both $\pi^\pm$ can be absorbed in flight. The rates for these modes are large, compared, for example, to (π^+, p). On the other hand, study of the reactions requires measurement of two particles in coincidence: either by photographic methods (emulsions, bubble chambers) or by counter techniques. Since the detection of neutrons is more difficult than that of protons, the π^+ studies (Eq. 42b) have generally been restricted to $(\pi^+, 2p)$. For π^- reactions, the time-of-flight approach has been used for (π^-, nn) and (π^-, np) modes.

Much of the interest in these reactions depends on the assumption that they are direct, or have a considerable direct component, meaning that the two nucleons ejected are the only ones essentially involved in the dynamics. To the extent that this is true, this mode is an interesting probe of nuclear structure, leading to two-hole states in the final nuclei (B^* in Eq. 42).

Let us first consider the kinematics of the reaction, which are illustrated in a number of ways in Fig. 7.9. For a given T_π, there are five independent kinematic variables; these are fixed by the momenta $\mathbf{p}_1$ and $\mathbf{p}_2$ of the two emitted nucleons, relative to the pion momentum $\mathbf{q}$, in the lab frame. Since rotation about $\mathbf{q}$ is not a change of kinematics, $\mathbf{p}_1$ and $\mathbf{p}_2$ specify only five variables. If $T_\pi = 0$, no $\mathbf{q}$ direction is fixed, and the number of variables reduces to three (see Fig. 7.9a). Thus a complete kinematic specification is obtained by measuring $\mathbf{p}_1$ and $\mathbf{p}_2$ (relative to $\mathbf{q}$); spin polarizations give additional information. With such measurement, the energy E^* of the final nuclear state is specified by energy conservation, which in the lab frame, may be written

$$E^* = \varepsilon_\pi(q) + E_A - T(p_1) - T(p_2) - T(K_r), \qquad [43]$$

where E_A is the target ground state energy, ε_π is the relativistic energy of the incoming pion, $T(p_{1,2})$ is the kinetic energy of each of the outgoing nucleons, and $T(K_r)$ is the recoil kinetic energy of the final nucleus. The recoil momentum $\mathbf{K}_r$ (Fig. 7.9b and c) is determined by momentum conservation, which in the lab is written

$$\mathbf{K}_r = -\mathbf{p}_1 - \mathbf{p}_2 + \mathbf{q} = -\mathbf{P} + \mathbf{q} = -\mathbf{K}, \qquad [44a]$$

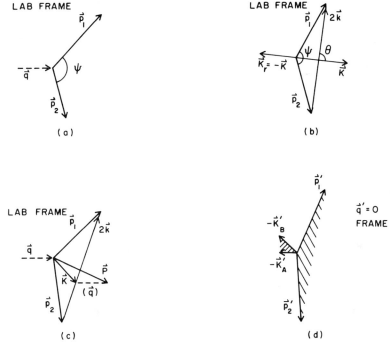

Fig. 7.9 Kinematics of $(\pi, 2N)$ reaction: (a) momenta in lab; (b) defining variables for absorption at rest; (c) for absorption in flight; (d) kinematics in pion rest frame to define Treiman-Yang angle, as in Eq. 46.

where

$$\mathbf{P} \equiv \mathbf{p}_1 + \mathbf{p}_2 \quad \text{and} \quad \mathbf{K} \equiv \mathbf{P} - \mathbf{q}. \qquad [44\mathbf{b}]$$

Two convenient variables are E^* (or the excitation energy $\Delta E^* = E^* - E_\mathbf{B}$ relative to the ground state of the final target) and the magnitude $K_r = |K|$. If we assume nonrelativistic kinematics for the nucleons, it is useful to introduce the relative momentum

$$\mathbf{k} \equiv \tfrac{1}{2}(\mathbf{p}_1 - \mathbf{p}_2). \qquad [44\mathbf{c}]$$

(For relativistic kinematics, the equivalent would be the momentum $\mathbf{p}_1$ in the two-nucleon c.m. frame). For stopped pions, $|\mathbf{k}|$ is fixed by ΔE^* and K through Eq. 43, since

$$T(p_1) + T(p_2) = \frac{p_1^2 + p_2^2}{2M} = \frac{1}{2M}\left(2K^2 + \frac{1}{2}k^2\right) \qquad [45]$$

and $K = P$ for $q = 0$. The remaining variable for stopped pions could then be an angle (see Fig. 7.9b): either the opening angle ψ ($\cos\psi = \hat{\mathbf{p}}_1 \cdot \hat{\mathbf{p}}_2$), which is directly measured in the lab frame, or the angle θ ($\cos\theta = \hat{\mathbf{k}} \cdot \hat{\mathbf{K}}$). For energetic pions, the magnitude k is independent of ΔE^* and K, leaving two angles (cosines) to be chosen (note that $\hat{\mathbf{q}} \cdot \hat{\mathbf{K}}$ is now determined, using Eqs. 43, 44b, and 45). An angle of possible interest is the *Treiman-Yang* (TY) angle (Tre–62, Sha–65), which is defined in the frame of reference in which the *pion* is at rest (see Fig. 7.9d) and in which we denote all momenta by primes. The target A has momentum $\mathbf{K}'_A$; the final target B recoils with momentum $\mathbf{K}'_B$ in this frame. The angle of interest, ϕ_{TY}, may be defined as the angle between the $\mathbf{K}'_A \mathbf{K}'_B$ and $\mathbf{p}'_1 \mathbf{p}'_2$ planes in Fig. 7.9d, or as

$$\cos\phi_{TY} = \frac{(\mathbf{K}'_A \times \mathbf{K}'_B)}{|\mathbf{K}'_A \times \mathbf{K}'_B|} \cdot \frac{(\mathbf{p}'_1 \times \mathbf{p}'_2)}{|\mathbf{p}'_1 \times \mathbf{p}'_2|} . \qquad [46]$$

(This angle is not defined for stopped pions.) The Treiman-Yang angle ϕ_{TY} for the case of energetic pions, and the angle θ for the case of stopped pions, both carry information about the angular momentum transferred in the $(\pi, 2N)$ reaction under certain model dependent assumptions, which we discuss below (Sha–65, Kol–67). The distributions in these angles are therefore useful in testing these model assumptions.

A direct reaction description of (π, NN) assumes that the pion interacts primarily with the two nucleons in the target that are ultimately emitted. The interaction of the pion with other nucleons is assumed, at most, to involve the optical distortion of the incoming wave. Similarly, interactions or exchange effects of the emitted nucleons with others in the target are also neglected, again excepting the possibility of scattering of the outgoing waves on the final target (in state B*). The simplest case ignores the initial and final distortions: thus the initial pion and emitted nucleon states are treated as plane waves of momenta $\mathbf{q}, \mathbf{p}_1$, and $\mathbf{p}_2$, respectively. The transition amplitude then takes the simple form, as in the plane-wave impulse approximation,

$$\langle B^*, \mathbf{p}_1, \mathbf{p}_2 | T | \mathbf{q}, A \rangle \simeq \int \int \frac{d\mathbf{p}'_1 d\mathbf{p}'_2}{(2\pi)^6} \langle \mathbf{p}_1, \mathbf{p}_2 | T | \mathbf{q}, \mathbf{p}'_1, \mathbf{p}'_2 \rangle \langle B, \mathbf{p}'_1, \mathbf{p}'_2 | A \rangle,$$

$$[47]$$

where the last factor is the probability (overlap) amplitude for decomposing state A into final state B with two plane wave nucleons: It is analogous to the overlap integral in Eq. 40 for (π, p). In this approximation, the pion momentum is transferred to the absorbing pair of nucleons; hence, using Eq. 44b, we obtain

$$\mathbf{p}'_1 + \mathbf{p}'_2 = \mathbf{p}_1 + \mathbf{p}_2 - \mathbf{q} = \mathbf{K}. \qquad [48]$$

It is possible to rewrite Eq. 47 as

$$\langle B, \mathbf{p}_1, \mathbf{p}_2 | T | \mathbf{q}, A \rangle \simeq \int \frac{d\mathbf{k}'}{(2\pi)^3} \langle \mathbf{k} | T | \mathbf{q}, \mathbf{k}' \rangle \langle B, \mathbf{K}, \mathbf{k}' | A \rangle, \qquad [49]$$

where $\mathbf{K}$ is fixed by the external kinematics through Eq. 48. The final nucleus B is a "spectator" in the initial target state, with momentum $\mathbf{K}_r = -\mathbf{K}$.

In the impulse forms, Eqs. 47 and 49, the nuclear structure of the states A and B appears in the overlap amplitude. For example, for a shell model in which the emitted particles are removed from definite initial orbits, the overlap amplitude takes the form

$$\langle B, \mathbf{p}_1', \mathbf{p}_2' | A \rangle = \sum_{\alpha, \beta} \mathcal{S}_{\alpha\beta}(B, A) u_\alpha(\mathbf{p}_1') u_\beta(\mathbf{p}_2'), \qquad [50]$$

where the orbitals α and β are denoted by $u(\mathbf{p})$. The angular momenta and isospins of the two nucleons must also be coupled to those of states A and B; this is indicated by the sum in Eq. 50. For the evaluation of the integral in Eq. 49, it is useful to reexpress Eq. 50 in terms of the relative and c.m. momenta, $\mathbf{k}'$ and $\mathbf{K}'$, for the harmonic oscillator shell model; this can be done in closed form, using the Talmi-Moshinsky transformation (Bro–60). Correlations between the two nucleons in the initial target state can be included explicitly in the overlap (Eq. 50), for example, by a multiplicative term, depending on the relative momentum $\mathbf{k}'$ (and spins, e.g., for tensor correlations), as we discuss below.

There is still considerable freedom in the treatment of the two-nucleon transition amplitude $\langle \mathbf{p}_1, \mathbf{p}_2 | T | \mathbf{q}_1, \mathbf{p}_1', \mathbf{p}_2' \rangle$. In the strictest plane wave Born approximation, this term would be given simply by the one-body absorption current, operating on each nucleon separately:

$$\langle \mathbf{p}_1, \mathbf{p}_2 | T | \mathbf{q}, \mathbf{p}_1', \mathbf{p}_2' \rangle \simeq (2\pi)^3 \langle \mathbf{p}_1 | j(\mathbf{q}) | \mathbf{p}'_1 \rangle \delta(\mathbf{p}_2 - \mathbf{p}_2')$$
$$+ (2\pi)^3 \langle \mathbf{p}_2 | j(\mathbf{q}) | \mathbf{p}_2' \rangle \delta(\mathbf{p}_1 - \mathbf{p}_1') \qquad \text{(PWIA)}.$$

$$[51]$$

This is analogous to Eq. 41. However, even though we have omitted the scattering of the incoming pion and outgoing nucleons on the initial and final targets, respectively, we may still keep track of the scattering of the pion *by the two nucleons*, and of these nucleons by each other. To do this, we must interpret the two nucleon amplitude in Eqs. 47 and 49 as the t-matrix for absorption of a plane wave meson by an interacting nucleon pair, initially in a plane wave state $(\mathbf{p}_1, \mathbf{p}_2)$ or $(\mathbf{K}, \mathbf{k}')$. This amplitude is similar to that for the absorption of a meson by a free pair of nucleons (as in Section 3.5), with the differences that energy is not conserved by the $\pi + N + N$ system, because of the nuclear aspects in Eq. 43, and that there is no initial nucleon-nucleon interaction in the present interpretation of $\langle k|T|q, k'\rangle$, since that has already been included in Eq. 50.

One can carry the relation of the amplitude $\langle k|T|q, k'\rangle$ even closer to that for free $\pi + N + N \rightarrow N + N$, by extracting the initial two-nucleon correlation from the initial state A, as follows. We assume for simplicity that the correlation is given by a state-independent function $g(k)$ (independent of α and β in Eq. 50); thus we can write the correlated overlap amplitude, in terms of K and k', as

$$\langle B, K, k'|A\rangle \cong \int dk'' g(k' - k'')\langle B, K, k''|\tilde{A}\rangle \qquad [52]$$

where $\tilde{A}$ refers to an uncorrelated state—for example, the shell model state that appears in Eq. 50. (The convoluted form in Eq. 52 corresponds to a *product* wave function in configuration space $\psi \sim g(r)\tilde{\psi}(r)$.) If the correlation function $g(k)$ refers to short-range behavior in space (e.g., hard core potentials), then it will vary with k'' more slowly than will the shell-model factor $\langle B, K, k''|\tilde{A}\rangle$, so that the integral in Eq. 52 can be approximated by

$$\langle B, K, k'|A\rangle \simeq g(k') \int dk'' \langle B, K, k''|\tilde{A}\rangle$$

$$\equiv g(k') F_{B\tilde{A}}(K), \qquad [53]$$

where the last factor F is an average nuclear structure factor for removing an uncorrelated pair with c.m. momentum K. Now the absorption amplitude in Eq. 49 takes the (approximate) form

$$\langle B, p_1, p_2|T|q, A\rangle \simeq F_{B\tilde{A}}(K) \int \frac{dk'}{(2\pi)^3} \langle k|T|q, k'\rangle g(k'). \qquad [54]$$

Equation 54 gives a simple and compact expression for the pair absorption model in the impulse approximation. The nuclear structure (without pair correlations) appears in the structure factor $F(K)$, and influences the c.m. motion of the emitted nucleon pair through Eq. 48. The transition amplitude in relative momenta, given by the integral in Eq. 54, now represents the t-matrix for absorption of a meson by a *correlated* pair of nucleons. The factoring of Eqs. 53 and 54 is essentially a short-range approximation, appropriate for s-wave correlations. Note that this gives no *angular correlation* between k and K, that is, in the angle θ in Fig. 7.9b. Higher angular momentum requires more careful treatment of Eq. 52.

The term under the integral in Eq. 54 may still be quite complicated microscopically, since it has all the possible interactions of the pion plus two nucleons, as in Chapter 3. Multiple scattering of the pion, as well as the correlation of the nucleons in the initial and final (and intermediate!) state, may all play a role. Calculation in a detailed model can be as complicated as for the reaction $\pi + d \rightarrow p + p$ in Section 3.5. One qualitative feature that emerges is that the integral in Eq. 54 is expected to be a slower function of the relative momentum $\mathbf{k}$ than the nuclear structure factor $F(\mathbf{K})$ is of $\mathbf{K}$. This is because all the short-range interactions (with large momentum transfer) have been kept for relative motion of the pair, but not for the c.m. motion, which is determined by the average shell model potential. Conservation of energy relates $\mathbf{k}$ and $\mathbf{K}$ through Eqs. 43, 44b, and 45:

$$2k^2 + \tfrac{1}{2}(\mathbf{K} + \mathbf{q})^2 = 2M\left[\varepsilon_\pi(q) + E_A - E^* - T(K)\right]. \qquad [55]$$

If we assume that $F(\mathbf{K})$ dominates Eq. 54 with a "range" of $|K| \lesssim 150$ MeV/c, then Eq. 55 leads to an estimate of

$$|k| \gtrsim 450 \text{ MeV}/c \qquad [56]$$

for $q = 0$, and increasing with $|q|$. This means that $|2k| \gg K$; that is, the relative momentum of the emitted pair is generally larger than the recoil momentum. For stopped pions, this leads to a large opening angle ψ between the emitted particles (see Fig. 7.9b).

These qualitative features of the direct two-nucleon absorption, illustrated by the simplified form of Eq. 54, have been demonstrated in a number of $(\pi, 2N)$ experiments, both with stopped π^- and with low energy ($\lesssim 100$ MeV) π^+, with light nuclear targets, typically ^{6}Li, ^{12}C, or ^{16}O. For these low pion energies, the neglect of the nuclear scattering of the pion is probably not serious. The neglect of the scattering of the outgoing nucleons at these energies may be more important; however, for very light targets, the impulse approximation in Eq. 47 may still apply. We illustrate the dominance of low values of K for these reactions in Fig. 7.10a and b. The angular distribution in θ (Fig. 7.9b) has been measured for some cases of (π^-, nn) (Bas-77). The Treiman-Yang distribution given in Eq. 46 has been measured for low energy $(\pi^+, 2p)$ reactions. For both of these angles, isotropic distributions have been found for simple cases where no angular momentum transfer is expected, as for ^{6}Li$(\pi, 2N)$ ^{4}He (ground state). The expected result is then isotropy, as given by the zero-range approximation in Eq. 54; this is found experimentally, as illustrated in Fig. 7.10c (Bas-77).

The simple features of the approximate form in Eq. 54 can also be obtained by an effective absorption operator, as we discussed earlier in

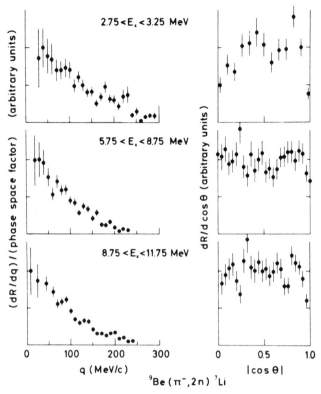

Fig. 7.10 Distributions measured in (π^-, 2n) for stopped π^- (from Bas–77). The rates as a function of q (which is equivalent to our K in Fig. 7.9b) are shown for different excitation energies of the final target. Distributions in $\cos\theta$ (as in Fig. 7.9b) are on the right.

Section 7.2, Eqs. 35 and 36. All the complications of the π-scattering on the nucleon pair, and the pair interactions, are now lumped together into the coefficients in Eq. 36, where the short range approximation that led to Eqs. 53 and 54 has become explicitly a *zero-range* assumption. The integral in Eq. 54 becomes a simple function of pion momentum

$$\int \frac{d\mathbf{k}'}{(2\pi)^3} \langle \mathbf{k}|T|\mathbf{q}, \mathbf{k}'\rangle g(\mathbf{k}') \to \alpha(q); \qquad [57]$$

the dependence on $\mathbf{k}$ has disappeared in the zero-range limit. (We have suppressed explicit dependence on spin-isospin.) In this limit, it is tempting to identify the functions $\alpha(q)$ with the amplitudes for absorption on two free nucleons, $\pi + 2N \to 2N$—for example, $\pi^+ + d \to pp$ (for $T=0$ target nucleon pair). However, this requires not only the assumption of zero range, in which the off-shell dependence on k is suppressed, but also the assumption that the 2N correlation $g(\mathbf{k}')$ is the same for nucleon pairs in nuclei as for free pairs.

On the other hand, one need not be restricted to these zero-range, on-shell approximations; one can directly attack the pair absorption problem by microscopic calculation, beginning with Eq. 49. Aside from the method of treating the initial pair wave function (e.g., by Eq. 52), one needs the absorption operator $\mathcal{C}(\omega_q)$, the relative wave function for the outgoing NN pair, and some approximation for the scattering of the incoming pion. There is a considerable literature of microscopic calculations of this sort, that is, of the pair absorption amplitude (e.g., Kol–68, Kop–71, Par–71). Generally, the detailed results show sensitivity to the assumptions or models involved for a number of elements that go into the calculation, which were just mentioned above. Therefore, there is no complete agreement on the relative importance of short-range correlations in the target, final-state interactions of the outgoing pair, and absorption with and without scattering of the pion. To some exent, this uncertainty does not change the qualitative aspects of the (π, NN) reaction, which were discussed earlier and which follow from the general properties of the amplitudes—for example, in Eq. 54, in the impulse approximation for pair absorption. To the extent that this general approximation is valid, the nuclear structure aspects of the (π, NN) reaction, as reflected in the overlap amplitude in Eq. 50 or structure factor $F_{B\bar{A}}(\mathbf{K})$ of Eq. 54, are not strongly affected by the details of the $\pi NN \rightarrow NN$ amplitude.

How good is the impulse approximation we have been discussing? For low energy or for stopped pions, the neglect of distortion of the incoming pion wave by the target may not be serious, but it can also be included relatively easily through an optical potential. For higher energy, for example, near the $\Delta(1232)$ resonance, this DW approximation may be quite inadequate for reasons we have discussed in Section 7.2, in the text above Eq. 29. The scattering of the outgoing nucleons on the final nucleus could easily be a serious problem; the nucleon-nucleus interaction at these energies (50–150 MeV) is quite strong. Strong distortion of the outgoing nucleons would change the dependence of the absorption on the kinematic parameters, for example, as given by Eq. 54 for no distortion. There have been attempts to treat the final state $(N + N + B)$ as a three-body scattering problem (Mor–73, Gar–74). These calculations indicate, that even for light nuclei (e.g., ^{12}C), the distortion effects may be considerable and may require the inclusion of both NN and NB interactions.

Finally, in discussing further corrections, one should distinguish between calculations of cross sections or rates for *inclusive* versus *exclusive* reactions. If one is interested in a (π, NN) reaction to a specific final state

B* as in Eq. 42 (exclusive), then it is appropriate to include *optical* absorption for the incoming pion, or outgoing nucleons, since this accounts for loss of flux to other channels. If one is interested in all final states B* that accompany two fast nucleons (inclusive), then *optical* absorption should be omitted. If one were interested in what further particles might be emitted (e.g., from N + B* collisions), it would be appropriate to apply a statistical calculation to the final channel, as we have discussed at the end of Section 7.2 (see Eq. 37).

7.3c Other Aspects of Absorptive Reactions

There are many other types of nuclear absorption reactions than the two we have examined in detail in this section; there are also a number of theoretical questions of reaction theory and nuclear structure that we have barely discussed. Without trying to survey the entire subject, we conclude this section with a brief discussion of several interesting problems in meson absorption.

1. Clusters of nucleons.

Two separate issues are involved in nucleon clusters: absorption of mesons by clusters in the target, and emission of clusters from the nucleus as a result of absorption. In the second case, these clusters are bound states of several nucleons—d, ^{3}H, ^{3}He, and ^{4}He—that may be detected singly or in coincidence with other clusters or nucleons. In the first case, one is thinking of a correlated substructure of the target nucleus, which may be a stable cluster (d, ^{4}He, etc.), or may only have the symmetries and quantum numbers of such clusters, whether stable or unstable (see Fig. 7.1). For example, in discussing the (π, NN) reaction in Section 7.3b, we followed an argument that led to Eqs. 54 and 57, in which the pair absorption amplitude was given in the form

$$T \simeq F(\mathbf{K})\alpha(q). \qquad [58]$$

This can be interpreted as absorption by a 2N cluster, whose momentum wave function in the target is given by $F(\mathbf{K})$. If the cluster has $J = 1$ and $T = 0$, it can be thought of as a "quasi-deuteron" in the target. If, in addition, the internal structure of the 2N cluster is similar to that of the deuteron, at least in its short-range structure [e.g., $g(\mathbf{k})$], then the cluster absorption amplitude $\alpha(q)$ may be similar to that for a free deuteron given by $\pi^+ d \rightarrow pp$ (quasi-deuteron model). However, a 2N cluster with $J = 0$ and $T = 1$ may also absorb mesons, with an amplitude of the form of Eq. 58, but does not correspond to a bound 2N state. The use of the word cluster here is somewhat arbitrary, but does indicate the quantum numbers and the degree of correlation of the group of nucleons.

The alpha (^{4}He) cluster is of interest in nuclear physics, both for light $A = 4n$ nuclei (^{12}C, ^{16}O, ^{20}Ne, etc.) that show some aspects of being composed of alphas, or nuclei, including these but also heavier A, that have large probabilities (widths) for α-emission or α-transfer reactions. The question arises as to whether the α-clusters in these cases absorb pions directly. Presumably this would mean an absorption amplitude of the form of Eq. 58, where $F(K)$ is an α-structure factor, and $\alpha(q)$ is the amplitude for pion absorption by ^{4}He. For this assumption, all products of the absorption by ^{4}He should be emitted, for example, for π^+:

$$\pi^+ + {}^4\text{He} \rightarrow {}^3\text{He} + \text{p}$$

$$\text{d} + 2\text{p}$$

$$\text{n} + 3\text{p}. \qquad \qquad [59]$$

This possibility shows that *one* source of emitted clusters in pion absorption is the fragments (in this case, ^{3}He and d) of the absorbing cluster. (See the analysis using the "pole approximation" of Kol–66.)

From this point of view, the question of clusters is a question of nuclear structure; it does not necessarily follow that there is a special mechanism for absorption by a cluster, which could not be given by a microscopic absorption theory. For example, for absorption of stopped π^- by ^{4}He, the pair impulse model of Section 7.3b gives moderately successful quantitative results for the total rate (Kol–68). It is quite possible for higher-energy π^+—for example, for $T_\pi \sim 200$ MeV—that the pair model would not work for ^{4}He, and that all four nucleons are dynamically involved in the absorption.

Last, it should be noted that the emission of clusters, which has been seen experimentally, does not *require* a cluster absorption model, as in Eq. 59. One could imagine that the emitted cluster has shared the pion energy with another nucleon or cluster in the absorption process. Otherwise, clusters can be knocked out: for example, $\pi + 2\text{N} \rightarrow \text{N} + \text{N}'$ in the target, and $\text{N}' + \alpha \rightarrow \alpha + \text{N}''$, where the α is knocked out. Similarly, clusters may be emitted statistically (boil-off).

2. Radiative absorption and photoproduction.

We mentioned the (π, γ) absorptive reaction given in Eq. 3 in Section 7.1a; we now also note the inverse reaction (γ, π), which is known as photoproduction. Both reactions have been studied experimentally and theoretically. In many ways this reaction differs from the other absorption modes, principally in that there is an on-energy-shell reaction possible for one nucleon:

$$\pi^+ + n \leftrightarrow p + \gamma$$
$$\pi^- + p \leftrightarrow n + \gamma \qquad [60]$$
$$\pi^0 + \left\{ \begin{array}{c} p \\ n \end{array} \right\} \leftrightarrow \left\{ \begin{array}{c} p \\ n \end{array} \right\} + \gamma,$$

unlike the case for the strong interaction, $\pi + N \leftrightarrow N$. The difference from the nonradiative absorptions is not only that an on-shell impulse approximation becomes possible here, but also that it may well apply to a good accuracy, since the momentum transfer to the nucleon is no longer required to be large, because the photon may carry away both energy and momentum. The reaction is therefore similar to an inelastic scattering reaction, the particular details depending on the spin, isospin, and angular momentum structure of the amplitudes.

In the impulse approximation, the amplitude is assumed to depend on that for $\pi N \leftrightarrow N\gamma$, which may be taken (on-energy-shell) from γp photoproduction cross sections. The amplitude R is characterized as a function of the vectors $\mathbf{k}$ and $\mathbf{q}$, the momenta of the photon and meson; σ, the nucleon spin; and ε, the photon polarization. For example, one may write (e.g., Del–75)

$$R = R_1 \sigma \cdot \varepsilon + R_2 (\sigma \cdot \mathbf{q})(\sigma \cdot \mathbf{k} \times \varepsilon) + R_3 (\sigma \cdot \mathbf{k})(\mathbf{q} \cdot \varepsilon) + R_4 (\sigma \cdot \mathbf{q})(\mathbf{q} \cdot \varepsilon),$$

$$[61]$$

where the R_i depend on scalars (s, t or E, q^2, note $\varepsilon \cdot \mathbf{k} = 0$ for the usual gauge). The leading term as $q \rightarrow 0$ is $R_1 \sigma \cdot \varepsilon$, sometimes called the "catastrophic" term* (s-wave π), and contributes only for charged pions: $R_1 \equiv 0$ for π^0. The p-wave terms are known to reflect the energy dependence of the $\Delta(1232)$. The particular spin (and isospin) structure of Eq. 61 makes it of interest for exciting particular nuclear states (spin-flip). The similarity of some of the terms in Eq. 61 to those of the $M1$ electromagnetic transition, or Gamow-Teller beta transition, make it of interest to compare the (π, γ) and (γ, π) reactions with inelastic electron scattering (e, e') and μ^--capture reactions on nuclei.

The corrections to the impulse approximation are also of interest here (Del–75). The problem is closely related to the renormalization of axial currents in nuclei (PCAC) or to the mesonic correction to electromagnetic interactions in nuclei (Eri–72).

---

*The somewhat excessive term "catastrophic" refers to a coincidence of the point of π absorption with that of γ emission.

3. K⁻ –absorption.

We have already discussed the (K^-, π^-) reaction in Section 5.4 as a strangeness-exchange scattering reaction. Although the K^- is absorbed, the pion carries away momentum and energy, such that for suitable T_{K^-}, the momentum transfer to the target may be small and the final target state a low (10–20 MeV) excited state of a hypernucleus $(K^- + A \rightarrow \pi^- + {}_\Lambda A^*)$. In kinematical respects the (K^-, π^-) reaction is similar to the (π, γ). For higher (or lower) K^- energy, the momentum and energy transfers may be greater, and the process becomes more *inelastic*. Finally, there may be no outgoing meson, which means true absorption. As in the case of pions, true absorption requires at least two nucleons, with some interactions among them to balance momentum and energy. In principle, the reactions could proceed as illustrated in Fig. 7.11, where we show one fast baryon emitted for simplicity. In (*a*), the K^- is directly absorbed by one nucleon, emitting a hyperon. The vertex for this process is not as well known as in the pion case. Also, this diagram requires even higher momentum transfer than the π case. In (*b*), we show the K^- scattering from the emitted particle, which is then absorbed by the residual target. In (*c*), the K^- has converted to a π^- on the outgoing particle and is absorbed (as a π^-) on the residual target. (Cf. Figs. 7.6. and 7.7*a*). These reactions have received very little attention. (See, however, Bur–62, Com–64, and Fow–66.)

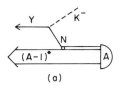

(a)

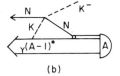

(b)

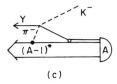

(c)

Fig. 7.11 Possible processes for K^- absorption, with no outgoing mesons.

Let us summarize our theoretical understanding of meson absorption reactions. General kinematical arguments, which we presented in Section 7.1, lead us to expect that these processes are more complicated than the scattering reactions discussed in Chapter 5. The microscopic theory, which has so far only been applied to the emission of one or two fast nucleons, does not fully explore this complexity. However, we have already seen how many aspects of the meson physics *and* the nuclear physics become involved, even in the one- and two-nucleon reactions: Multiple scattering, with off-energy-shell propagation of the meson, the absorption vertex, and the correlations among nucleons are all elements of the problem about which we have, at best, incomplete theoretical knowledge. This is in sharp contrast to the prominent effects seen in elastic scattering of mesons (Section 5.1) in which diffraction is dominant. Simple properties of the πN system (its on-energy-shell t-matrix) and of the nucleus [the nucleon density distribution $\rho(r)$] already give us a good qualitative picture of what is going on.

In spite of all the uncertainties involved in the theory of absorption, some of the general features do seem to be correctly given by the microscopic approach, when compared to the few experiments available for light nuclear targets. For example, the kinematic aspects of the $(\pi, 2N)$ reaction, in which the summed-momentum distribution of the emitted pair reflects the ground state (spectator) distribution and in which angular distribution reflects the short range of the interaction region in the relative coordinate, do seem to be supported by the data (e.g., Fig. 7.10).

The nuclear structure aspects of absorption reactions have barely been utilized. To the extent that direct reactions do dominate—which is not fully established—they provide a tool for the study of two-hole nuclear states. As in the case of the $(p, 2p)$ reaction, which is used to study one-hole states, the greater interest may be in locating the deep-lying hole structure, which is otherwise inaccessible in nuclear physics. The study of valence shells would also be interesting, making use of the selection rules possible in the reaction (e.g., Sha–65, Kol–67). The use of absorption for study of clusters in nuclei has barely begun.

Last, with the possible exception of statistical studies of the reactions, there is no good theory that gives the total absorption rates or cross sections. These are also not well known experimentally. As we discussed in Chapter 6, the coupling of absorption to scattering reactions is one of the open problems in multiple scattering theory, whose solution will have its impact on our understanding of all meson reactions with nuclei.

References

The subject has been reviewed in a number of articles or conference reports: Kol–69, Kol–71, Eis–75b, Huf–75a, Nob–76, and Wal–77, most of which concentrate on pions. For kaon absorption, see Dov–73c.

8 Meson Interactions and Nuclear Structure

In this chapter we give a brief survey of the role that mesonic degrees of freedom play in the structure of nuclei. We look for specific effects of the interactions of mesons and nuclei that show up in the dynamics of nuclear states, such as the ground state or low-lying states studied in conventional nuclear spectroscopy, and in the transitions among such states, induced by electromagnetic or weak interactions. In most of this book we have been concerned with the interaction of *real* mesons with nuclei, in nuclear reactions for which the mesons are projectiles that may be scattered and detected in space far away from a nuclear target. Now we deal only with virtual mesons, which are encountered in the interior of stable nuclei, since energy conservation does not permit them to escape to large distance.

In discussing nuclear reactions induced by π or K mesons we have found that these particles, as projectiles, can excite within a nuclear target excitations like ρ-mesons or N*-resonances, which may be either real or virtual, depending on the available energy brought in. In stable nuclear states, such excitations are always virtual. We include them in our discussion of mesonic effects in nuclear states. Our basic concern is to identify that part of the physics of nuclear structure that relates specifically to the excitation of virtual mesons, N*, and so forth, for nucleons put together into a stable nucleus, leaving out mesonic aspects of the nucleons when considered as isolated particles. Since we are talking only about virtual processes, it is more difficult to find direct experimental evidence for them than for real mesonic processes, such as scattering reactions.

The subject we are dealing with is a large one, in which there has been growing interest in recent years. However, since the problems involve the strong interactions of elementary particles in a nuclear system, progress has been somewhat limited and only certain aspects have been treated in any depth. For that reason, we only introduce the reader to the subject matter and send him to the literature for specific details. A timely general reference is the compilation *Mesons in Nuclei* (Rho-79).

8.1 Outline of Theory

The theoretical problems we deal with are quite similar to those encountered in Chapter 6, except that we now consider the low energy states of the nuclear system, that is, well below the threshold for emitting mesons. For conventional nuclear physics the degrees of freedom are taken to be those of the constituent nucleons, and a Schroedinger equation with hermitian Hamiltonian is assumed to give the dynamics. Similarly, electromagnetic and weak transitions are assumed to be governed by operators on the individual nucleons. One could approach the problem of including meson degrees of freedom by starting afresh with an equation of motion for nucleons interacting with quantum fields, such as for the π, ρ, ω, and similar mesons. One would then set about finding, for example, the ground states of that combined system, with total baryon number A, and charge Z, which would represent the ground state of the nucleus with those quantum numbers (A, Z). To get the rest of conventional nuclear structure physics, we would have to calculate the excitation spectrum and transition probabilities for electromagnetic or weak interactions from the field theoretic problem. However, this direct approach is not really tractable, except for highly simplified model cases. There have been some attempts to use a simplified field treatment to study particular aspects of the ground state of nuclei, in particular saturation of the energy and stability of the state (e.g., against pion condensation, which we discuss below in Section 8.2c; see, e.g., Wal–74).

Instead of a general field theoretic approach to nuclear structure, we try to reformulate the problem in terms of an effective Schroedinger theory and make use of some of the techniques discussed in Chapter 6, where we took a similar approach for scattering reactions. We look particularly for those aspects of the meson degrees of freedom that do *not* lead to the conventional results of nuclear structure theory, since it is in terms of these non-Schroedinger aspects that we expect to learn new things about mesons and nuclei.

8.1a Effective Schroedinger Equation

Let us assume that there is a Hamiltonian H that includes all meson-nucleus interactions, and which defines a Schroedinger equation

$$H \Psi_\lambda = E_\lambda \Psi_\lambda \qquad [1]$$

for the particle-stable states of the nucleus. The state vectors Ψ_λ include nucleon, meson, and N* degrees of freedom. The idea of the effective

Schroedinger equation is to give an equation equivalent to Eq. 1 for some specified set of states, but in terms of fewer degrees of freedom. We then write

$$\tilde{H}\Phi_\lambda = E_\lambda\Phi_\lambda \qquad\qquad [2]$$

for some set $\{\lambda\}$, with the same energies E_λ as for Eq. 1, but with a simpler set of state vectors Φ_λ, which depend, for example, only on the nucleon degrees of freedom. The operator $\tilde{H}$ that gives the same spectrum E_λ for the given set of states Φ_λ as H does for the set Ψ_λ is the effective Hamiltonian. The first problem is to produce the operator $\tilde{H}$, from a theory of interacting mesons and nucleons.

A number of methods of approach have been suggested; few have been developed to the point of giving tools for quantitative theoretical work. One technique is based on projection operators, which select the *model space* of state vectors Φ_λ of Eq. 2 from the full space of vectors Ψ_λ of Eq. 1 by projection: $P\Psi_\lambda = \Phi_\lambda$. This method, introduced by Okubo (Oku–54) for the study of Tamm–Dancoff and perturbation theory for fields, is quite similar to the coupled-channels formulation of scattering theory (Fes–58, 62) that was discussed in Section 6.2a, where we used it to write an effective Schroedinger equation for meson-nucleus scattering (Eq. 6.21). If we follow the steps that took us from Eq. 6.18 to Eq. 6.21 for the present case of a bound state, we find that Eq. 2 takes the form

$$\tilde{H}(E_\lambda)\Phi_\lambda = E_\lambda\Phi_\lambda \qquad\qquad [3a]$$

with an energy-dependent effective Hamiltonian given by

$$\tilde{H}(E) = H_{PP} + H_{PQ}(E - H_{QQ})^{-1}H_{QP}. \qquad\qquad [3b]$$

If the form of $\tilde{H}(E)$ is known, Eq. 3a may be solved self-consistently for the E_λ and Φ_λ.*

An alternative method leads to Eq. 2, with an energy-independent definition of $\tilde{H}$. Again, one projects the full eigenstates onto the model space: $P\Psi_\lambda = \Phi_\lambda$, with an orthogonal projection $(1 - P)\Psi_\lambda = X_\lambda$. Then one

---

*The solutions Φ_λ of this equation are not, in general, orthogonal; in addition, there may be more solutions to Eq. 3 than the dimension of the model space. These peculiar properties have been studied in many-body theory, and methods are available for restricting the number of solutions and for handling the problems of nonorthogonal model state vectors (Blo–58, Bra–67).

performs a unitary transformation,

$$\begin{pmatrix} \Phi_\lambda \\ X_\lambda \end{pmatrix} = U \begin{pmatrix} \Phi'_\lambda \\ X'_\lambda \end{pmatrix}, \qquad [4a]$$

such that the Schroedinger equation for the primed vectors uncouple:

$$\tilde{H} \Phi'_\lambda = E_\lambda \Phi'_\lambda \qquad [4b]$$

with $\tilde{H}$ the transformed Hamiltonian (effective) in the new model space.*

The formal methods of Eqs. 3 and 4 define not only the effective Hamiltonians and model states, but also the effective operators for transitions (e.g., electromagnetic or beta-decay). We discuss practical methods for obtaining these operators in the following subsections. However, we ought first to consider how the equations might be used if we were able to obtain the effective operators. One approach is that which concentrates on nucleon dynamics by eliminating the meson degrees of freedom entirely. The model space then is chosen to have only nucleons, and might be even more restricted, for example, to a finite vector space, as for the shell model. It is the job of meson theory to calculate the appropriate nuclear interactions, in terms of the effective Hamiltonians in Eqs. 3 or 4. It is the job of nuclear theory to solve the nuclear Schroedinger problem, or restricted shell model problem, by the standard techniques of that subject. However, we see below that interactions and transition operators that come out of the suppression of the meson degrees of freedom are not of the conventional form used in nuclear physics in that there are interactions of three or more particles simultaneously and the transition operators also have many-body effects.

A different view is that leaving out all the meson degrees of the effective dynamics leaves out some interesting physical effects. For example, if there are virtual N* excitations in normal nuclei, it would be useful to have them in the wave functions and to try to see their effects experimentally. Second, the nucleon-only picture of stable nuclear states may not always work. For example, if pion condensates are possible (see Section 8.2c), the effective Eqs. 3 and 4 become unstable, and multiple solutions—signaling competing phases—appear. Here it makes more sense to include the *average* pion field in the dynamical equations.

These considerations suggest that it may be useful to enlarge the model spaces by including some of the extra degrees of freedom explicitly. For

---

*The transformation in Eq. 4a is not unique; conditions that must be satisfied have been given by Oku–54, who also gives some examples. The solutions of Eq. 4b are, however, orthogonal.

example, suppose we consider $\Delta(1232)$ excitation in nuclei. We could define the projection operator P to include a (finite) number of Δ-excitations:

$$P = P_0 + P_1 + P_2 + \dots . \qquad [5]$$

Then the effective Eqs. 3a or 4b become *coupled channel* equations for these modes of excitation.

8.1b Diagrammatic Expansion of the Effective Interaction

Let us consider how to construct effective interactions to lowest order in the meson-nucleon couplings. We suppose that we are dealing with nucleons coupled to various meson fields $(\pi, \rho, \omega, \dots, \text{etc.})$, through a Hamiltonian (Eq. 1) of the form

$$H = H_0 + H_I \qquad [6]$$

where H_0 is the kinetic energy operator of the bare nucleons and the meson fields, and H_I gives the coupling of mesons to nucleons. We may assume that H_I is of the Yukawa form in Eqs. 2.60 and 6.1—that is, linear in each field

$$H_I = \sum_k \int dx j_k(x) \phi_k(x) \qquad [7]$$

where k labels the meson type. Each term in Eq. 7 has a coupling strength determined by a coupling constant g_k (which we always take as the renormalized constant, since we only evaluate Eq. 7 between "dressed" or renormalized nucleon states (see Eqs. 2.65 and 2.67)).

Now we can see how this Hamiltonian in Eqs. 6 and 7 contributes to the effective Hamiltonian $\tilde{H}$, where we consider the projected form (Eq. 3b), with P defined to project onto a model space of dressed nucleons only. Then the first term of Eq. 3b,

$$H_{PP} \equiv PHP \cong \sum_{j=1} M \left(1 + \frac{p_j^2}{2M^2} \right), \qquad [8]$$

gives the kinetic energy of A dressed nucleons that do not interact. The interaction among these nucleons comes from the second term in Eq. 3b, through mesons that are excited and de-excited by the terms

$$H_{QP} = QH_I P \quad \text{and} \quad H_{PQ} = PH_I Q,$$

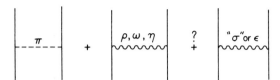

Fig. 8.1 One-meson exchange contributions to the two-nucleon interaction.

and which propagate among the nucleons with the propagator $(E - H_{QQ})^{-1}$. The lowest order contributions of this term are the one-meson-exchange (or one-boson-exchange: OBE) terms of the type we discussed in Section 2.5 in connection with the NN interaction; these are illustrated in Fig. 8.1. With the usual static approximation for the nucleons, the interaction term of Eq. 3b takes the form of a Yukawa potential:

$$\sum_{i<j} V(r_{ij}), \qquad V(r) \propto g_k^2 \frac{e^{-m_k r}}{r}, \qquad \qquad [9]$$

as given in Eq. 2.110 for π-exchange.

Higher-order processes certainly also contribute to the effective interaction, involving multiple exchange of mesons. For example, in Fig. 8.2 we show several kinds of two-meson-exchange terms that are likely to enter. The figure suggests that in order to calculate the effective interaction systematically, it is useful to have a diagrammatic expansion procedure to keep track of different types of contributions. In Section 2.5, where we considered multiple meson exchange in the NN interaction, we considered

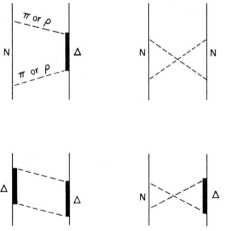

Fig. 8.2 Two-meson exchange contributions to the two-nucleon interaction.

scattering processes, and the diagrams were scattering graphs of the Feynman type. What we need for the effective interaction expansion are something like the Feynman-Goldstone graphs that are used in many-body theory (see, e.g., Bra–67, and Fet–71). To handle the energy dependence of $\tilde{H}(E)$, and the resultant orthogonality problems, one can introduce *folded diagrams* (Bra–67, Joh–71). We do not enter into the rather complicated subject of diagram rules, which are quite technical and which have not been worked out completely for the present case of nucleons plus mesons. (See, however, Joh–76 on diagrams for meson exchange potentials.)

In some of the examples shown in Fig. 8.2, a Δ-resonance is excited or de-excited by the exchanged mesons. These are likely to be important processes, since the coupling of nucleons and mesons to the Δ is strong. We discuss the consequences of these excitations in Sections 8.2a and 8.2b, where other N* resonances are also considered. If we want to consider multiple excitation of Δ-resonances in the nucleus, it is useful to extend the model space as discussed in Section 8.1a (see Eq. 5) to include the Δs explicitly, using coupled channel equations. For these we need, among other things, the transition interactions $H_{nm} = P_n H P_m$ between subspaces with different numbers (n, m) of Δ-excitations. These in turn may be obtained as transition *potentials*, generated by meson exchange as in V_{OBE} in Eq. 9. We show the lowest-order diagrams in Fig. 8.3. In the static limit these potentials take the generalized Yukawa form, in momentum space

$$V_k^{tr}(\mathbf{q}) = \frac{\Gamma_1 \cdot \Gamma_2}{\mathbf{q}^2 + m_k^2}, \qquad [10]$$

where m_k is the mass of the exchanged meson (k), and the vertex functions Γ depend on the Δ-N-meson coupling (spin and isospin).

One of the effects of multiple exchange of mesons, other than the excitation of Δ or other N* modes in the nucleus, is the generation of *many-body* interactions among the nucleons. We illustrate some simple contributions to a three-body interaction in Fig. 8.4. We discuss these again in Section 8.2b.

8.1c Transitions

Consider transitions between nuclear states through the electromagnetic or weak (beta-decay) interactions. Let us assume we know the interaction operator M for the meson-nuclear system, which induces transitions between states $\Psi_\lambda \to \Psi_\mu$ of Eq. 1. We want to find the proper expression for the effective transition operator that will give the same amplitude within

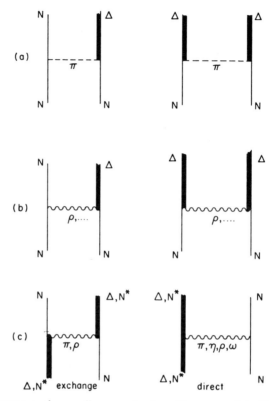

Fig. 8.3 One-meson exchange diagrams for transition potentials, Eq. 10, involving nucleons, N* and Δ-resonances.

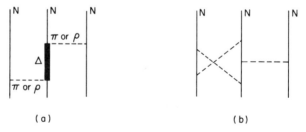

Fig. 8.4 Meson-exchange contributions to three-nucleon interaction: (*a*) Δ-excitation; (*b*) three-meson term.

the model spaces. If we use the projected states of Eq. 3, we want to find M such that the transition amplitudes in the two spaces agree; that is,

$$M_{\mu\lambda} = \frac{\langle \Psi_\mu | M | \Psi_\lambda \rangle}{(\langle \Psi_\mu | \Psi_\mu \rangle \langle \Psi_\lambda | \Psi_\lambda \rangle)^{1/2}} = \frac{\langle \Phi_\mu | M | \Phi_\lambda \rangle}{(\langle \Phi_\mu | \Phi_\mu \rangle \langle \Phi_\lambda | \Phi_\lambda \rangle)^{1/2}} . \qquad [11]$$

We may obtain an expression for M, using the formal solution of Eq. 1 in terms of Φ_λ (see Eq. 6.20):

$$\Psi_\lambda = \left[P + Q(E_\lambda - H_{QQ})^{-1} H_{QP} \right] \Phi_\lambda . \qquad [12]$$

Substituting this in Eq. 11, we find

$$N_\mu N_\lambda M = M_{PP} + M_{PQ}(E_\lambda - H_{QQ})^{-1} H_{QP}$$

$$+ H_{PQ}(E_\mu - H_{QQ})^{-1} M_{QP}$$

$$+ H_{PQ}(E_\mu - H_{QQ})^{-1} M_{QQ}(E_\lambda - H_{QQ})^{-1} H_{QP} \qquad [13]$$

where $M_{PP} \equiv PMP$, and so forth, and where N_μ gives the ratio of normalizations of Ψ_μ and $\Phi_\mu = P\Psi_\mu$:

$$N_\mu^2 = \frac{\langle \Psi_\mu | \Psi_\mu \rangle}{\langle \Phi_\mu | \Phi_\mu \rangle} . \qquad [14a]$$

It is generally more convenient to normalize the model states to unity, so that

$$N_\mu^2 = 1 + \langle \Phi_\mu | H_{PQ}(E - H_{QQ})^{-2} H_{QP} | \Phi_\mu \rangle . \qquad [14b]$$

The transition amplitude is given by

$$M_{\mu\lambda} = \langle \Phi_\mu | M | \Phi_\lambda \rangle . \qquad [15]$$

Note that the effective operator M depends on the states μ and λ through the energies E_μ and E_λ in Eq. 13, as well as through the normalizations N_μ and N_λ. It is clear that M is closely related to the effective interaction $\tilde{H}(E)$ of Eq. 3; presumably approximation schemes should be applied consistently to both operators. This is not generally done at present, since nuclear model states are not calculated from effective interactions based on meson theory. An exception is the OBE approximation, which may be used for nuclear forces and transition operators to the same order (see, e.g., Sti-70b).

Diagrammatic expansions may be developed for the effective operator M, based on methods of many-body theory, similar to those used for the effective Hamiltonian. Again we give no details which may, however, be found in Bra–67 and Ell–77. Alternative expressions for effective transition operators and diagram expansions for these may also be developed for the unitary transform method of Eq. 4 (see, e.g., Gar–76). These have the advantage of defining effective operators independent of model states. The connection of these methods to each other, and to the Feynman diagram expansions of S-matrix theory, are of interest, but they do not seem to have been worked out completely. This last would be particularly useful, since, as for the relativistic theory of meson scattering, the coupling of meson fields by the electromagnetic and weak interactions is studied in a covariant formulation (see, e.g., Che–71).

We illustrate the different contributions to the operator M by giving examples of the individual parts of Eq. 13 in terms of diagrams shown in Fig. 8.5. The operator M is indicated as an external wavy line, attached to nucleon, meson, or Δ lines. In (a) we have the term M_{PP}, which acts as a one nucleon operator on the model space nucleons. The second term of

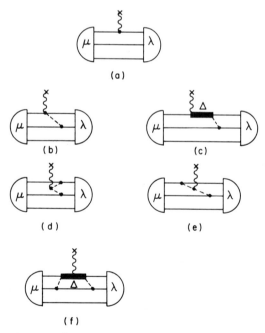

Fig. 8.5 Diagrams for contributions to the nuclear transition due to an external field, corresponding to the matrix $M_{\mu\lambda}$ of Eq. 11.

Eq. 13 is represented in Fig. 8.5 in (b), (c), and (d), in which M_{PQ} connects the model space to the excited (Q) space of mesons and Δs. We note that these terms are all two-nucleon operators. The third term of Eq. 13 involves diagrams obtained from Fig. 8.5b, c, d, and e by left-right reflection. The last term in Eq. 13 has contributions like that of (f), in which M_{QQ} couples to a Δ. Conventionally, all the diagrams of Fig. 8.5b–f are considered mesic corrections to the single-nucleon operator of (a): M_{PP}. These terms all involve exchanges of mesons with other nucleons; they are referred to as *mesonic* or *exchange current* effects. Additional corrections come from the expansion of the N_μ and N_λ in diagrams (see Bra–67 and Ell–77).

8.1d Nuclear Currents: Their Uses and Properties

For nuclear transitions involving electromagnetic or weak interactions, the effective operator required in the matrix element of Eq. 15 is most conveniently expressed in terms of nuclear currents. These are defined and their uses and properties are very briefly sketched in the present section. As we see, there is a surprisingly close connection between one member of this family of currents—the weak axial vector current—and pionic processes; hence some of the physics that this book describes may be directly pertinent to nuclear weak interactions.

We start by defining the well-known* electromagnetic and weak interaction Lagrangian densities

$$\mathcal{L}_{em}(x) = e J_\mu^{em}(x) A_\mu(x) \qquad [16]$$

and

$$\mathcal{L}_{weak}(x) = -\frac{G_F}{2\sqrt{2}} \left[J_\mu^{weak}(x) \tilde{J}_\mu^{weak}(x) + \tilde{J}_\mu^{weak}(x) J_\mu^{weak}(x) \right], \qquad [17]$$

where $J_\mu^{em}(x)$ is the electromagnetic four-current, $A_\mu(x)$ is the four-potential, and e is the absolute value of the electron charge, $e^2/4\pi \cong 1/137$. For the weak interaction, $J_\mu^{weak}(x)$ is the four-current and $G_F \cong 1.03 \times 10^{-5}/M_P^2$

---

*A far more complete description of the meson theory of nuclear currents is given in Che–79, while details of the calculation of nuclear transitions governed by $\mathcal{L}_{em}$ and $\mathcal{L}_{weak}$ may be found in Eis–70. A general reference on weak interactions is Mar–69b. All of these references use the same metric and conventions for γ-matrices as employed here, thus facilitating comparison.

is the Fermi coupling constant. The metric we use here for four-vector contractions is Euclidean, $\delta_{\mu\nu}(\mu,\nu=1,2,3,4)$; the vectors have imaginary time components $x_\mu = (\mathbf{x}, x_4 = it)$, $p_\mu = (\mathbf{p}, p_4 = iE)$, and $\partial_\mu = \partial/\partial x_\mu = (\nabla, \partial/\partial x_4 = -i\partial/\partial t)$; and we take $\tilde{V}_\mu = (-1)^{\delta_{\mu 4}}V_\mu^\dagger$.

Each of the currents splits into a leptonic and a hadronic part,

$$J_\mu^{em}(x) = l_\mu^{em}(x) + j_\mu(x), \qquad [18]$$

and

$$J_\mu^{weak}(x) = l_\mu^{weak}(x) + J_\mu(x), \qquad [19]$$

where the leptonic forms are

$$l_\mu^{em}(x) = -i(\bar{u}_e\gamma_\mu u_e + \bar{u}_\mu\gamma_\mu u_\mu) \qquad [20]$$

and

$$l_\mu^{weak}(x) = i\left[\bar{u}_{\nu_e}\gamma_\mu(1+\gamma_5)u_e + \bar{u}_{\nu_\mu}\gamma_\mu(1+\gamma_5)u_\mu\right]. \qquad [21]$$

The spinors here are labeled by the particles to which they pertain, and the Dirac γ-matrices are hermitian and satisfy

$$\{\gamma_\mu, \gamma_\nu\} = 2\delta_{\mu\nu}; \qquad [22a]$$

for subsequent use we note also

$$\slashed{V} = V_\mu\gamma_\mu, \quad \sigma_{\mu\nu} = \frac{1}{2i}(\gamma_\mu\gamma_\nu - \gamma_\nu\gamma_\mu), \quad \gamma_5 = \gamma_1\gamma_2\gamma_3\gamma_4. \qquad [22b]$$

The Lagrangian densities and currents of Eqs. 16–21 are constructed so as to incorporate the known symmetry properties of the respective interactions, such as Lorentz invariance, electromagnetic current conservation, and muon-quantum-number conservation. The weak leptonic current $l_\mu^{weak}(x)$ is written so as to be in conformity with the V − A theory of weak interactions (Mar–69b), and is thus the sum (or difference, depending on the order of the γ-matrices) of a vector (γ_μ) and an axial vector ($\gamma_\mu\gamma_5 = -\gamma_5\gamma_\mu$) part.

The hadronic currents are naturally modified by the strong interactions from the simple forms of Eqs. 20 and 21, but we may, of course, expect that the electromagnetic current $j_\mu(x)$ will again be a four-vector, while the weak hadronic current will be a sum of vector and axial vector terms:

$$J_\mu(x) = J_\mu^V(x) + J_\mu^A(x). \qquad [23]$$

The electromagnetic current may also be divided into isoscalar and isovector parts, as through

$$j_\mu(x) \sim \bar{u}_N O_\mu \frac{1+\tau_3}{2} u_N \sim j_\mu^s(x) + j_\mu^v(x), \qquad [24]$$

where the explicit illustration is for a single-nucleon current. The weak current is taken to be isovector in character, $J_\mu(x) \sim \bar{u}_N O_\mu \tau_\pm u_N$, where O_μ is a generic vector (or axial vector) operator. The electromagnetic current is, as already noted, conserved,

$$\partial_\mu j_\mu(x) = 0. \qquad [25a]$$

This current conservation in the nuclear context has direct implications for the matter of mesic exchange currents. This may be seen by recasting it in the form

$$\nabla \cdot \mathbf{j} + i[\tilde{H}, \rho] = 0, \qquad [25b]$$

where $\tilde{H}$ is the effective nuclear Hamiltonian discussed in Sections 8.1a and 8.1b. When mesic degrees of freedom have been removed from $\tilde{H}$ their influence is still felt through the appearance of charge exchange potentials involving the Heisenberg isospin-exchange operator $\Sigma_{i<j} V_{ij}^H P_\tau^{ij}$, with $P_\tau^{ij} = \frac{1}{2}(1 + \vec{\tau}(i) \cdot \vec{\tau}(j))$. These, in the commutator* of Eq. 25b, requires the appearance of an exchange current—in this case pertaining to nucleon variables—such that

$$\nabla \cdot \mathbf{j}^{\text{exch}} = -i \left[\sum_{i<j} V_{ij}^H P_\tau^{ij}, \rho \right]. \qquad [25c]$$

This exchange current then contributes to many electromagnetic processes, as we see in Section 8.2d, such as static magnetic moments and magnetic multipole transitions. General forms of exchange diagrams for such effects are seen in Fig. 8.5. The exchange effects do not enter for electric multipoles in processes where the limit of long wavelengths (or low Fourier components) is applicable. This is because such cases are governed by the charge density only (Siegert's theorem; see Boh–69 and Eis–70), and conservation of charge guarantees that even in the presence of nuclear collisions the aggregate charge, which is what is seen in the long wavelength limit, is preserved.

Under the hypothesis of conserved vector currents (CVC), for which there is now much corroborative evidence (Mar–69b, Eis–70), it is

*Explicit nonlocality in $\tilde{H}$ also contributes to the commutator and therefore to the exchange current.

assumed that—just like the electromagnetic current—the weak vector current is also conserved, and is, in fact, merely a rotation in isospin space of the electromagnetic current:

$$\partial_\mu J_\mu^V(x) = 0, \qquad\qquad [26]$$

and

$$J_\mu^{V\pm}(x) = \left[\vec{J}_\mu^V(x)\right]^\pm, \qquad j_\mu^v(x) = \left[\vec{J}_\mu^V(x)\right]^3 \qquad [27]$$

for a general vector, isovector current $\vec{J}_\mu^V(x)$, which, again for a single-nucleon case, would have the structure

$$\vec{J}_\mu^V(x) \sim \bar{u}_N O_\mu \vec{\tau} u_N. \qquad\qquad [28]$$

These isovector currents $\vec{J}_\mu^V(x)$ and the analogous isovector (and in space axial vector) currents $\vec{J}_\mu^A(x)$ then generate a current algebra,

$$\delta(x_0 - y_0)\left[J_0^{Vi}(x), J_\mu^{Vj}(y)\right] = i\varepsilon_{ijk}\,\delta^{(4)}(x-y)J_\mu^{Vk}(x),$$

$$\delta(x_0 - y_0)\left[J_0^{Vi}(x), J_\mu^{Aj}(y)\right] = i\varepsilon_{ijk}\,\delta^{(4)}(x-y)J_\mu^{Ak}(x),$$

$$\delta(x_0 - y_0)\left[J_0^{Ai}(x), J_\mu^{Aj}(y)\right] = i\varepsilon_{ijk}\,\delta^{(4)}(x-y)J_\mu^{Vk}(x), \qquad [29]$$

whose consequences we do not have the space to explore here (see, for example, Che–79).

Explicit general constructs for these various currents are easily given for *single-particle* states by exploiting symmetry arguments. Thus, for a single-nucleon case with four-momenta p and p',

$$\langle N(p')|\vec{J}_\mu^V(0)|N(p)\rangle = i\frac{\vec{\tau}}{2}\left[F_1^v(k^2)\gamma_\mu - \frac{F_2^v(k^2)}{2M}\sigma_{\mu\nu}k_\nu\right], \qquad [30]$$

$$\langle N(p')|j_\mu^s(0)|N(p)\rangle = \frac{i}{2}\left[F_1^s(k^2)\gamma_\mu - \frac{F_2^s(k^2)}{2M}\sigma_{\mu\nu}k_\nu\right], \qquad [31]$$

$$\langle N(p')|\vec{J}_\mu^A(0)|N(p)\rangle = i\frac{\vec{\tau}}{2}\left[g_A(k^2)\gamma_\mu\gamma_5 - ig_P(k^2)k_\mu\gamma_5\right], \qquad [32]$$

where $k = p' - p$ and the form factors relate to the coupling constants through

$$F_1^v(0) = F_1^s(0) = 1, \quad F_2^v(0) = \kappa_v = 3.70, \quad F_2^s(0) = \kappa_s = 0.12, \quad g_A(0) = g_A = 1.23.$$

$$[33]$$

For initial and final pion states, with isospin indices l and m, the vector current for coupling to a photon, say, is

$$\langle \pi^m(p')|[\vec{J}_\mu^V(0)]^j|\pi^l(p)\rangle = -iF_\pi^v(k^2)\epsilon_{mlj}(p+p')_\mu, \qquad [34]$$

with $F_\pi^v(0)=1$, while if the pion decays (or is produced),

$$\langle \pi^l(k)|[\vec{J}_\mu^A(0)]^j|0\rangle = -i\delta_{lj}\frac{f_\pi}{\sqrt{2}\,m_\pi^2}k_\mu, \qquad [35]$$

with $f_\pi = 0.946\,m_\pi^3$ from the measured decay $\pi \to \mu\nu$.

Unlike the vector current, the axial current is not conserved. [Indeed, if it were Eq. 35 would require $k_\mu\langle\pi^l(k)|J_\mu^{Aj}|0\rangle = -i\delta_{lj}(f_\pi k^2/\sqrt{2}\,m_\pi^2)=0$ or $f_\pi = 0$, whence the pion could not decay through $\pi \to \mu\nu$.] But the hypothesis of partially conserved axial currents (PCAC), for which there is also much evidence (Mar–69b), states that the matrix elements for the four-divergence of the axial current $\partial_\mu \vec{J}_\mu^A(x)$ are dominated by a pion pole term (or, equivalently, may be extrapolated smoothly away from that pole at $k^2 = -m_\pi^2$), written in operator form as

$$\partial_\mu \vec{J}_\mu^A(x) = \frac{f_\pi}{\sqrt{2}}\vec{\phi}(x), \qquad [36]$$

where $\vec{\phi}$ is the pion field and f_π is the constant of Eq. 35. For example, for single-nucleon states of four momentum p and p' the relevant matrix elements may be written

$$\langle N'(p')|\vec{J}_\mu^A(0)|N(p)\rangle = \vec{P}_\mu^A(k) + \langle N'(p')|\vec{J}_\mu^{A'}(0)|N(p)\rangle \qquad [37]$$

where $\vec{P}_\mu^A \sim (k^2 + m_\pi^2)^{-1}$ contains a pion pole and the term with $\vec{J}_\mu^{A'}$ does not. The pole in $\vec{P}_\mu^A$ is hidden in Eq. 32 in the induced pseudoscalar coupling constant $g_P(k^2)$ which is dominated by pion exchange, and which, from PCAC, is expected to satisfy $g_P(k^2) \cong 2Mg_A/(k^2 + m_\pi^2)$. For the pole-free part, PCAC states

$$k_\mu\langle N'(p')|\vec{J}_\mu^{A'}(0)|N(p)\rangle = i\frac{k^2 + m_\pi^2}{m_\pi^2}\frac{f_\pi}{\sqrt{2}}\langle N'(p')|\vec{\phi}(0)|N(p)\rangle. \qquad [38]$$

The basic expression in Eq. 36 used with Eq. 32 yields, in the limit $k \to 0$, the Goldberger-Treiman relation

$$f_\pi = \frac{\sqrt{2}\,Mg_A m_\pi^2}{g} = 0.87\,m_\pi^3, \qquad [39]$$

where g is the πNN coupling constant ($g^2/4\pi \sim 16$). This compares satisfactorily with the empirical value quoted after Eq. 35.

For nuclear applications, we of course require expressions for the current operators between states—including, in general, excited states—of the many-nucleon system, and one can no longer use symmetry arguments to produce expressions sufficiently simple to be useful for analysis. Moreover, one also wishes to be able to link the nuclear expression to known properties of the individual nucleon. Therefore the usual procedure has been to take single-nucleon results of the form of Eqs. 30–32, and posit, at least initially, that the operator for the many-nucleon system is a sum over the single-nucleon operators (impulse approximation). Furthermore, it is necessary—in the absence of a complete, relativistic theory for the many-nucleon system—to use the Foldy-Wouthuysen transformation or its equivalent to generate nonrelativistic forms for the single-nucleon operators. When two-particle correction terms are considered in order to incorporate mesic exchange effects as illustrated in Fig. 8.5, the procedure is then usually to calculate these as Feynman diagrams and take nonrelativistic limits to produce two-body operators for use with conventional nuclear wave functions (extensive details are given in Che–71, Che–79, and Had–75). In the course of this work, some guidance in the calculation of the diagrams may be obtained from the principles noted above (for example, the dynamics embodied in PCAC, or, for that matter, in vector meson dominance, and the general requirement of gauge invariance or current conservation as in Eq. 25). Other input as to the relative importance of various possible diagrams or the values of coupling constants and cutoff parameters must be phenomenological or even partially conjectural. We illustrate this process further in the following section.

8.2 Examples of Mesic Features in Nuclear Structure

The phenomena whereby virtual mesons may influence nuclear structure considerations have traditionally been placed under a variety of headings, depending on which feature of the mesic role was being stressed. These have included the notion that mesic exchange between nucleons bound in nuclei might lead to excitation of baryon resonances, which are then thought of as admixed in the nuclear states (see Fig. 8.3), or that mesic exchanges that were not merely ladders of OBE's might lead to three-body forces (as in Fig. 8.4). They have also involved the possibility of many virtual mesons contributing cooperatively in the nuclear state, as in pion condensates or abnormal nuclear matter, and the direct influence of mesic exchange on electromagnetic or weak interactions in what is usually referred to as exchange current contributions.* Since all of these features are manifestations of virtual meson presence in the nuclei, they are naturally intertwined and overlapping to some degree. We illustrate here each of them briefly, especially in terms of their pertinence to aspects of meson-nucleus interactions that we have already encountered in this book.

8.2a Isobar Configuration Admixtures in Nuclei

It is natural first to study manifestations of isobar admixtures in nuclei in the $A = 2$ system and, in particular, in its bound state, the deuteron. This state has isospin zero and so we are limited essentially to NN* or $\Delta\Delta$ configurations (see Table 2.3). The study of isobar admixtures in nuclear wave functions in fact had its inception when Kerman and Kisslinger (Ker–69) suggested that a small component of NN(1688) in the deuteron may help to explain the backward peak in scattering of protons on deuterons at about 1 GeV. Subsequent estimates, which are unfortunately very dependent on the cutoff functions used, suggest that the dominant isobar configuration admixtures are $\Delta(1232)\Delta(1232)$ and N(939)N(1470) with probabilities of very roughly 1% and 0.2%, respectively. The configuration N(939)N(1520) may also lie in about this range. Although these numbers are much smaller than the dominant S-state N(939)N(939) component of the deuteron—and indeed rather smaller than the D-state NN part as well—the isobars tend to enhance the higher-momentum components of the deuteron wave function, whence their importance for p-d backward scattering (or for that matter meson-deuteron scattering at high momentum transfer!). This is due to two effects: First, the higher mass of the isobar resonance means a larger value for the parameter $\alpha^2 = 2M_{red}|E|$ that governs the fall-off of the wave function in momentum space. Here M_{red} is the reduced mass for the pertinent two-baryon system and E is the binding energy. Second, and more important, the higher spins of the isobar resonances permit higher values of the orbital angular momentum in the two-baryon states. Thus the ordinary N(939) has spin $\frac{1}{2}$ and thus allows $L = 0$ or 2 that together with the spins can reach the $J^\pi = 1^+$ deuteron state. But a $\Delta(1232)\Delta(1232)$ configuration with two spin - $\frac{3}{2}$ particles, coupling to total spin as large as 3, can achieve $L = 4$. This higher orbital angular momentum implies a wave function in configuration space that is more severely pinched between the correspondingly higher centrifugal barrier and the long-range attraction of the potential. The greater restriction in configuration space then leads to increased support at higher momentum

---

*Exchange currents may, of course, also effect hadronic processes, including pion scattering itself; see, for example, Wil–75 and Ger–79.

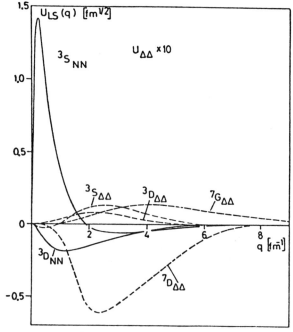

Fig. 8.6 Momentum space deuteron wave functions for the usual S- and D-state NN components and $\Delta\Delta$ admixtures totaling 0.97% probability (from Web–77).

values in the space of the Fourier transform. Examples of these effects are shown in some momentum-space deuteron wave functions in Fig. 8.6, where the large effects of $\Delta\Delta$ admixture at and above $q\sim6$ fm$^{-1}\sim1,200$ MeV/c are clear.

Even for low- or zero-momentum-transfer processes, isobar admixtures in the deuteron may have significant effects. Thus, for example, a 1% $\Delta(1232)\Delta(1232)$ component may increase the deuteron magnetic moment by as much as 5%, although again these results are dependent on the cutoff used. The effect of such an admixture on the deuteron quadrupole moment is smaller ($\sim1.5\%$ reduction) and arises mainly from reducing the NN probability by the amount given over to $\Delta\Delta$, the two isobars themselves contribute little since, as we have noted, their relative wave function does not extend out very far in configuration space while the quadrupole operator involves r^2 and thus suppresses small r contributions. Of course, at high momentum transfers in electromagnetic interactions of the deuteron, as in elastic electron scattering, the isobar configurations cause a large effect, on the order of 100% for the magnetic form factor at $q\gtrsim4$ fm^{-1}.

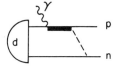

Fig. 8.7 A Δ-isobar contribution to $n + p \rightarrow d + \gamma$.

A striking manifestation of baryon resonances in the electromagnetic interactions of the deuteron occurs in radiative neutron capture $n + p \rightarrow d + \gamma$. Here a 10% discrepancy exists for the capture cross section for thermal neutrons, between experiment and theory based on $N(939)N(939)$ components only. Of this, some 6–7% is supplied by mesic exchange effects, and the remaining 3–4% by the isobar effect illustrated in Fig. 8.7 (see Section 8.2d and Ris–72). Figure 8.7 has an obvious analog for the process $\pi^+ + d \leftrightarrow p + p$, discussed in Chapter 3, where the Δ is also important.

A major aspect of isobar effects in the $A = 2$ system has to do with its role in the nucleon-nucleon force.* Here one would like to develop a picture of the force as mediated by the exchange of single bosons, as shown in Fig. 8.1 and commented on briefly in Table 1.1. Unfortunately, to obtain a fit to the NN data one requires scalar mesons $\sigma_{0,1}$ with $T = 0$ and 1 for the intermediate range part of the force with mass $m_\sigma \sim 550$ MeV, and no sharp resonance of this sort is seen in the $\pi\pi$ system. It may be that part of the required σ-meson is in fact a simulation of two-pion-exchange (TPE) effects. In particular, the TPE box diagrams of Fig. 8.8, which contain one or two intermediate Δ-isobars, have been shown to account for a good part of the σ-exchange effect, while the $\pi\rho$-exchange may produce some of the short-range strong repulsion in the NN interaction (Dur–77; see also the reviews by Gre–76, Web–78, and Eis–77).

With this understanding, a fairly clear picture has begun to emerge whereby the NN force is mediated primarily by π, ρ, and ω exchanges. The η-meson appears to couple somewhat weakly to the nucleon, and some σ-meson component may yet be required to represent nonresonant two-pion exchange. The π provides the longest range part of the force, which therefore has considerable tensor character. The intermediate range arises appreciably from TPE with intermediate isobars, and also from ρ-exchange. The ρ, as a vector meson, has both vector (convective) and tensor (magnetization) coupling of which the tensor form is expected to dominate for this isovector particle. Thus it also produces a tensor component in the NN force, but with opposite sign to that of the pion, and therefore causes a partial cancellation. The ω as an isoscalar, vector meson

---

*See also Section 2.5.

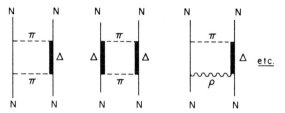

Fig. 8.8 Box diagrams with $\pi\pi$ or $\pi\rho$ exchange and internal $\Delta(1232)$ isobars.

couples primarily through the vector form and produces a strong short-range repulsion in all spin and isospin channels (the analog here of the statement that like charges repel in the case of the zero-mass vector boson —the photon). This is presumably partly responsible for the "hard core" of the NN force, with some of the rest coming from $\pi\rho$ exchange. Heavier masses may affect the situation at still shorter ranges, but this for the moment remains *terra incognita*.

This picture carries with it, of course, complications and further questions in considering the NN force: What is the effective range for transition potentials whose iteration is supposed to produce the intermediate-range effects of the box graphs of Fig. 8.8? How important are the crossed graphs of Fig. 8.9? Can the resulting NN force be taken as local and static? And, especially, what will the consequences of such an NN interaction be when used in nuclear structure calculations?

Last, in the context of the two-baryon system their remains the intriguing possibility of "bound" states in $N\Delta$ or $\Delta\Delta$ systems. Such states, whose presence is suggested by some theoretical work and has been hinted at in some experiments, would have their energies lowered with respect to the sum of the on-shell energies for the particles involved. They would also have reduced decay widths for the isobars involved. Should convincing evidence be found for the existence of such effects it would have considerable impact on our handling of pion scattering in nuclei near the 3,3 region where it would presumably require very careful attention to binding effects in the multiple-scattering treatment. Needless to say, with the

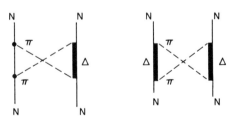

Fig. 8.9 Crossed diagrams for consideration in constructing the nucleon-nucleon interaction.

precise mechanisms of the NN force still in question it is difficult to give a definitive analysis of this binding issue so that these interrelated issues must be regarded as still open.

In considering systems with $A > 2$ we again encounter the same range of phenomena we have already noted, as well as significant new effects. For example, as we have seen, it now becomes possible to have isobar configuration admixtures in the nuclear ground state containing a single Δ which is ruled out for the deuteron. In addition to the electromagnetic effects we have mentioned, the isobar admixtures can now influence the Coulomb energy. Furthermore, one can study the effects of the isobars in weak interaction processes. Isobars also play a role in many-body forces, and in pion condensation, which are discussed in the following sections.

The direct observation of isobar configuration admixtures has been considered repeatedly (see especially Web–75 and Kis–79, as well as Are–72, Gre–76, Web–78, and Eis–77). The general idea is to perform an experiment in which the preexisting isobar, which is far off its mass shell, is nudged on shell by some probe while the other nuclear constituents participate merely as relatively passive spectators. The complication lies in the existence of "background" mechanisms in which Δs are actually produced in the reaction or various hadronic rescattering effects simulate the signal expected from the preexisting isobar. As a result, no definitive direct observation of isobar admixtures has yet been claimed. Nonetheless, their inevitability as part of the theoretical description of strongly interacting many-body systems seems clear, and we must expect to consider their effects in addressing the short-range, high-momentum phenomena that are the central issue in medium-energy nuclear physics.

8.2b Many-Body Forces

One of the direct consequences of mesic degrees of freedom in nuclei is the occurrence of many-body interactions among the nucleons. This comes about because processes involving exchange of several mesons may involve more than two nucleons, as indicated earlier, in Fig. 8.4. Since the multiple exchange of mesons contributes significantly to the two-body NN interaction, it is reasonable to expect that the many-nucleon interactions induced may not be small. How large or small these interactions actually are is not well established either by experimental evidence or convincing theoretical calculation. With many-body interactions, the effective Hamiltonian $\tilde{H}$ of Eqs. 2 and 3 for a nucleus with A nucleons, may be written

$$\tilde{H} = \sum_i^A \left(M + \frac{p_i^2}{2M} \right) + \sum_{i<j}^A V_{ij} + \sum_{i<j<k}^A V_{ijk} + \dots \qquad [40]$$

One might expect to find differences in the predictions of such a Hamiltonian in ordinary nuclear physics, compared to the conventionally assumed forms, in which only the kinetic energy and two-nucleon potential V_{ij} appear. The binding energies of ^{3}H and ^{3}He and of nuclear matter are places where one might try to estimate how much room there is for unconventional forces, based on the degree of success of conventional two-body potentials in these cases. Given the uncertainties of present theoretical calculations, there is room for a contribution of the order of 1 MeV/nucleon to the binding energy due to three-body forces. (See, e.g., the reviews of Del–72 for $A = 3$ and Day–78 for nuclear matter.) A small amount of evidence from nuclear spectra (e.g., Kol–73b) also allows for some specifically many-body interactions. The interpretation of all these cases is very elusive, however, since in general it is impossible to separate, in the study of energies, the effect of a many-nucleon interaction from that of a many-nucleon correlation induced, say, by two-nucleon forces.

The theoretical evidence that there *ought* to be many-body interactions seems much simpler. One way of looking at the problem is as follows. We assume that there are two-body interactions mediated by exchange of single mesons, as in Fig. 8.1 (OBE). For a pair of nucleons in a nucleus, the exchanged meson, say, a pion, may scatter from an intermediate nucleon on its way between the interacting pair. This three-body effect is illustrated in Fig. 8.10, where we have also indicated some simple contributions to this process based on our understanding of πN scattering. In (*a*), the intermediate nucleon is excited to a resonant N* intermediate state; in (*b*) a vector meson is exchanged between the π and N_2. Such processes have been calculated for nuclear matter: for excitation of the $\Delta(1232)$, see Gre–73 and Bla–75a (see also the review in Gre–76); for diagrams like Fig. 8.10b, with exchanges of π, ω, and $\varepsilon(\sigma)$, refer to Ued–77 and Nym–77. There is no complete agreement about how important these terms are, in part because of the many theoretical uncertainties and difficulties in the actual calculations. First of all, the diagrams of Fig. 8.10 are analogous to πN scattering diagrams, but in fact the pion is far off-energy-shell. This requires extrapolation of πN scattering to the unphysical value of $\omega_\pi = 0$. Second, the calculation of the potentials corresponding to these diagrams is technically not trivial, and third, the contribution of these potentials to the nuclear matter binding energy requires the use of correlated wave functions for the nucleons, which are also theoretically uncertain. In addition to this, whenever a given diagram, such as that of Fig. 8.10a is important, there are usually related diagrams, like that of Fig. 8.11, with contributions of comparable magnitude. Similarly, there may be many-body forces for $n > 3$ (see Bla–75b, c). This suggests that if, for example, the Δ contributes in an important way to the binding of nuclei, it may be

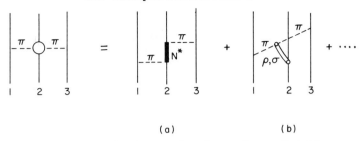

(a)   (b)

Fig. 8.10 Contributions to the three-nucleon interaction mediated by π-exchange between particles 1 and 3, with intermediate scattering on particle 2.

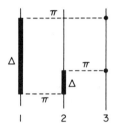

Fig. 8.11 Three-nucleon interaction involving multiple $-\Delta$ excitation.

necessary to treat the dynamic problem by a coupled channel method, in which the Δ is explicitly included in the dynamics, as in Section 8.1a. (See also Hol–77a, b.)

8.2c Pion Condensation

Pion condensation concerns the ground state of a nuclear system interacting with the pion field, that is, through exchange of pions. The most dramatic theoretical examples are for infinite nuclear matter. For $N = Z$ this is the asymptotic extension of "ordinary" stable nuclei (without Coulomb interaction) in nucleon number $A \rightarrow \infty$, at fixed nucleon density; for $N \gg Z$ this is neutron matter, appropriate to the discussion of neutron stars as astrophysical (macroscopic) objects. Now we consider two possible phases of infinite nuclear matter: *normal* and *pion condensate*. Normal nuclear matter has the characteristics of a degenerate Fermi gas: a uniform distribution of nucleons, with zero spin density (and zero isospin density for $N = Z$). The average pion field $\langle \phi(\mathbf{r}) \rangle$ in this nuclear system is zero, since a uniform source cannot produce a pseudoscalar field. There may, however, be nonzero fluctuations of the pion field: $\langle \phi^2(\mathbf{r}) \rangle \neq 0$.

The phase with a pion condensate is quite different: the density of nucleons may be nonuniform, in particular the spin-isospin density,* and the average pion field $\langle \phi(\mathbf{r}) \rangle$ may take on nonzero values as a function of $\mathbf{r}$ in the nonuniform medium (condensate). This involves a considerable rearrangement of the nucleons, and creation of mesons from the vacuum, relative to the normal phase, at the same average nucleon density. For a given density, either phase could represent the ground state of the system, depending on the balance between the attractive and repulsive interactions among the nucleons and mesons, as well as the meson kinetic energy required to produce the condensate. Therefore, one could have a (theoretical) phase transition by changing the average nucleon density (e.g., by external pressure) or the strength of the πN interaction (e.g., through the coupling constants). Most theories conclude that for very dilute systems, the ground state is normal, and the phase transition sets in with increasing density or coupling strength. The question of what that density is for nuclear matter or neutron matter, given known meson-nucleon interactions, is of considerable interest. The subject has been recently reviewed at length in Bro–76a and Mig–78, to which articles the reader is referred for a fuller account and for further bibliography.

We are interested here in making contact between the phenomenon of pion condensation and the general subject matter of this chapter, and of meson interactions with nuclei. First let us consider nuclear matter from the point of view of the effective Schroedinger equation (Eq. 2) of Section 8.1a, in which the degrees of freedom are those of the nucleons alone. It is possible for Eq. 2 to have two solutions, one spatially uniform, corresponding to normal nuclear matter, and one with spin-isospin waves. The former could be the ground state for low density or weak coupling, and the second for high density or strong coupling. But where is the pion condensate in this picture, from which the meson degrees of freedom have been removed? In part, it may be said to be in the effective Hamiltonian $\tilde{H}$, or its expectation value $\langle \tilde{H} \rangle$, which has a different spatial distribution in the two states: normal and spin-wave. But there is more to the story.

Let us consider the case of a finite nucleus, rather than nuclear matter, which we assume to have $N = Z$, even-even. Part of the energy spectrum of such a nucleus is illustrated in Fig. 8.12a: the ground state has $J^\pi = 0^+$, which is a "normal" ground state, with isospin $T = 0$. We show one $J^\pi = 0^-, T = 1$ excited state, which we assume to be a simple collective state, based on spin-isospin dipole excitation of the 0^+ ground state. A typical excitation energy is of the order of ~ 10 MeV, as illustrated. So far, the description is consistent with an effective Schroedinger equation with

*That is, standing or traveling spin-isospin waves.

nucleons only; in the simplest approximations, the ground state is an independent particle state like the Fermi gas, and the 0^- state is a simple particle-hole excitation of the ground state. We may also consider another excited $0^-, T = 1$ state of the nuclear system, which we obtain by adding a π^0 of zero kinetic energy to the ground state $(0^+, T = 0)$ of the nucleus; this is a scattering state, the lowest of a continuum starting at an excitation energy equal to the pion mass (see Fig. 8.12a). Now, because of the interaction of the pion with the nucleus, these two $0^-, T = 1$ states will actually be mixed: the dipole excitation will be coupled to the π^0 continuum state. We illustrate this in Fig. 8.12b, where the nuclear 0^- state decomposes into the ground state plus a π^0. The state at ~ 10 MeV is partly pionic, and the continuum state(s) are partly nuclear. In terms of the effective Schroedinger equation (Eqs. 2 and 3) these show up as *multiple solutions*, with different E_λ, for the same quantum numbers: $\{\lambda\} = J^\pi, T = 0^-, 1$. Here the energy dependence of the effective interaction $\tilde{H}(E)$ recovers the pionic excitation of the nucleus, which has been left out of the degrees of freedom of the effective vector space. A "phase transition" can be induced by increasing the πN coupling constant f^2, as shown in Fig. 8.12c. With increasing f^2, the 0^- state may drop below the normal 0^+ state, replacing it as the ground state. Because of the increased mixing with the π^0 continuum, which produces the "level repulsion" to drive down the 0^- state, the pionic component of the lower 0^- state is also increased. This is also part of the condensation phenomenon for this finite system. (See Bar–73b and Mig–78, p. 158, for a discussion of the shift of 0^- levels in nuclei and its relation to condensation.)

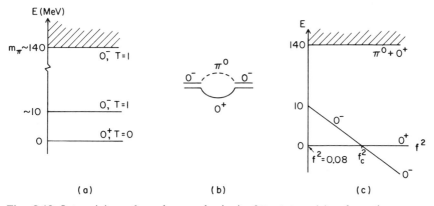

Fig. 8.12 Intermixing of nuclear and pionic 0^- states: (*a*) schematic energy spectrum of an even-even nucleus with $N = Z$; (*b*) mixing of 0^- states; (*c*) onset of pion condensation with increasing πN coupling constant f^2.

The mixing of the pionic and nuclear spin-isospin excitations can also be investigated for the case of infinite nuclear matter $(N = Z)$ by considering the propagation of a pion through the system. This is given by the Green function or propagator of Eq. 6.47, for a uniform medium

$$G(\omega, k) = \left[\omega^2 - \omega_k^2 - \Pi(\omega, k) + i\eta \right]^{-1}, \qquad \eta \to 0^+, \qquad \omega_k^2 = k^2 + m_\pi^2.$$

$$[41]$$

We may characterize an excitation of momentum $\mathbf{k}$ as having an energy of ω given by the singularities of $G(\omega, k)$, or equivalently, by solutions of the dispersion equation

$$\omega^2 = \omega_k^2 + \Pi(\omega, k). \qquad\qquad [42]$$

For the case of the coupling of a pion to a single spin-isospin wave in the nuclear medium (of the same momentum $\mathbf{k}$), there will be two solutions to Eq. 42 for each value of k^2. This is illustrated in Fig. 8.13 for two values of the nuclear density. [Here we follow Mig–78, who uses a simple first-order model for the polarization function (or optical potential) $\Pi(\omega, k)$.] We assume the πN interaction is negligible for $k^2 \to 0$, so that the upper branch has $\omega^2 \to m_\pi^2$, and corresponds to a pion excitation of the medium (the imaginary part of $\Pi(\omega, k)$ is ignored here). The lower branch has $\omega^2 \to 0$ and corresponds to a nuclear excitation mode (spin-isospin wave). For small k^2 ($< m_\pi^2$) both branches rise with increasing kinetic energy. With increasing k^2 the πN interaction increases, causing the two modes to intermix and the two branches to "repel" each other. This repulsion is greater for greater nuclear density; for sufficiently high density (Fig. 8.13b) the repulsion may cause the lower branch to reach negative values of ω^2 for some k^2. This is a sign that the system is unstable to the excitation of pions, and that condensation has set in. For this density the ground state is no longer

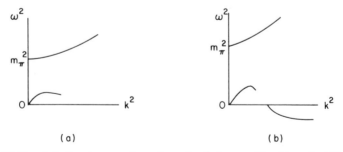

(a)   (b)

Fig. 8.13 Sketch indicating two branches of solutions to Eq. 42, as in Mig–78.

uniform, so that the use of Eqs. 41 and 42, is no longer valid. The onset of pion condensation with increasing nuclear density occurs at the lowest density for which $\omega^2 = 0$ ($k^2 > 0$). If other excitations of nuclear matter also couple to the spin-isospin wave, and/or to the propagating pion, there will be additional branches for the solutions of the dispersion equation given as Eq. 42. (A Δ-hole excitation is a simple example.) Again, we refer the reader to Bro–76a and Mig–78 for further details and discussion.

8.2d Electromagnetic Transitions and Meson Exchange Currents

One of the first reactions for which convincing theoretical evidence was provided concerning the contribution of exchange currents was neutron radiative capture at threshold $n + p \rightarrow d + \gamma$ with thermal neutrons. The experimental cross section for this process is 334.2 ± 0.5 mb, whereas calculations based on nucleon degrees of freedom only, with standard single-particle magnetic moment operators for this $M1$ process, give 302.5 ± 4.0 mb. An explanation for this 9.5% discrepancy was provided by Riska and Brown (Ris–72), who calculated the meson exchange diagrams shown in Fig. 8.14. The first of these (a) corresponds to the $\overline{NN}$ pair process in the intermediate state, and for low-energy purposes due to the high intermediate mass state can be taken as a point interaction (often called a seagull term, or, in the older literature, catastrophic term; see Eq. 7.61 and thereafter). This is evaluated by the usual "minimal coupling" replacement for the electromagnetic field in the πNN coupling term (see, for example,

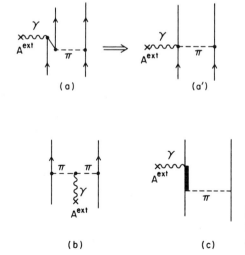

Fig. 8.14 Meson exchange graphs that contribute to radiative capture of thermal neutrons, $n + p \rightarrow d + \gamma$.

Eq. 2.65)

$$\frac{f}{m_\pi}\,\boldsymbol{\sigma}\cdot\mathbf{k}\to\frac{f}{m_\pi}\,\boldsymbol{\sigma}\cdot(\mathbf{k}-e\mathbf{A}),\qquad\qquad[43]$$

where the last term gives the form of the four-particle $\pi NN\gamma$ vertex. The second term (b) in Fig. 8.14 involves direct electromagnetic coupling of the photon to the pion exchange current of Eq. 34. The effects of these diagrams in the deuteron system are then calculated by means of two-body operators generated from the nonrelativistic limit of the graphs.

The correction due to exchange terms is taken relative to the conventional single-particle (sp) result in the form

$$\delta=\frac{\langle{}^3S_1+{}^3D_1|M^{\mathrm{exch}}|{}^1S_0\rangle}{\langle{}^3S_1|M^{\mathrm{sp}}|{}^1S_0\rangle},\qquad\qquad[44]$$

where the $M1$ nature of the low-energy capture has been incorporated explicitly. The exchange contribution for the transition from the ${}^1S_0$ continuum state to the bound state receives almost as much strength from the deuteron ${}^3D_1$ component as from the much larger ${}^3S_1$ part; in fact $\delta^{S\to S}=1.90\%$ and $\delta^{S\to D}=1.38\%$. To these must be added the contribution of Fig. 8.14c, which exists only for the ${}^1S_0\to{}^3D_1$ component and amounts to $\delta^{S\to D}(\Delta)=1.88\%$. The total effect is then to increase the cross section to ≈302.5 mb$\times(1+1.90\%+1.38\%+1.88\%)^2=335$ mb. This value is in quite good agreement with the experimental one; a slight improvement on it may be obtained if a renormalization of the ${}^1S_0$ state is made for the NΔ component in it. As noted in Section 8.2a, the appearance of the correction of Fig. 8.14c may be seen as evidence for Δ-admixture in the nuclear state.

If deuteron photodisintegration $\gamma+d\to p+n$ is considered, one can study the relative contribution of the exchange currents with energy; at about $E^\gamma_{\mathrm{lab}}\sim100$ MeV the combined effects of these plus isobar configurations become fairly large ($\sim$20–30%), though the latter dominate appreciably. The photodisintegration process is also richer in that it allows for studies of the consequences of meson currents in angular distributions and polarizations.*

A natural extension of the study of mesic currents in the processes $n+p\rightleftarrows d+\gamma$ is to consider the electron scattering reactions $e+d\to e+d$ or $e+d\to n+p+e$. At very high momentum transfer, the nucleon impulse approximation for elastic scattering may have dropped sufficiently to allow

*See Had–75, Gre–76, Web–78, and Kis–79, where the role of exchange currents (and baryon resonance admixtures) in a variety of other reactions is also discussed.

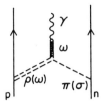

Fig. 8.15 A meson exchange graph expected to contribute appreciably to elastic electron-deuteron scattering at high momentum transfer. The three-meson vertex may involve $\rho\omega\pi$ or $\omega\omega\sigma$.

exchange current effects to show up clearly. With this in mind, Chemtob, Moniz, and Rho (and later Des Planques and Kisslinger—see Rho–76 and Kis–79) calculated the meson exchange graph of Fig. 8.15, with the result shown in Fig. 8.16. When form factors at the meson-nucleon vertices of Fig. 8.15 are included, the graph does not change the impulse approximation result overly dramatically, so that it is difficult to pinpoint the mesic current contribution. It is interesting to note that the observed fall-off of the cross section can also be given by a quark model for the deuteron (as made up of six quarks; this issue is reviewed in Rho–76).

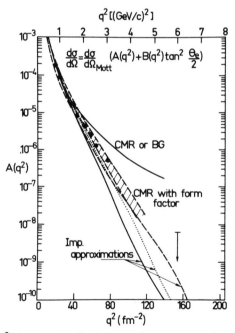

Fig. 8.16 The $A(q^2)$ form factor in electron-deuteron cross sections with (CMR) or without (impulse approximation) the graphs of Fig. 8.15. When the graphs are included, the effects of further incorporating meson-nucleon form factors are also shown.

8.2e Weak Interactions: Axial Coupling Renormalization and the Pion-Nucleus Optical Potential

We have seen in Section 8.1d that the PCAC relation of Eq. 36 implies an intimate connection involving the axial current taken between two hadronic states and the pion field between those two states. This suggests that we may use our knowledge of pion-nucleus reactions to learn something about nuclear weak interactions. A particularly attractive example of such a process is provided by the study of the quenching of the axial-vector coupling constant in nuclei that may be related to the Lorentz-Lorenz effect in the pion-nucleus optical potential, as first pointed out by M. Ericson (Eri–71b and Eri–73b) and later developed in a somewhat different formalism in Rho–74 (whose treatment we follow here) and in Oht–74.

The point of departure is the PCAC relation that we quote from Eq. 36,

$$\partial_\mu \vec{J}_\mu^A(x) = \frac{f_\pi}{\sqrt{2}}\,\vec{\phi}(x), \qquad f_\pi = 0.946\,m_\pi^3, \qquad [45]$$

between the axial current $\vec{J}_\mu^A$ and the pion field $\vec{\phi}$. We then use the Klein-Gordon equation for the pion field

$$(\Box - m_\pi^2)\vec{\phi}(x) = \vec{j}_\pi(x) \qquad [46]$$

in terms of the source $\vec{j}_\pi(x)$, which for a single-nucleon case in the nonrelativistic limit is (cf. Eqs. 2.61 and 2.62)

$$\vec{j}_\pi \sim i\frac{g}{2M}\,\vec{\tau}\,\sigma\cdot\mathbf{k}, \qquad [47]$$

where $\mathbf{k}$ is the pion three-momentum. Taking the matrix elements of Eqs. 45 and 46 between general (nuclear) states i and f of momenta p_i and p_f, with $k = p_i - p_f$ and $k^2 = \mathbf{k}^2 - k_0^2$, we have

$$ik_\mu\langle f|J_\mu^{A\pm}|i\rangle = \frac{f_\pi}{k^2 + m_\pi^2}\langle f|j_\pi^\pm|i\rangle \qquad [48]$$

for the charge raising or lowering components, $+$ or $-$, which will be relevant for β^- or β^+ decay in J_μ^A and π^+ or π^- absorption in j_π. For β decay, the four-momentum transfer is small, and the "soft pion" limit $k_\mu \to 0$ is relevant. First taking the fourth component to zero, and then $\mathbf{k} \to 0$, we have

$$i\mathbf{k}\cdot\langle f|\mathbf{J}^{A\pm}|i\rangle = \frac{f_\pi}{m_\pi^2}\langle f|j_\pi^\pm|i\rangle. \qquad [49]$$

This may then be expanded in a Taylor series in $\mathbf{k}$ to obtain

$$i\mathbf{k} \cdot \langle f|\mathbf{J}^{A\pm}|i\rangle|_{\mathbf{k}=0} = \frac{f_\pi}{m_\pi^2}\mathbf{k} \cdot \left(\nabla_{\mathbf{k}}\langle f|j_\pi^\pm|i\rangle\right)_{\mathbf{k}=0}, \qquad [50]$$

to order k, where we have used the fact that the matrix element of the pseudoscalar pion source operator vanishes for zero momentum transfer. From Eq. 50 we identify

$$i\langle f|\mathbf{J}^{A\pm}|i\rangle\Big|_{\mathbf{k}=0} = \frac{f_\pi}{m_\pi^2}\nabla_{\mathbf{k}}\langle f|j_\pi^\pm|i\rangle\Big|_{\mathbf{k}=0}, \qquad [51]$$

and from this we conclude that if we are able to calculate the complete virtual pion-nucleus vertex, then we have information on the corresponding exact axial vector matrix element.

By now we have accumulated much information on the ingredients that affect the pion-nucleus interaction and the πAA vertex, and we may anticipate that the nuclear absorption vertex for the virtual pion will involve distortion of the pion in a nuclear optical potential, followed by the actual absorption on a nucleon, mediated by the operator of Eq. 47. Rho (Rho-74) supplies details for this reasoning and also argues that the main effect of relevance in the pion-nucleus optical potential for the virtual meson with $\omega = k_0 \approx 0$ is the Lorentz-Lorenz feature discussed in Section 5.1e. The coupling of Eq. 47 yields a single-nucleon operator term on the right-hand side in Eq. 51 of the form

$$i\frac{f_\pi}{m_\pi^2}\frac{g}{2M}\vec{\tau}\sigma = \frac{i}{\sqrt{2}}g_A\vec{\tau}\sigma, \qquad [52]$$

where we have used the Goldberger-Treiman relation in Eq. 39.

The consequence of the pion-nucleus distortion on the right-hand side in the full nuclear relation of Eq. 51 is to modify the effective axial strength in the nuclear case. If the primary distortive effect derives from the full, unsuppressed Lorentz-Lorenz feature, this takes the form of a renormalized axial coupling constant

$$g_A^{\text{eff}} = \frac{g_A}{1+\alpha}, \qquad \alpha = \frac{4\pi}{3}c_0 A\rho. \qquad [53]$$

Here α is the factor in the Lorentz-Lorenz denominator and has the value $\alpha \approx 0.3$ in nuclear matter (see Eq. 5.48 and immediately thereafter). Since $g_A \cong 1.23$, as noted in Eq. 33, this implies an effective axial coupling in nuclei that is quenched by some 30%, or $g_A^{\text{eff}} \approx 0.77 g_A \approx 0.95$. This value is roughly appropriate for very heavy nuclei, while the smaller effective density in lighter nuclei may imply $\alpha \sim 0.16$ and $g_A^{\text{eff}} \sim 0.86 g_A \sim 1.1$. These last values would seem to compare favorably with the roughly 10% quenching in g_A found in β- decay in light nuclei, thus providing some support for a picture in which features of nuclear weak interactions may be elucidated through our understanding of pion-nucleus interactions.

References

The section of the book in which a reference is cited is given in brackets [ ].

Aar-68   R. Aaron, R. D. Amado, and J. E. Young, *Phys. Rev.*, **174**, 2022 (1968) [3.4, 6.1]

Afa-65   B. P. Afanasev and V. I. Oustroumov, *Yad. Fiz.* (USSR) **1**, 647 (1965); *Sov. J. Nucl. Phys.* **1**, 463 (1965) [7.1].

Afn-74   I. R. Afnan and A. W. Thomas, *Phys. Rev.*, **C10**, 109 (1974) [3.3, 3.5]

Aga-73   D. Agassi and A. Gal, *Ann. Phys.*, **75**, 56 (1973) [5.1]

Aga-75a   D. Agassi, A. Gal, and V. Mandelzweig, *Ann. Phys.*, **91**, 194 (1975) [5.1]

Aga-75b   D. Agassi and A. Gal, *Ann. Phys.*, **94**, 184 (1975) [4.3, 5.1]

Aga-75c   D. Agassi, H. A. Weidenmüller, and G. Montzouranis, *Phys. Rep.*, **22C**, 147 (1975) [7.2]

Aga-77   D. Agassi, C. M. Ko, and H. A. Weidenmüller, *Ann. Phys.*, **107**, 140 (1977) [7.2]

Ahm-65   A. Ahmadzadeh and J. A. Tjon, *Phys. Rev.*, **139**, B1085 (1965) [3.2]

Alb-76   M. Alberg, E. M. Henley, and L. Wilets, *Ann. Phys.*, **96**, 43 (1976) [5.3]

Ale-65   V. A. Alessandrini and R. L. Omnes, *Phys. Rev.*, **139**, B167 (1965) [3.4]

Ale-67   G. Alexander, Ed., *High Energy Physics and Nuclear Structure*, Rehovot, 1967, (North-Holland, Amsterdam, 1967) [5]

All-72   B. W. Allardyce *et al.*, *Phys. Lett.*, **41B**, 577 (1972) [5.1]

All-73   B. W. Allardyce *et al.*, *Nucl. Phys.*, **A209**, 1 (1973) [5.1]

Alm-72   S. Almehed and C. Lovelace, *Nucl. Phys.*, **B40**, 157 (1972) [2.2]

Als-79   J. Alster and J. Warszawski, *Phys. Rep.*, **52**, 87 (1979) [5.4]

And-66   F. Ando and D. Feldman, *Suppl. Prog. Theoret. Phys.*, **37 & 38**, 524 (1966) [3.4]

Are-72   H. Arenhövel and H. J. Weber, *Springer Tracts in Modern Physics*, **65**, 58 (1972) [8.2]

Aue-72   N. Auerbach, J. Hüfner, A. K. Kerman, and C. M. Shakin, *Revs. Mod. Phys.*, **44**, 48 (1972) [5.4]

Aue-77   N. Auerbach, *Phys. Rev. Lett.*, **38**, 804 (1977) [6.4]

Aul-78   E. G. Auld *et al.*, *Phys. Rev. Lett.*, **41**, 462 (1978) [7.3]

379

Aus–70   N. Austern, *Direct Nuclear Reaction Theories*, (Wiley, New York, 1970) [6, 6.2]

Aus–77   N. Austern, F. Tabakin, and M. Silver, *Am. J. Phys.*, **45**, 361 (1977) [4]

Axe–76   D. Axen *et al.*, *Nucl. Phys.*, **A256**, 387 (1976) [3.3]

Bac–72   G. Backenstoss *et al.*, *Phys. Lett.*, **38B**, 181 (1972) [5.3]

Bai–74   J. Bailey *et al.*, *Phys. Lett.*, **50B**, 403 (1974) [3.3]

Baj–75   K. K. Bajaj and Y. Nogami, *Phys. Rev. Lett.*, **34**, 701 (1975) [2.4]

Bal–69   R. Balian and E. Brezin, *Nuovo Cimento*, **61B**, 403 (1969) [3.2]

Bar–68   P. Bareyre, C. Bricman, and G. Villet, *Phys. Rev.*, **165**, 1730 (1968) [2.2]

Bar–71   W. A. Bardeen and E. W. Torigoe, *Phys. Rev.*, **C3**, 1785 (1971) [5.3]

Bar–73a   S. Barshay, V. Rostokin, and G. Vagradov, *Nucl. Phys.*, **B59**, 189 (1973) [5.1]

Bar–73b   S. Barshay and G. E. Brown, *Phys. Lett.*, **47B**, 107 (1973) [8.2]

Bar–76   P. D. Barnes, R. A. Eisenstein, and L. S. Kisslinger, Eds., *Meson-Nuclear Physics*-1976, (American Institute of Physics, New York, 1976) [5] See also Dov–76b, Eis–76a, Gal–76b, Kis–76b, Mil–76, Mon–76, Tab–76, Tho–76b, and Wal–76.

Bas–77   B. Bassaleck, W. D. Klotz, F. Takeutchi, H. Ullrich, and M. Furic, *Phys. Rev.*, **C16**, 1526 (1977) [7.3]

Bat–73   C. J. Batty, G. T. A. Squier, and G. K. Turner, *Nucl. Phys.*, **B67**, 492 (1973) [5.2]

Bay–75   G. Baym and G. E. Brown, *Nucl. Phys.*, **A247**, 395 (1975); **A262**, 539 (1976) [5.1, 6.4]

Bec–71   F. Becker *et al.*, Eds., *Proceedings of the International Seminar on Pi-Meson Nucleus Interactions* (Strasbourg, 1971) [5] See also Bin–71b.

Beg–61   M. A. B. Bèg, *Ann. Phys.*, **13**, 110 (1961) [5.1].

Bei–73   J. Beiner, *Nucl. Phys.*, **B53**, 349 (1973) [5.2]

Ben–76   J. J. Benayoun, C. Gignoux, and J. Gillespie, *Nucl. Phys.*, **A274**, 525 (1976) [4.3]

Bet–73   H. A. Bethe, *Phys. Rev. Lett.*, **30**, 105 (1973) [5.1]

Bet–78   H. A. Bethe and M. B. Johnson, *Nucl. Phys.*, **A305**, 418 (1978) [4.3]

Bin–70   F. Binon *et al.*, *Nucl. Phys.*, **B17**, 168 (1970) [5.1, 5.5]

Bin–71a   F. Binon *et al.*, *Nucl. Phys.*, **B33**, 42 (1971) [5.2]

Bin–71b   F. Binon, in Bec–71, p. II–25 [5.1]

Bin–75   F. Binon *et al.*, *Phys. Rev. Lett.*, **35**, 145 (1975) [5.1]

Bir–75   A. Birnbaum and H. Feshbach, private communication (1975); A. Birnbaum, M.I.T. doctoral thesis (1975) [5.1]

Bjo–64   J. D. Bjorken and S. D. Drell, *Relativistic Quantum Mechanics* (McGraw-Hill, New York, 1964) [2, 2.1, 2.4, 2.5]

Bjo–65   J. D. Bjorken and S. D. Drell, *Relativistic Quantum Fields* (McGraw-Hill, New York, 1965) [2, 2.4, 3.4, 6, 6.2]

Bla–52   J. M. Blatt and V. F. Weisskopf, *Theoretical Nuclear Physics* (Wiley, New York, 1952) [3.5, 4.1]

Bla–66   R. Blankenbecler and R. Sugar, *Phys. Rev.*, **142**, 1051 (1966) [3.4]

Bla–75a   D. W. E. Blatt and B. H. J. MacKellar, *Phys. Rev.*, **C11**, 614 (1975) [8.2]

Bla–75b   D. W. E. Blatt and B. H. J. MacKellar, *Phys. Rev.*, **C11**, 2040 (1975) [8.2]

Bla–75c   D. W. E. Blatt and B. H. J. MacKellar, *Phys. Rev.*, **C12**, 637 (1975) [8.2]

Blo–58   C. Bloch and J. Horowitz, *Nucl. Phys.*, **8**, 91 (1958) [8.1]

Boh–69   A. Bohr and B. R. Mottelson, *Nuclear Structure* (Benjamin, New York, 1969) [8.1]

Bol–74   D. Bollé and T. A. Osborn, *Phys. Rev.*, **C9**, 441 (1974) [3.2]

Bra–67   B. Brandow, *Revs. Mod. Phys.*, **39**, 771 (1967) [8.1]

Bra–69   B. H. Bransden, in *High Energy Physics*, E. H. S. Burhop, Ed., vol. 3 (Academic, New York, 1969) p. 1 [2, 2.4]

Bra–73   B. H. Bransden and R. G. Moorehouse, *The Pion-Nucleon System* (Princeton University Press, Princeton, 1973) [2, 2.2, 2.4]

Bra–75   D. D. Brayshaw, *Phys. Rev.*, **C11**, 1196 (1975) [3.4]

Bra–77   M. Brack, D. O. Riska, and W. Weise, *Nucl. Phys.*, **A287**, 425 (1977) [3.5]

Bre–63   J. J. Brehm and J. Sucher, *Ann. Phys.*, **25**, 1 (1963) [3.4]

Bro–60   T. A. Brody and M. Moshinsky, *Tables of Transformation Brackets*, (Instituto de Fisica, Mexico, 1960) [7.3]

Bro–69   M. M. Broido, *Rep. Prog. Phys.*, **32**, 493 (1969) [2.1, 3.4]

Bro–71   G. E. Brown, *Unified Theory of Nuclear Models and Forces*, 3rd rev. ed. (North-Holland, Amsterdam, 1971) [6.4]

Bro–75   G. E. Brown and W. Weise, *Phys. Rep.*, **22C**, 279 (1975) [2.4, 5.1, 6, 6.3, 6.4]

Bro–76a   G. E. Brown and W. Weise, *Phys. Rep.*, **27C**, 1 (1976) [5.1, 8.2]

Bro–76b   G. E. Brown and A. D. Jackson, *The Nucleon-Nucleon Interaction* (North-Holland, Amsterdam, 1976) [2, 2.5]

Bru–53   K. A. Brueckner, *Phys. Rev.*, **89**, 943 (1953) [3.3]

Bug–73   D. V. Bugg, A. A. Carter, J. R. Carter, *Phys. Lett.*, **44B**, 278 (1973) [2.2, 2.4]

Bur–62   E. H. S. Burhop, A. K. Common, and K. Higgins, *Nucl. Phys.*, **39**, 644 (1962) [7.3]

Cam–73   J. B. Cammarata and M. K. Banerjee, *Phys. Rev. Lett.*, **31**, 610 (1973) [6.1]

Cam–75   J. B. Cammarata and M. K. Banerjee, *Phys. Rev.*, **C12**, 1595 (1975) [6.3]

Car–70   C. Carlson, *Phys. Rev.*, **C2**, 1224 (1970) [3.3]

Car–73   J. R. Carter, D. V. Bugg, and A. A. Carter, *Nucl. Phys.*, **B58**, 378 (1973) [2.2]

382 References

Car–76	A. S. Carroll et al., Phys. Rev., **C14**, 635 (1976) [5.2]
Cas–56	L. Castillejo, R. H. Dalitz, and F. Dyson, Phys. Rev., **101**, 453 (1956) [2.4, 6.3]
Cas–71	P. J. Castleberry, L. Coulson, R. C. Minehart, and K. O. H. Ziock, Phys. Lett., **34B**, 57 (1971) [7.1]
Cel–74	L. S. Celenza, M. K. Liou, L. C. Liu, and C. M. Shakin, Phys. Rev., **C10**, 398 (1974) [6.3]
Cel–75	L. S. Celenza, L. C. Liu, and C. M. Shakin, Phys. Rev., **C11**, 1593 (1975); **C12**, 721 (1975) [6.1]
Cer–63	International Conference on High Energy Physics and Nuclear Structure, Geneva, 1963, CERN 63-28 (CERN, Geneva) [5]
Cha–63	R. Chand, Ann. Phys., **22**, 438 (1963) [3.3]
Cha–66	J. Chahoud, G. Russo, and F. Selleri, Nuovo Cimento, **45**, 38 (1966) [3.4]
Che–56	G. F. Chew and F. E. Low, Phys. Rev., **101**, 1570 (1956) [2, 2.2, 2.4, 5.1]
Che–71	M. Chemtob and M. Rho, Nucl. Phys., **A163**, 1 (1971) [8.1]
Che–79	M. Chemtob, in Rho–79 [8.1]
Clo–74	A. S. Clough et al., in Tib–74, p. 213 [5.1, 5.2]
Com–64	A. K. Common and K. Higgins, Nucl. Phys., **60**, 465 (1964) [7.3]
Coo–76	M. D. Cooper and M. B. Johnson, Nucl. Phys., **A260**, 352 (1976) [5.2]
Coo–77	M. D. Cooper, M. B. Johnson, and G. B. West, Nucl. Phys., **A292**, 350 (1977) [5.2]
Cro–69	K. M. Crowe, A. Fainberg, J. Miller, and A. S. L. Parsons, Phys. Rev., **180**, 1349 (1969) [5.1]
Cut–58	R. E. Cutkosky, Phys. Rev., **112**, 1027 (1958) [3.4]
Czy–71	W. Czyż, in Advances in Nuclear Physics, M. Baranger and E. Vogt, Eds., Vol. 4 (Plenum, New York, 1971) p. 61 [4]
Czy–75	W. Czyż, Interactions of High Energy Particles with Nuclei, National Bureau of Standards Monograph 139 (National Bureau of Standards, Washington, 1975) [4]
Dah–73	S. Dahlgren, P. Grafström, B. Höistad, and A. Åsberg, Phys. Lett., **47B**, 439 (1973) [7.3]
Dal–60	R. H. Dalitz and S. F. Tuan, Ann. Phys., **10**, 307 (1960) [2.2]
Dal–76a	R. H. Dalitz and A. Gal, Phys. Lett., **64B**, 19 (1976) [5.4]
Dal–76b	R. H. Dalitz and A. Gal, Phys. Rev. Lett., **36**, 362 (1976) [5.4]
Day–78	B. D. Day, Revs. Mod. Phys., **50**, 495 (1978) [8.2]
Ded–76	J. P. Dedonder and C. Schmit, Phys. Lett., **65B**, 131 (1976) [4.3]
Def–66	T. deForest and J. D. Walecka, Adv. Phys., **15**, 1 (1966) [5.4]
Del–72	L. M. Delves, in Advances in Nuclear Physics, M. Baranger and E. Vogt, Eds., Vol. 5 (Plenum, New York, 1972) p. 1 [8.2]
Del–75	J. Delorme, M. Ericson, and G. Fäldt, Nucl. Phys., **A240**, 493 (1975) [7.3]
Del–76	J. Delorme and M. Ericson, Phys. Lett., **60**, 451 (1976) [5.1]

Del–77   E. Delacroix and F. Gross, *Phys. Lett.*, **66B**, 337 (1977) [7.2]

Dev–70   S. Devons, Ed., *High-Energy Physics and Nuclear Structure* (Plenum, New York, 1970) [5]

Dil–77   M. Dillig, R. Händel, and M. G. Huber, in Loc–77, contributed paper [5.1]

Dov–71   C. B. Dover, J. Hüfner, and R. H. Lemmer, *Ann. Phys.*, **66**, 248 (1971) [5.1, 6.3]

Dov–73a   C. B. Dover and R. H. Lemmer, *Phys. Rev.*, **C7**, 2312 (1973) [5.1, 6.3]

Dov–73b   C. B. Dover, *Ann. Phys.*, **79**, 441 (1973) [6.3]

Dov–73c   C. B. Dover, in *Proceedings of Summer Study Meeting on Nuclear and Hypernuclear Physics with Kaon Beams*, H. Palevsky, Ed. (Brookhaven National Laboratory Informal Report, 1973) p. 264 [7]

Dov–74   C. B. Dover, D. J. Ernst, R. A. Friedenberg, and R. M. Thaler, *Phys. Rev. Lett.*, **33**, 728 (1974) [2.4]

Dov–76a   C. B. Dover, and R. H. Lemmer, *Phys. Rev.*, **C14**, 2211 (1976) [5.1]

Dov–76b   C. B. Dover, in Bar–76, p. 249 [5.2]

Dro–75   B. J. Dropesky *et al.*, *Phys. Rev. Lett.*, **34**, 821 (1975) [5.6]

Dub–75   S. Dubnička and O. Dumbrajs, *Phys. Rep.*, **19C**, 141 (1975) [5.2]

Dur–77   J. W. Durso, M. Saarela, G. E. Brown, and A. D. Jackson, *Nucl. Phys.*, **A287**, 445 (1977) [8.2]

Dyt–79   S. A. Dytman *et al.*, *Phys. Rev.*, **C19**, 971 (1979) [5.1]

Dzh–72   V. P. Dzhelepov, Ed., *Proceedings of the Sixth International Conference on High Energy Physics and Nuclear Structure* (Dubna, 1972) [5] See also Kol–71.

Edw–71   G. W. Edwards and E. Rost, *Phys. Rev. Lett.*, **26**, 785 (1971) [5.5]

Eis–70   J. M. Eisenberg and W. Greiner, *Excitation Mechanisms of the Nucleus-Nuclear Theory*, Vol. 2 (North-Holland, Amsterdam, 1970) [8.1]

Eis–72   J. M. Eisenberg, *Ann. Phys.*, **71**, 542 (1972) [4.4]

Eis–73a   J. M. Eisenberg and H. J. Weber, *Phys. Lett.*, **45B**, 110 (1973) and in Gib–73, p. 162 [5.1]

Eis–73b   J. M. Eisenberg, J. Hüfner, and E. J. Moniz, *Phys. Lett.*, **47B**, 381 (1973) [5.1]

Eis–74   J. M. Eisenberg, *Phys. Lett.*, **49B**, 224 (1974) [5.1]

Eis–75a   J. M. Eisenberg and A. Gal, *Phys. Lett.*, **58B**, 390 (1975) [5.4]

Eis–75b   J. M. Eisenberg, in Nag–75, p. 17 [5.6, 7, 7.1]

Eis–75c   J. M. Eisenberg, J. V. Noble, and H. J. Weber, *Phys. Rev.*, **C11**, 1048 (1975) [6.1]

Eis–76a   R. Eisenstein, in Bar–76, p. 55 [5.1]

Eis–76b   J. M. Eisenberg, *Phys. Rev.*, **C14**, 2343 (1976) [5.3]

Eis–77   J. M. Eisenberg and H. J. Weber, in Loc–77 [8.2]

Eis–79   J. M. Eisenberg, J. V. Noble, and H. J. Weber, *Phys. Rev.*, **C19**, 276 (1979) [7.3]

384 References

Eli–69	J. E. Elias *et al.*, *Phys. Rev.*, **177**, 2075 (1969) [3.1]
Ell–77	P. J. Ellis and E. Osnes, *Revs. Mod. Phys.*, **49**, 777 (1977) [8.1]
Eri–66a	M. Ericson and T. E. O. Ericson, *Ann. Phys.*, **36**, 323 (1966) [5.1, 5.3]
Eri–66b	T. E. O. Ericson, in *Proceedings of Williamsburg Conference on Intermediate Energy Physics, 1966*, H. O. Funsten *et al.*, Eds. (College of William and Mary, Williamsburg, 1966) p. 187 [5.1]
Eri–70a	T. E. O. Ericson and J. Hüfner, *Phys. Lett.*, **33B**, 601 (1970) [5.1]
Eri–70b	T. E. O. Ericson and M. P. Locher, *Nucl. Phys.*, **A148**, 1 (1970) [5.2]
Eri–71a	T. E. O. Ericson *et al.*, Eds., *Spring School on Pion Interactions at Low and Medium Energies*, Zuoz, 1971, (CERN, Geneva, 1971) [5] See also Wil–71.
Eri–71b	M. Ericson, *Ann. Phys.*, **63**, 562 (1971) [8.2]
Eri–72	M. Ericson and M. Rho, *Phys. Rep.*, **5C**, 58 (1972) [7.3]
Eri–73a	T. E. O. Ericson, in Gib–73, p. 113 [5.1]
Eri–73b	M. Ericson, A. Figureau, and C. Thévenet, *Phys. Lett.*, **45B**, 19 (1973) [8.2]
Eri–75	M. Ericson and M. Krell, *Nucl. Phys.*, **A241**, 487 (1975) [5.2]
Eri–78	T. E. O. Ericson, private communication (1978) [4.4]
Ern–78	D. J. Ernst and M. B. Johnson, *Phys. Rev.*, **C17**, 247 (1978) [2.4]
Fad–61a	L. D. Faddeev, *Sov. Phys. JETP*, **12**, 1014 (1961) [3.2]
Fad–61b	L. D. Faddeev, *Sov. Phys. Dokl.*, **6**, 384 (1961) [3.2]
Fad–63	L. D. Faddeev, *Sov. Phys. Dokl.*, **7**, 600 (1963) [3.2]
Fad–65	L. D. Faddeev, *Mathematical Aspects of the Three-Body Problem in the Quantum Scattering Theory* (Israel Program for Scientific Translations, Jerusalem, 1965) [3.2]
Fal–69	G. Fäldt, *Nucl. Phys.*, **B10**, 597 (1969) [3.4]
Fal–77	G. Fäldt, *Phys. Scr.*, **16**, 81 (1977) [4.3]
Fes–58	H. Feshbach, *Ann. Phys.*, **5**, 357 (1958) [2.3, 6, 6.2, 8.1]
Fes–62	H. Feshbach, *Ann. Phys.*, **19**, 287 (1962) [2.3, 6, 6.2, 8.1]
Fes–66	H. Feshbach and A. K. Kerman, in *Preludes in Theoretical Physics*, A. de-Shalit *et al.*, Eds. (North-Holland, Amsterdam, 1966), p. 260 [5.4]
Fes–67	H. Feshbach, A. K. Kerman, and R. Lemmer, *Ann. Phys.*, **41**, 230 (1967) [6, 6.2, 6.3, 6.4]
Fes–70	H. Feshbach and J. Hüfner, *Ann. Phys.*, **56**, 268 (1970) [5.1]
Fes–71	H. Feshbach, A. Gal, and J. Hüfner, *Ann. Phys.*, **66**, 20 (1971) [4.2, 5.1]
Fet–65	A. L. Fetter and K. M. Watson, in *Advances in Theoretical Physics*, K. A. Brueckner, Ed., Vol. 1, (Academic, New York, 1965) p. 115 [4]
Fet–71	A. L. Fetter and J. D. Walecka, *Quantum Theory of Many-Particle Systems* (McGraw-Hill, New York, 1971) [8.1]
Fol–69	L. L. Foldy and J. D. Walecka, *Ann. Phys.*, **54**, 447 (1969) [4.3, 5.1]
Fow–66	G. N. Fowler and P. N. Poulopoulos, *Nucl. Phys.*, **77**, 689 (1966) [7.3]

Fra–53a   N. C. Francis and K. M. Watson, *Am. J. Phys.*, **21**, 659 (1953) [4]

Fra–53b   N. C. Francis and K. M. Watson, *Phys. Rev.*, **92**, 291 (1953) [4]

Fra–56   R. M. Frank, J. L. Gammel, and K. M. Watson, *Phys. Rev.*, **101**, 891 (1956) [4, 4.3]

Fra–66   V. Franco and R. J. Glauber, *Phys. Rev.*, **142**, 1195 (1966) [4]

Fre–66   D. Z. Freedman, C. Lovelace, and J. M. Namysłowski, *Nuovo Cimento*, **43A**, 258 (1966) [3.4]

Fri–74   J. L. Friar, *Phys. Rev.*, **C10**, 955 (1974) [7.2]

Fri–75   W. Friedman, *Phys. Rev.*, **C12**, 1294 (1975) [6.4]

Fri–77   J. L. Friar, *Phys. Rev.*, **C15**, 1783 (1977) [7.2]

Gab–74   K. Gabathuler and C. Wilkin, *Nucl. Phys.*, **B70**, 215 (1974) [3.3]

Gal–75a   A. Gal, in *Advances in Nuclear Physics*, M. Baranger and E. Vogt, Eds., Vol. 8, (Plenum, New York, 1975) p. 1 [5.4]

Gal–75b   A. Gal, in Nag–75, p. 185 [5.4]

Gal–76a   A. Gal and J. M. Eisenberg, *Phys. Rev.*, **C14**, 1273 (1976) [5.4]

Gal–76b   A. Gal, in Bar–76, p. 694 [5.4]

Gar–74   H. Garcilazo and J. M. Eisenberg, *Nucl. Phys.*, **A220**, 13 (1974) [5.3, 7.2, 7.3]

Gar–76   M. Gari and H. Hyuga, *Z. Phys.*, **A277**, 291 (1976) [8.1]

Gar–78   H. Garcilazo, *Nucl. Phys.*, **A302**, 493 (1978) [4.3, 5.1]

Gas–66   S. Gasiorowicz, *Elementary Particle Physics* (Wiley, New York, 1966) [5.2]

Ger–79   J. F. Germond and C. Wilkin, in Rho–79, [8.2]

Gib–71   W. Gibbs, *Phys. Rev.*, **C3**, 1127 (1971) [3.1, 3.3]

Gib–73   W. R. Gibbs and B. F. Gibson, Eds., *Lectures from the LAMPF Summer School On the Theory of Pion-Nucleus Scattering*, LA-5443-C (National Technical Information Service, Washington, 1973) [4, 5] See also Eis–73a, Eri–73a, Hes–73, Nat–73a, Ros–73, Wal–73d.

Gib–76   W. R. Gibbs, B. F. Gibson, A. T. Hess, G. J. Stephenson, Jr., and W. B. Kaufmann, *Phys. Rev.*, **C13**, 2433 (1976) [4.2, 5.1, 7.2]

Gib–77   W. R. Gibbs, private communication (1977) [5.4]

Gil–64   L. Gilley et al., *Phys. Lett.*, **11**, 244 (1964) [5.4]

Gin–78   J. N. Ginocchio, *Phys. Rev.*, **C17**, 195 (1978) [7.1]

Gla–55   R. J. Glauber, *Phys. Rev.*, **100**, 242 (1955) [3.1, 4]

Gla–59   R. J. Glauber, in *Lectures in Theoretical Physics*, W. E. Brittin et al., Eds., Vol. 1, (Interscience, New York, 1959) p. 315 [3.1, 4]

Gla–67   R. J. Glauber, in *High Energy Physics and Nuclear Structure*, G. Alexander, Ed., (North-Holland, Amsterdam, 1967) p. 311 [3.1, 4]

Gla–70   R. J. Glauber, in *High-Energy Physics and Nuclear Structure*, S. Devons, Ed., (Plenum, New York, 1970) p. 207 [3.1, 4]

Gol–64   M. L. Goldberger and K. M. Watson, *Collision Theory* (Wiley, New

York, 1964) [1, 2, 2.1, 2.3, 4, 5.1, 5.6, 6.1, 6.2]

Gop-74   B. Goplen, W. Gibbs, and E. Lomon, *Phys. Rev. Lett.*, **32**, 1012 (1974) [3.5]

Gra-65   I. S. Gradshteyn and I. M. Ryzhik, *Table of Integrals, Series and Products* (Academic, New York, 1965) [4.1, 4.4]

Gre-73   A. M. Green, T. K. Dahlblom, and T. Kouki, *Nucl. Phys.*, **A209**, 52 (1973) [8.2]

Gre-76   A. M. Green, *Rep. Prog. Phys.*, **39**, 1109 (1976) [8.2]

Gro-74a   F. Gross, *Phys. Rev.*, **D10**, 223 (1974) [6.1]

Gro-74b   Z. Grossmann, F. Lenz, and M. P. Locher, *Ann. Phys.*, **84**, 348 (1974) [7.3]

Gua-71   R. Guardiola, *Nuovo Cimento*, **3A**, 747 (1971) [5.5]

Gue-64   R. F. Guertin and F. Stern, *Phys. Rev.*, **134**, A427 (1964) [5.1]

Gup-76   M. K. Gupta and G. E. Walker, *Nucl. Phys.*, **A256**, 444 (1976) [5.5]

Gur-75   S. A. Gurvitz, Y. Alexander, and A. S. Rinat, *Ann. Phys.*, **93**, 152 (1975) [5.1]

Gur-76   S. A. Gurvitz, Y. Alexander, and A. S. Rinat, *Ann. Phys.*, **98**, 346 (1976) [5.1]

Hac-78   F. Hachenberg and H. J. Pirner, *Ann. Phys.*, **112**, 401 (1978) [5.3]

Had-75   E. Hadjimichael, in *Effets Mésoniques dans les Noyaux, Comptes Rendus, Saclay, 1975*, J. R. Bellicard *et al.*, Eds. (Commissariat à l'Energie Atomique, Saclay, 1975), p. 143 [8.1, 8.2]

Haf-70   P. K. Haff and J. M. Eisenberg, *Phys. Lett.*, **33B**, 133 (1970) [5.3]

Ham-63   J. Hamilton and W. S. Woolcock, *Revs. Mod. Phys.*, **35**, 737 (1963) [2.2]

Ham-67   J. Hamilton, in *High Energy Physics*, E. H. S. Burhop, Ed., Vol. 1, (Academic, New York, 1967) p. 194 [2, 2.4]

Har-69   D. R. Harrington, *Phys. Rev.*, **184**, 1745 (1969) [3.1, 4.4]

Har-73   G. D. Harp, K. Chen, G. Friedlander, Z. Fraenkel, and J. M. Miller, *Phys. Rev.*, **C8**, 581 (1973) [7.1]

Hel-76   L. Heller, G. Bohannon, and F. Tabakin, *Phys. Rev.*, **C13**, 742 (1976) [2.1, 5.1, 6.1]

Hen-69   E. M. Henley, in *Isospin in Nuclear Physics*, D. H. Wilkinson, Ed. (North-Holland, Amsterdam, 1969) [2.2]

Hes-58   W. N. Hess, *Revs. Mod. Phys.*, **30**, 368 (1958) [2.2]

Hes-73   A. T. Hess and J. M. Eisenberg, *Phys. Lett.*, **47B**, 311 (1973) and in Gib-73, p. 177 [5.1]

Hes-75   A. T. Hess and J. M. Eisenberg, *Nucl. Phys.*, **A241**, 493 (1975) [5.5]

Het-65a   J. H. Hetherington and L. H. Schick, *Phys. Rev.*, **137B**, 935 (1965) [3.3]

Het-65b   J. H. Hetherington and L. H. Schick, *Phys. Rev.*, **138B**, 1411 (1965) [3.3]

Het-66   J. H. Hetherington and L. H. Schick, *Phys. Rev.*, **141**, 1314 (1966) [3.3]

Hew-69   P. W. Hewson, *Nucl. Phys.*, **A133**, 659 (1969) [5.6]

Hil-19   E. Hilb, *Math. Z.*, **5**, 17 (1919); **8**, 79 (1920), quoted in *La Theorie des Fonctions de Bessel*, G. Petiau (Centre National de la Recherche Scientifique, Paris, 1955), p. 297 [4.4]

Hir-77a   M. Hirata, F. Lenz, and K. Yazaki, *Ann. Phys.*, **108**, 116 (1977) [5.1, 6.4]

Hir-77b   M. Hirata, J. H. Koch, F. Lenz, and E. J. Moniz, *Phys. Lett.*, **70B**, 281 (1977) [6.4]

Ho-75   H. W. Ho, M. Alberg, and E. M. Henley, *Phys. Rev.*, **C12**, 217 (1975) [6.1]

Hoe-74   M. Hoenig and A. S. Rinat, *Phys. Rev.*, **C10**, 2102 (1974) [3.3, 5.1]

Hoh-64   G. Höhler, G. Ebel, and J. Giesecke, *Z. Phys.*, **180**, 430 (1964) [4.1, 5.5]

Hol-77a   K. Holinde and R. Machleidt, *Nucl. Phys.*, **A280**, 429 (1977) [8.2]

Hol-77b   K. Holinde, R. Machleidt, A. Faessler, and H. Müther, *Phys. Rev.*, **C15**, 1432 (1977) [8.2]

Huf-73   J. Hüfner, *Nucl. Phys.*, **B58**, 55 (1973) [5.1]

Huf-74   J. Hüfner, S. Y. Lee, and H. Weidenmüller, *Nucl. Phys.*, **A234**, 429 (1974) [5.4]

Huf-75a   J. Hüfner, *Phys. Rep.*, **21C**, 1 (1975) [1, 4, 5, 7, 7.1]

Huf-75b   J. Hüfner, in Nag-75, p. 1 [5.1]

Huf-75c   J. Hüfner, H. J. Pirner, and M. Thies, *Phys. Lett.*, **59B**, 215 (1975) [5.6]

Huf-78   J. Hüfner, *Ann. Phys.*, **115**, 43 (1978) [7.2]

Iac-74   F. Iachello and A. Lande, *Phys. Lett.*, **50B**, 313 (1974) [5.1]

Ing-74   R. L. Ingraham, *Lett. Nuovo Cimento*, **9**, 331 (1974) [6.3]

Iof-56   B. L. Ioffe, I. Ia. Pomeranchuk, and A. P. Rudik, *J. Exper. Theor. Phys. (USSR)* **31**, 712 (1956) [3.4]

Jac-62   J. D. Jackson, *Classical Electrodynamics* (Wiley, New York, 1962) [5.1]

Joh-71   M. B. Johnson and M. Baranger, *Ann. Phys.* **62**, 172 (1971) [8.1]

Joh-76   M. B. Johnson, *Ann. Phys.* **97**, 400 (1976) [8.1]

Joh-78   M. B. Johnson and B. D. Keister, *Nucl. Phys.*, **A305**, 461 (1978) [4.3]

Kal-64   G. Källén, *Elementary Particle Physics* (Addison-Wesley, Reading, 1964) [2, 2.4, 6.1]

Kar-74   B. R. Karlsson and E. M. Zeiger, *Phys. Rev.*, **D9**, 1761 (1974) [3.2]

Kei-74   B. D. Keister, *Nucl. Phys.*, **A235**, 520 (1974) [4.3]

Kei-76   B. D. Keister, *Nucl. Phys.*, **A271**, 342 (1976) [4.3]

Ker-59   A. K. Kerman, H. McManus, and R. M. Thaler, *Ann. Phys.*, **8**, 551 (1959) [4]

Ker-69   A. K. Kerman and L. S. Kisslinger, *Phys. Rev.*, **180**, 1483 (1969) [8.2]

Ker-71   A. K. Kerman and H. J. Lipkin, *Ann. Phys.*, **66**, 738 (1971) [5.4]

Kim-71   J. K. Kim, *Phys. Rev. Lett.*, **27**, 356 (1971) [2.2]

Kis-55   L. S. Kisslinger, *Phys. Rev.*, **98**, 761 (1955) [5.1]

Kis-73   L. S. Kisslinger and W. L. Wang, *Phys. Rev. Lett.*, **30**, 1071 (1973) [6.2]

388 References

Kis–76a L. S. Kisslinger and W. L. Wang, *Ann. Phys.*, **99**, 374 (1976) [6, 6.2, 6.4]

Kis–76b L. S. Kisslinger, in Bar–76, p. 159 [5.1, 6, 6.2, 6.4]

Kis–79 L. S. Kisslinger, in Rho–79, [8.2]

Kli–78 K. Klingenbeck, M. Dillig, and M. G. Huber, *Phys. Rev. Lett.*, **41**, 387 (1978) [5.1]

Koc–74 J. H. Koch and J. D. Walecka, *Nucl. Phys.*, **B72**, 283 (1974) [4.3]

Kol–66 V. M. Kolybasov, *Yad. Fiz.* (*USSR*), **3**, 729, 965 (1966); *Sov. J. Nucl. Phys.*, **3**, 535, 704 (1966) [7.2, 7.3]

Kol–67 D. S. Koltun, *Phys. Rev.*, **162**, 963 (1967) [7.3]

Kol–68 D. S. Koltun and A. Reitan, *Nucl. Phys.*, **B4**, 629 (1968) [7.2, 7.3]

Kol–69 D. S. Koltun, in *Advances in Nuclear Physics*, M. Baranger and E. Vogt, Eds., Vol. 3 (Plenum, New York, 1969) p. 71 [1, 4, 5, 5.4, 7, 7.1]

Kol–71 D. S. Koltun, in Dzh–72, p. 201 [7, 7.1]

Kol–73a V. M. Kolybasov and A. E. Kudryavtsev, *Zh. Ekspert. Teor. Fiz.* (*USSR*), **63**, 35 (1972); *Sov. Phys. JETP*, **36**, 18 (1973) [3.3]

Kol–73b D. S. Koltun, *Ann. Rev. Nucl. Sci.*, **23**, 163 (1973) [8.2]

Kol–74 D. S. Koltun and O. Nalcioğlu, *Phys. Lett.*, **51B**, 19 (1974) [5.5]

Kol–77 D. S. Koltun and F. Myhrer, *Z. Phys.*, **A283**, 397 (1977) [4.3, 5.3]

Kol–79 D. S. Koltun and D. M. Schneider, *Phys. Rev. Lett.*, **42**, 211 (1979) [5.6]

Kop–71 T. Kopaleishvili and J. Z. Machabeli, *Nucl. Phys.*, **A160**, 204 (1971) [7.2, 7.3]

Kre–69 M. Krell and T. E. O. Ericson, *Nucl. Phys.*, **B11**, 521 (1969) [5.1, 5.3]

Kre–70 M. Krell and S. Barmo, *Nucl. Phys.*, **B20**, 461 (1970) [5.1]

Kro–61 N. Kroll, unpublished, quoted in R. M. Edelstein, W. F. Baker, and J. Rainwater, *Phys. Rev.*, **122**, 252 (1961) [5.1]

Kwo–78 Y. R. Kwon and F. Tabakin, *Phys. Rev.*, **C18**, 932 (1978) [5.3]

Lam–73 E. Lambert and H. Feshbach, *Ann. Phys.*, **76**, 80 (1973) [5.1]

Lan–73 R. H. Landau, S. C. Phatak, and F. Tabakin, *Ann. Phys.*, **78**, 299 (1973) [5.1]

Lan–78 R. H. Landau and A. W. Thomas, *Nucl. Phys.*, **A302**, 461 (1978) [4.3]

Lay–61 W. M. Layson, *Nuovo Cimento*, **20**, 1207 (1961) [2.4]

Lee–71 H. Lee and H. McManus, *Nucl. Phys.*, **167**, 257 (1971) [5.5]

Lee–74 T. S. H. Lee and F. Tabakin, *Nucl. Phys.*, **A226**, 253 (1974) [5.5]

Lee–77a T. S. H. Lee and S. Chakravarti, *Phys. Rev.*, **C16**, 273 (1977) [5.1]

Lee–77b T. S. H. Lee, D. Kurath, and B. Zeidman, *Phys. Rev. Lett.*, **39**, 1307 (1977) [5.4]

Len–75 F. Lenz, *Ann. Phys.*, **95**, 348 (1975) [5.1, 6, 6.4]

Leo–76 M. Leon, *Nucl. Phys.*, **A260**, 461 (1976) [5.3]

Lev–77 E. Levin and J. M. Eisenberg, *Nucl. Phys.*, **A292**, 459 (1977) [5.1, 6.3]

Lip–65 H. Lipkin, *Phys. Rev. Lett.*, **14**, 18 (1965) [5.4]

References **389**

Liu–75   L. C. Liu and V. Franco, *Phys. Rev.*, **C11**, 760 (1975) [5.4]

Loc–71   M. P. Locher, O. Steinmann, and N. Straumann, *Nucl. Phys.*, **B27**, 598 (1971) [5.1]

Loc–77   *High-Energy Physics and Nuclear Structure*, Zurich, 1977, M. P. Locher, Ed. (Birkhäuser Verlag, Basel and Stuttgart, 1977) [5] See also Dil–77, Eis–77, Wal–77.

Loh–77   K. P. Lohs and V. B. Mandelzweig, *Z. Phys.*, **A283**, 51 (1977) [4.4,5.4]

Lon–74   J. T. Londergan, K. W. McVoy, and E. J. Moniz, *Ann. Phys.*, **86**, 147 (1974) [2.3]

Lon–75   J. T. Londergan and E. J. Moniz, *Ann. Phys.*, **94**, 83 (1975) [5.1,6.1]

Lov–64a   C. Lovelace, *Phys. Rev.*, **135**, B1225 (1964) [3.2]

Lov–64b   C. Lovelace, in *Strong Interactions in High Energy Physics*, R. G. Moorhouse, Ed. (Plenum, New York, 1964) p. 437 [3.2]

Low–55   F. Low, *Phys.Rev.*, **97**, 1392 (1955) [2.4]

Mac–72   R. Mach, *Phys. Lett.*, **40B**, 46 (1972) [4.3,5.1]

Mac–73   R. Mach, *Nucl. Phys.*, **A205**, 56 (1973) [4.3,5.1]

Mai–71   J. P. Maillet, C. Schmit, and J. P. Dedonder, *Lett. Nuovo Cimento*, **1**, 191 (1971) [5.1]

Mai–76   J. P. Maillet, J. P. Dedonder, and C. Schmit, *Nucl. Phys.*, **A271**, 253 (1976) [4.3]

Man–76   V. B. Mandelzweig, H. Garcilazo, and J. M. Eisenberg, *Nucl. Phys.*, **A256**, 461 (1976) [3.2,3.3,3.4]

Mar–69a   B. R. Martin and M. Sakitt, *Phys. Rev.*, **183**, 1345, 1352 (1969) [2.2]

Mar–69b   R. Marshak, Riazuddin, and C. P. Ryan, *Theory of Weak Interactions in Particle Physics* (Wiley-Interscience, New York, 1969) [8.1]

Mar–75   B. R. Martin, *Nucl. Phys.*, **B94**, 413 (1975) [2.2]

Mar–77a   B. R. Martin and M. K. Pidcock, *Nucl. Phys.*, **B126**, 285 (1977) [2.2]

Mar–77b   T. Marks *et al.*, *Phys. Rev. Lett.*, **38**, 148 (1977) [5.4]

McV–77   K. W. McVoy, *Nucl. Phys.*, **A276**, 491 (1977) [5.1,5.2]

Mig–78   A. B. Migdal, *Revs. Mod. Phys.*, **50**, 107 (1978) [8.2]

Mil–74a   G. A. Miller, *Phys. Rev.*, **C10**, 1242 (1974) [5.1]

Mil–74b   L. D. Miller, *Phys. Rev.*, **C9**, 537 (1974) [6.1]

Mil–76   G. A. Miller, in Bar–76, p. 684 [5.4]

Mil–77   G. A. Miller, *Phys. Rev.*, **C16**, 2325 (1977) [6.3]

Mil–78   L. D. Miller and H. J. Weber, *Phys. Rev.*, **C17**, 219 (1978) [7.2,7.3]

Miz–75   T. Mizutani, doctoral dissertation, Rochester, University of Rochester Technical Report UR–533 (1975) [6.2,6.3]

Miz–77   T. Mizutani and D. S. Koltun, *Ann. Phys.*, **109**, 1 (1977) [3.3, 3.5, 6.2]

Mon–73   E. J. Moniz, *Phys. Rev.*, **C7**, 1750 (1973) [5.1]

Mon–76   E. J. Moniz, in Bar–76, p. 105 [5.1]

390 References

Mon-78   E. J. Moniz, in *NATO Advanced Study Institute on Theoretical Methods in Intermediate Energy and Heavy-Ion Physics*, Madison, Wisconsin, June, 1978 [5.1]

Moo-69   R. G. Moorehouse, *Ann. Rev. Nucl. Sci.*, **19**, 301 (1969) [2, 2.1, 2.4]

Mor-53   P. M. Morse and H. Feshbach, *Methods of Theoretical Physics* (McGraw-Hill, New York, 1953) [4.1]

Mor-60   M. J. Moravcsik, *Ann. Rev. Nucl. Sci.*, **10**, 324 (1960) [2.5]

Mor-73   J. W. Morris and H. J. Weber, *Ann. Phys.*, **79**, 34 (1973) [5.3, 7.2, 7.3]

Moy-69   L. Moyer and D. S. Koltun, *Phys. Rev.*, **182**, 999 (1969) [5.3]

Mut-75   G. S. Mutchler *et al.*, *Phys. Rev.*, **C11**, 1873 (1975) [5.2]

Myh-73a F. Myhrer, *Phys. Lett.*, **45B**, 96 (1973) [3.3]

Myh-73b F. Myhrer and D. S. Koltun, *Phys. Lett.*, **46B**, 322 (1973) [3.3, 5.1]

Myh-74   F. Myhrer, *Nucl. Phys.*, **80**, 491 (1974) [3.3]

Myh-75   F. Myhrer and D. S. Koltun, *Nucl. Phys.*, **B86**, 441 (1975) [3.3]

Nag-74   M. A. Nagarajan and W. L. Wang, *Phys. Rev.*, **C10**, 2125 (1974) [6.4]

Nag-75   D. E. Nagle *et al.*, Eds., *High-Energy Physics and Nuclear Structure*, Santa Fe and Los Alamos, 1975, (American Institute of Physics, New York, 1975) [5] See also Eis-75, Gal-75b, Huf-75b, Pov-75, Tau-75.

Nam-68   J. M. Namysłowski, *Nuovo Cimento*, **57A**, 355 (1968) [3.4]

Nat-73a P. Nath and S. S. Kere, in Gib-73, p. 226 [2.4]

Nat-73b N. R. Nath, H. J. Weber, and J. M. Eisenberg, *Phys. Rev.*, **C8**, 2488 (1973) [5.1]

Nob-76   J. V. Noble, in Bar-76, p. 221, and review article preprint [7, 7.1, 7.3]

Nob-78   J. V. Noble, *Phys. Rev.*, **C17**, 2151 (1978) [7.3]

Nor-71   J. H. Norem, *Nucl. Phys.*, **B33**, 512 (1971) [3.3]

Nym-77   E. M. Nyman, *Nucl. Phys.*, **A285**, 368 (1977) [8.2]

Oht-74   K. Ohta and M. Wakamatsu, *Phys. Lett.*, **51B**, 325 (1974) [8.2]

Oku-54   S. Okubo, *Prog. Theor. Phys.*, **12**, 603 (1954) [2.3, 8.1]

Par-71   C. Park and J. P. Rickett, *Phys. Rev.*, **C3**, 1926 (1971) [7.2, 7.3]

Par-78   Particle Data Group, *Phys. Lett.*, **75B**, 1 (1978) [1, 2, 2.2, 2.4]

Pen-63   H. N. Pendleton, *Phys. Rev.*, **131**, 1833 (1963) [3.4]

Pet-73   N. M. Petrov and V. V. Peresypkin, *Phys. Lett.*, **44B**, 321 (1973) [3.3]

Pov-75   B. Povh, in Nag-75, p. 173 [5.4]

Pov-76   B. Povh, *Rep. Prog. Phys.*, **39**, 823 (1976) [5.4]

Pre-76   B. M. Preedom *et al.*, *Phys. Lett.*, **65B**, 31 (1976) [3.5]

Rei-72   A. Reitan, *Nucl. Phys.*, **B50**, 166 (1972) [7.3]

Rem-75   E. A. Remler, *Ann. Phys.*, **95**, 455 (1975) [7.2]

Rev-73   J. Revai, *Nucl. Phys.*, **A205**, 20 (1973) [4.3]

Rho-74   M. Rho, *Nucl. Phys.*, **A231**, 493 (1974) [8.2]

Rho-76   M. Rho, in Bar-76, p. 146 [8.2]

Rho-79   M. Rho and D. H. Wilkinson, Eds., *Mesons in Nuclei* (North-Holland, Amsterdam, 1979) [8] See Che-79, Ger-79, and Kis-79.

Ric–70   C. Richard-Serre et al., *Nucl. Phys.*, **B20**, 413 (1970) [3.5]

Rin–77a   A. S. Rinat and A. W. Thomas, *Nucl. Phys.*, **A282**, 365 (1977) [3.3, 3.4]

Rin–77b   A. Rinat, *Nucl. Phys.*, **A287**, 399 (1977) [3.3, 3.5]

Ris–72   D. O. Riska and G. E. Brown, *Phys. Lett.*, **38B**, 193 (1972) [8.2]

Riv–77   J. M. Rivera and H. Garcilazo, *Nucl. Phys.*, **A285**, 505 (1977) [3.3, 3.4]

Rog–71   C. Rogers and C. Wilkin, *Lett. Nuovo Cimento*, **1**, 575 (1971) [5.5]

Rop–65   L. D. Roper, R. M. Wright, and B. T. Feld, *Phys. Rev.*, **138B**, 190 (1965) [2.2]

Ros–73   E. Rost, in Gib–73, p. 135 [5.5]

Sak–67   J. J. Sakurai, *Advanced Quantum Mechanics* (Addison-Wesley, Reading, 1967) [2, 2.5]

Sal–57   G. Saltzman and F. Saltzman, *Phys. Rev.*, **108**, 1619 (1957) [2.4]

Sal–74   M. Salomon, TRIUMF report TRI-74-2 (1974) [2.2]

Sam–72a   V. K. Samaranayake and W. S. Woolcock, *Nucl. Phys.* **45B**, 205 (1972) [2.2]

Sam–72b   V. K. Samaranayake and W. S. Woolcock, *Nucl. Phys.* **49B**, 128 (1972) [2.2]

Saw–72   R. F. Sawyer, *Astrophys. J.*, **176**, 205 (1972) [5.1]

Sch–61   S. S. Schweber, *Introduction to Relativistic Quantum Field Theory* (Row and Peterson, Evanston, 1961) [2, 2.4]

Sch–67   D. Schiff and J. Tran Thanh Van, *Nucl. Phys.*, **B3**, 671 (1967) [3.4]

Sch–68a   D. Schiff and J. Tran Thanh Van, *Nucl. Phys.*, **B5**, 529 (1968) [3.4]

Sch–68b   L. I. Schiff, *Quantum Mechanics* (McGraw-Hill, New York, 1968) [4.1, 4.2]

Sch–70   C. Schmit, *Lett. Nuovo Cimento*, **4**, 454 (1970) [5.1]

Sch–71   L. Scherk, *Can. J. Phys.*, **49**, 306 (1971) [5.1]

Sch–72a   C. Schmit, *Nucl. Phys.*, **A197**, 449 (1972) [5.1]

Sch–72b   F. Scheck and C. Wilkin, *Nucl. Phys.*, **B49**, 541 (1972) [5.1]

Sco–72   M. L. Scott et al., *Phys. Rev. Lett.*, **28**, 1209 (1972) [5.2]

Sed–77   J. E. Sedlak and W. A. Friedman, *Phys. Rev.*, **C16**, 2306 (1977) [5.1]

Sek–73   R. Seki, contributed paper, Fifth International Conference on High-Energy Physics and Nuclear Structure, Uppsala, 1973, p. 69 [5.1]

Sha–65   I. S. Shapiro, V. M. Kolybasov, and J. P. August, *Nucl. Phys.*, **61**, 353 (1965) [7.3]

Sha–67   I. S. Shapiro, in *Interactions of High Energy Particles with Nuclei*, ("Enrico Fermi," course 38), T. E. O. Ericson, Ed. (Academic, New York, 1967) [7.2, 7.3]

Sha–69   *Pion-Nucleon Scattering*, G. L. Shaw and D. Y. Wong, eds., (Wiley-Interscience, N. Y. 1969) [2]

Sha–76   Y. Shamai et al., *Phys. Rev. Lett.*, **36**, 82 (1976) [5.4]

Sil–77   R. R. Silbar, J. N. Ginocchio and M. M. Sternheim, *Phys. Rev.*, **C15**, 371 (1977) [5.6]

Slo–72   I. R. Sloan and J. C. Aarons, *Nucl. Phys.*, **A198**, 321 (1972) [3.2]

Spu–75 J. Spuller and D. F. Measday, *Phys. Rev.*, **D12**, 3550 (1975) [3.4,3.5]

Ste–64 V. J. Stenger *et al.*, *Phys. Rev.*, **134B**, 1111 (1964) [2.2]

Ste–70 M. M. Sternheim and E. H. Auerbach, *Phys. Rev. Lett.*, **25**, 1500 (1970) [5.1]

Ste–74 M. M. Sternheim and R. R. Silbar, *Ann. Rev. Nucl. Sci.*, **24**, 549 (1974) [5,5.1]

Ste–75 M. M. Sternheim and R. R. Silbar, *Phys. Rev. Lett.*, **34**, 824 (1975) [5.6]

Sti–70a M. Stingl and A. S. Rinat, *Nucl. Phys.*, **A154**, 613 (1970) [3.2]

Sti–70b P. Stichel and F. Werner, *Nucl. Phys.*, **A145**, 257 (1970) [8.1]

Tab–76 F. Tabakin, in Bar–76, p. 38 [4.3,5.1]

Tan–77 P. C. Tandy, E. F. Redish, and D. Bollé, *Phys. Rev.*, **C16**, 1924 (1977) [5.6]

Tau–75 L. Tauscher, in Nag–75, p. 541 [5.3]

Tay–65 J. G. Taylor, *Nuovo Cimento Supp.*, **1**, 857 (1965) [6.3]

Tay–66 J. G. Taylor, *Phys. Rev.*, **150**, 1321 (1966) [6.3]

Tho–76a A. W. Thomas, *Nucl. Phys.*, **A258**, 417 (1976) [3.3]

Tho–76b A. W. Thomas, in Bar–76, p. 375 [3.4]

Tho–77 A. W. Thomas, Ed., *Modern Three-Hadron Physics* (Springer, Berlin, 1977) [3,3.2]

Tho–78 A. W. Thomas and F. Myhrer, TRIUMF preprint TRI-PP-78-7 (1978) [2.4]

Tib–69 G. Tibell, *Phys. Lett.*, **28B**, 638 (1969) [5.5]

Tib–74 G. Tibell, Ed., *High-Energy Physics and Nuclear Structure*, Uppsala, 1973 (North-Holland, Amsterdam, 1974) [5] See also Clo–74 and Zai–74.

Tow–75 D. Tow and J. M. Eisenberg, *Nucl. Phys.*, **A237**, 441 (1975) [5.4]

Tre–62 S. B. Treiman and C. N. Yang, *Phys. Rev. Lett.*, **8**, 140 (1962) [7.3]

Ued–77 T. Ueda, T. Sawada, and S. Takagi, *Nucl. Phys.*, **A285**, 429 (1977) [8.2]

Ull–74 J. J. Ullo and H. Feshbach, *Ann. Phys.*, **82**, 156 (1974) [5.1]

Ure–66 J. L. Uretsky, *Phys. Rev.*, **147**, 906 (1966) [5.3]

Vai–64 A. O. Vaisenberg, E. D. Kolganova, and N. V. Rabin, *Zh. Ekspert. Teor. Fiz.* (*USSR*), **47**, 1262 (1964): *Sov. Phys. JETP*, **20**, 854 (1965) [7.1]

Vai–70 A. O. Vaisenberg, N. V. Rabin, and V. F. Kuichev, *Yad. Fiz.* (*USSR*), **11**, 48 (1970); *Sov. J. Nucl. Phys.*, **11**, 26 (1970) [7.1]

Vas–65 K. V. Vasavada, *Ann. Phys.*, **34**, 191 (1965) [3.4]

Vei–70 V. R. Veirs and R. A. Burnstein, *Phys. Rev.*, **D1**, 1883 (1970) [3.5]

Wal–73a S. J. Wallace, *Ann. Phys.*, **78**, 190 (1973) [4.4]

Wal–73b S. J. Wallace, *Phys. Rev.*, **C8**, 2043 (1973) [4.4]

Wal–73c S. J. Wallace, *Phys. Rev.*, **D8**, 1846 (1973) [4.4]

Wal–73d G. Walker, in Gib–73, p. 72 [5.1]

Wal–74   J. D. Walecka, *Ann. Phys.*, **83**, 491 (1974) [8.1]

Wal–76   G. E. Walker, in Bar–76, p. 674 [5.5]

Wal–77   H. K. Walter, in Loc–77, p. 225 [7, 7.1]

War–76   J. B. Warren, Ed., *Nuclear and Particle Physics at Intermediate Energies*, Brentwood, 1975 (Plenum, New York, 1976) [5]

War–77   J. Warszawski and N. Auerbach, *Nucl. Phys.*, **A276**, 402 (1977) [5.4]

War–78a J. Warszawski, J. M. Eisenberg, and A. Gal, *Nucl. Phys.*, **A312**, 253 (1978) [5.1]

War–78b J. Warszawski, A. Gal, and J. M. Eisenberg, *Nucl. Phys.*, **A294**, 321 (1978) [5.4]

Wat–53   K. M. Watson, *Phys. Rev.*, **89**, 575 (1953) [4]

Wat–57   K. M. Watson, *Phys. Rev.*, **105**, 1388 (1957) [4]

Wat–58   K. M. Watson, *Revs. Mod. Phys.*, **30**, 565 (1958) [4]

Web–75   H. J. Weber, in *Interaction Studies in Nuclei*, H. Jochim and B. Ziegler, Eds. (North-Holland, Amsterdam, 1975) p. 749 [8.2]

Web–76   H. J. Weber, *Fortsch. Phys.*, **24**, 1 (1976) [5.5]

Web–78   H. J. Weber and H. Arenhövel, *Phys. Rept.*, **36C**, 277 (1978) [8.2]

Wei–64   S. Weinberg, *Phys. Rev.*, **133**, B232 (1964) [3.2]

Wei–66   S. Weinberg, *Phys. Rev. Lett.*, **17**, 616 (1966) [2.4]

Wei–72   U. Weiss, *Nucl. Phys.*, **B44**, 573 (1972) [3.4]

Wei–77   W. Weise, *Nucl. Phys.*, **A278**, 402 (1977) [6.4]

Wic–55   G. C. Wick, *Revs. Mod. Phys.*, **27**, 339 (1955) [2, 2.4]

Wil–70   C. Wilkin, *Lett. Nuovo Cimento*, **4**, 491 (1970) [5.1]

Wil–71   C. Wilkin, in Eri–71a, p. 289 [5.1]

Wil–73   C. Wilkin et al., *Nucl. Phys.*, **B62**, 61 (1973) [5.2]

Wil–74   C. Wilkin, *Nucl. Phys.*, **A220**, 621 (1974) [5.5]

Wil–75   C. Wilkin, in *Effets Mésoniques dans les Noyaux, Comptes Rendus, Saclay, 1975*, J. R. Bellicard et al., Eds. (Commissariat a l'Energie Atomique, Saclay, 1975) p. 167 [8.2]

Wol–75   R. M. Woloshyn, E. J. Moniz, and R. Aaron, *Phys. Rev.*, **C12**, 909 (1975) [3.3]

Wol–76   R. M. Woloshyn, E. J. Moniz, and R. Aaron, *Phys. Rev.*, **C13**, 286 (1976) [3.4]

Won–75   C. W. Wong and S. K. Young, *Phys. Rev.*, **C12**, 1301 (1975) [4.4]

Wyc–71   S. Wycech, *Nucl. Phys.*, **B28**, 541 (1971) [5.3]

Yak–67   O. A. Yakubovskii, *J. Nucl. Phys. (USSR)*, **5**, 1312 (1967); *Sov. J. Nucl. Phys.*, **5**, 937 (1967) [3.2]

Yuk–35   H. Yukawa, *Proc. Phys. Math. Soc. Japan*, **17**, 48 (1935) [2.5]

Zai–74   M. Zaider et al., in Tib–74, p. 219 [5.4]

Index

Absorption, 1, 2, 36, 89, 91, 106-111, 117-
    119, 158, 225, 231, 257, 261, 284,
    289, 292, 304, 305-346
  optical, *see* Optical absorption
  true, *see* True absorption
Absorption cross section, 119, 123, 225,
    227
Absorption of kaons, 108
Absorption radius, 121
Absorption vertex, 346
Almost local potential, 30
Alpha-clusters, 176, 207, 226, 325, 343
Analyticity, 228
Angle transformation, 205
Angular momentum, recoupling of, 49
Angular momentum barrier, 245
Annihilation, *see* True absorption
Annihilation operators, 34-35
Anomalous cuts, 230
Anticorrelations, *see* Correlations
Antinucleon, 41
Antiparticles, 229, 231
Antiquarks, 24
Antisymmetrization, 105, 133, 135, 143-
    148, 151, 153, 159, 168, 172, 207
Argand diagrams of partial wave ampli-
    tudes, 22, 23, 200
Associated production, 308
Asymptotic states, 8, 12, 13, 27, 28, 275,
    279, 320
Average field interaction, 301
Axial coupling renormalization, 376-378
Axial vector current, 357, 358, 361, 376

Backward propagation, 268
Baryon number, 5, 348

Baryon resonance, 1, 5, 24, 261, 275, 347,
    353, 362
Baryon resonance admixtures in nuclei,
    see Isobar configuration admixtures in
    nuclei
Baryons, 1
Bessel functions, cylindrical, 122, 123,
    178
Beta-transitions, 307, 344, 350, 353, 376,
    378
Bethe-Salpeter equation, 16, 99 101, 268,
    294, 295
Binding effects, 171, 205, 208, 213, 237,
    296
Binding energy, 171, 368
Blankenbecler-Sugar equation, 100-102
Bilinear covariants, 17
Bohr orbits, 233
Born series, 34
Born terms, 32, 38-42, 45-48, 52-57, 102,
    110, 230, 264, 273, 279, 288, 291, 293
Boson, 1, 2, 124
Bound states, 74
  in N$\triangle$ or $\triangle\triangle$ systems, 366
Breit-Wigner resonance shape, 107
Brueckner theory of nuclear matter,
    168
Bubble chambers, 334

Capture cross section for thermal
    neutrons, 365
Cascade, atomic, 233
  nuclear, 312, 313, 326
"Catastrophic" term, 344, 373
Cauchy's theorem, 228
Causality, 228

Causal propagator, 282
Centrifugal effects, 10, 363
Channel coupling, 219, 251, 261-304, 351, 369
Channels, 8
Charge conjugation, 265
Charge distribution of nuclei, 5
Charge exchange in absorption, 327, 333
Charge exchange potentials, 359
Charge exchange scattering, 4, 12, 18, 63, 89, 106, 241-247
Charge symmetry, 225
Chew-Low effective range formula, 20, 51
Chew-Low equation, 48
Chew-Low static model, 42, 46-56
Chew-Low theory, 42-54, 60, 95, 103, 105, 109, 197, 213, 265, 268, 273, 278, 290, 293, 304, 318
Clausius-Massotti equation, 214
Closed channels, 275
Closure, 67, 128, 144, 160, 172, 173, 206, 292
Cluster expansion, *see* Correlations
Clustering problem, 102
Clusters:
  in nuclei, 275, 311, 342, 343
  in pion absorption, 342, 343
Coherence, 126, 137, 148, 175, 192, 242, 243, 249, 250, 319, 321, 327
Collective states, 248
Compactness, 75, 77, 80
Complex potential, 150
Compton scattering, 38
Condon-Shortley phase convention, 35
Conserved vector currents (CVC), 359, 360
Continuity equation, 119
Convective coupling, 365
Convective terms, 254
Convergence, 90, 92, 154, 177, 206, 236
Correlation expansion, 167, 194
Correlation length, 163, 164, 176, 217
Correlations, 3, 126, 139, 160-167, 172, 186-188, 194, 205-207, 209, 215-226, 247, 251, 296, 304, 323, 337, 341
  isovector, 245
Coulomb effects, 18, 228, 230, 246
Coulomb energy, 367
Coulomb interaction, 63, 234

Coulomb-nuclear interference, 18, 228, 233, 246
Coupled-channels, 219, 251, 261, 304, 351, 369
Coupled equations, 287
Coupled integral equations, 87
Coupling constants, 38, 42, 47, 58, 233, 351, 362, 377
Covariant formulation, 356
Covariants, bilinear, 17
Covariant theory, 16, 41, 315
Creation operators, 34-35
Crossing, 5, 38-41, 45-52, 102, 104, 195, 229, 230, 262, 265-267, 300, 304, 366
Crossing matrix, 50
Crossing symmetry, 39, 45, 265-267
Cross section, 8
Current algebra, 56, 360
Current conservation, 358, 359, 362
Current operators, 35, 36, 55, 262, 291, 318, 331, 357-362
Cutoff parameters, 51, 52, 362, 363
Cyclic notation, 75, 81

Delta-hole model, 210, 300-304
Delta resonance, 20-23, 52, 71, 116, 164, 171, 191, 366, 368
  shifting, position, 198, 199
Detailed balance, 327
Deuteron photodisintegration, 374
Diagonal nuclear matrix element, 136, 138, 139, 148
Diagrammatic expansion of effective interaction, 351, 356
Dielectric constant, 214
Diffraction, 118-123, 178, 181, 188, 221, 245, 346
Dirac matrices, 17, 358
Dirac spinors, 17, 358
Direct reactions, 326, 334
Discontinuity function, 101, 103
Dispersion equation, 198, 372
Dispersion relations, 24, 42, 51, 53, 101-105, 208, 228, 266
Dispersive effects, 236, 238
Distorted wave Born approximation (DWBA), 327-333
Distorted wave impulse approximation (DWIA), 150, 175, 191, 227, 241-256, 314, 333. *See also* Wave distortion

Doorway states, 211, 274, 289, 296-
   300, 304
Double analog state, 241, 246
Double charge exchange, 63, 151, 231,
   241, 246
Double counting, 105, 176,
   219
Double hypenuclei, 248
Double scattering, 68, 70, 71, 92, 98, 247,
   264
Double strangeness exchange, 241, 247,
   248
Dressed nucleon states, 351
Driving term, 80, 103, 142, 291. *See
   also* Born terms
DWIA, see Distorted wave impulse
   approximation
Dynamic mixing effects in mesic atoms,
   239

Effective absorption operator, 332
Effective charges, 319
Effective field corrections, 172
Effective Hamiltonian, 349, 350, 359, 367-
   371
Effective interaction, 32-34, 58, 277, 278,
   284, 287, 290, 351, 352
Effective many-body vertices, 294
Effective operator, 316, 357
Effective radius method, 198
Effective Schroedinger equation, 261, 277-
   287, 319, 348, 369
Effective transition operators, 350, 353,
   355
Eikonal approximation, 70, 162, 165, 179,
   181, 188, 243, 244, 258
Elastic scattering of pions, 190-227
Electric multipoles, 255
Electrodynamics, 1, 35
Electromagnetic current, 357-360
Electromagnetic effects, 367
Electromagnetic interactions, 63, 318, 344-
   364
Electromagnetic potential, 357
Electromagnetic transitions, 307, 373-375
Electron charge, 357
Electron scattering, 5, 67, 136, 225, 249,
   271, 364, 374
Emission of mesons, *see* Production

Emulsions, nuclear, 334
Energy conservation, 15
Energy resolution, 4
ϵ-meson, 2, 56, 58, 59, 368
Equivalent local potential, 239
η-meson, 2, 365
Even isospin terms, 11, 230
Exchange currents, 318, 357-363, 373-375
Exchange diagrams, 54-57
Exchange forces, 2
Exchange of mesons, multiple, 353
Exchange of quantum numbers in
   scattering, 240-251
Excitons, 325
Exclusion principle, *see* Pauli principle
Exclusive reactions, 256-258
External wave, 115
Extinction of wave, 115

Faddeev theory, 72-99, 102-110, 126-131,
   268, 321
Fermi averaging, *see* Kinematical
   averaging
Fermi coupling constant, 357, 358
Fermi gas model, 249, 309, 312, 325, 371
Fermi momentum, 156, 236, 248, 249
Fermi motion, 201
Feynman diagrams, 36, 41, 99, 100, 112,
   274, 283, 294, 301, 315, 353, 356,
   362
Feynman-Goldstone diagrams, 301, 353
Field theory, 112, 261-304
Final-state interaction, 259
First-order optical potential, *see* Optical
   potential, first-order
Fission, 306
Fixed-scatterer approximation, 67, 71, 92,
   98, 125, 126, 155, 172-175, 226
Folded diagrams, 353
Foldy-Wouthuysen transformation, 318,
   362
Formalism hierarchy for optical potential,
   141, 142, 146, 159, 164, 205, 206
Form factor for scattering, 67, 69, 71, 92
Form factor for vertex, *see* Vertex function,
Four-body formalism, 85
Fourier-Bessel transform, 180
Frozen nucleus approximation, *see*
   Fixed-scatterer approximation

Galilei invariance, 14, 36, 73, 74, 81, 156,
   201, 203, 255, 270, 271, 317, 318, 331
γ-matrices, see Dirac matrices
Gamow-Teller beta-transitions, 344
Geometrical considerations, 120, 123, 192
Giant dipole resonance, 254
Glauber multiple-scattering theory, 70, 71,
   98, 127 , 163, 177-189, 221-224, 233,
   245, 246, 251
Goldberger-Treiman relation, 361, 377
Granularity, 126, 251
Green function, see Propagator
Green operator, see Propagator

Hadron currents, 358
Hadronic atoms, 239
Hadronic currents, 358
Hadronic matter distribution,
   see Nuclear matter distribution
Half off-energy-shell scattering, 39, 44-48,
   73
Hamiltonian methods, 274-290
Hard-core interactions, 1, 134, 164, 168,
   207, 220, 365
Hartree-Fock theory, 270
Hartree interaction, 301
Healing, 129
Heisenberg field operator, 282
Heisenberg isospin-exchange operator, 359
Heisenberg representation, 265
Hierarchy of formalisms for optical
   potential, see Formalism hierarchy
   for optical potential
Higher-order distorted wave impulse
   approximation, 245
Higher orders in the optical potential, 138-
   142, 159-168
High-momentum phenomena, 5, 367, 374
Hilb formula, 178
Hole states, 210, 300
Hypernuclear Hamiltonian, 248
Hypernuclei, 1, 247, 307, 345
Hypernucleon, see Hyperon
Hyperon, 1, 25, 60, 64, 106, 241, 247,
   249, 307, 308, 345

Impact coordinate, see Impact parameter
Impact parameter, 115, 120, 123, 178,
   182

Impulse approximation, 18, 66, 105, 150,
   158, 159, 172, 215, 226, 236, 239,
   292, 316, 318, 344
   two-nucleon, 332
Inclusive reactions, 257
Index of refraction, 113-118, 124, 199, 256
Inelastic election scattering, 249, 252, 254, 344
Inelasticity, 10, 11, 19, 200
Inelastic pion scattering, 251-255
Inelastic proton scattering, 249
Inelastic scattering, 22, 25, 63, 106, 147-
   154, 307, 310, 319
In field, 282
Ingoing waves, 27, 47
Inhomogeneous term, see Driving term
Integral equation, 28-31, 50, 86, 131
Internal energy, 73, 81
Internal momentum, 82, 124, 218
Internal pion momentum, 198
Internucleon separation distance, 237
Invariance, see Galilei invariance; Lorentz
   invariance; Parity inversion invariance;
   Translational invariance
Invariant total energy, 100
Inverse-square radius, 71
Isobar configuration admixtures in nuclei,
   2, 363-367, 374
Isobar doorway model, 299
Isobaric analog resonance, 241, 242
Isoscalar, 11
Isoscalar current, 359
Isospin, 6-11, 34-39, 55, 83, 241, 242
Isospin density, 157, 369
Isospin invariance, 18, 62, 241, 246
Isospin term, antisymmetric, 11, 230
Isospin term symmetric, 11, 230
Isotensor term, 230
Isovector, 11
Isovector correlations, 245
Isovector current, 359

Jacobian of coordinate transformation, 73, 81
Jastrow correlations, 218

Kaon, 1, 8, 12, 60, 271, 284, 306, 347
Kaon absorption, 345, 346
Kaon-deuteron scattering, 87, 91-95, 103,
   268
Kaon-deuteron scattering length, 95

Kaonic atoms, 27, 95, 234, 238-240
Kaon-nucleon interaction, 7, 17, 24-27, 56,
    57, 61, 93, 197, 239, 295
Kaon-nucleus scattering, 287, 299
Kinematical averaging, 155, 194, 205, 239
Kinematic variables, 81
Kisslinger potential, 30, 192-198, 202-205,
    211, 217, 218, 226, 330
Klein-Gordon equation, 195, 234, 269,
    272, 283, 376
Klein-Gordon propagator, 294
K-matrix, 26
K-meson, see Kaon
Knockout reactions, 127, 192, 255-259
K*-meson, 271

Ladder approximation, 99, 102
Lagrangian densities, 357, 358
Λ-particle, 1
Λ(1405)-resonance, 95, 238, 251, 273
Left-hand cut, 103
Legendre polynomial, 10, 178
Lepton currents, 358
Level broadening in mesic atoms, 234
Level shifts in mesic atoms, 233, 238
Linearized propagator, 162, 181
Linked diagrams, 80
Lippmann-Schwinger equation, 27-34, 72-
    74
Local density approximation, 239
Local field corrections, see Reflections
Locality, 30, 32, 55, 74, 125, 181, 184,
    187, 195, 197, 211, 218, 239, 366
London-Heitler method, 105
Lorentz invariance, 23, 203, 271, 315-
    318, 358
Lorentz-Lorentz effect, 124, 164, 213-221,
    238, 304, 376, 377
Lorentz transformation, 16, 202, 270
Low equation, 46, 279, 290, 292
Lowest-order optical potential,
    see Optical potential, first-order

Magnetic transitions, 254, 255, 344, 364,
    373
Magnetization coupling, see Tensor
    coupling
Many-body forces, 267, 353, 367-369
Many-body problem, 129

Many-body theory, 349, 353, 356
Many-channel systems, 33
Mean free path, 113-118, 206, 257
Medium-energy nuclear physics, 3, 112,
    113, 118, 121, 159, 176, 207, 367
Mesic atoms, 233-240, 318
Mesic degrees of freedom, 2, 176, 261,
    347, 348, 350, 367
Mesic exchange currents, see Exchange
    currents
Mesic scattering, 2
Meson-deuteron scattt

Meson-deuteron scattering, 18, 62-105, 134, 363
Meson exchange, 57
    multiple, 353
Meson factories, 4, 246
Meson-nucleon interaction, 13, 64, 168, 171
Meson-nucleon scattering, 83
Meson-nucleus scattering, 126, 140, 155,
    190-260, 349
Meson production, see Production
Metric, 358
Minimal coupling replacement, 373
Model space, 350, 356
Model states, 350, 355, 356
Molecular polarizability, 214
Monte Carlo methods, 134, 207, 245
Moshinsky-Talmi transformation, 337
Multiple scattering theory, 64-71, 86-98, 102-
    106, 112, 189, 190-260, 261-304
Multistep processes, 330, 332
Muon capture, 344
Muon quantum number conservation, 358

Natural variables, 84
N/D method, 53
Negative frequencies, 229
Neutron density, 3, 226, 227, 235, 243,
    246
Neutron radiative capture, 365
Neutron radii, 225, 226
Neutrons, 18
Nondiagonal nuclear matrix elements,
    136, 137, 148, 151
Nonlocality, 30, 155, 164, 196, 197, 210,
    218, 238-240, 254, 299, 359
Nonorthogonality, 349
Nonoverlapping scatterers, 219, 224, 226

Nonrelativistic formulations, 87, 177
Nonstatic effects, 211
Nonstatic potentials, 33
Notational conventions, 5
Nuclear currents, *see* Current operators
Nuclear density, 155, 156
Nuclear emulsions, 334
Nuclear force, *see* Nucleon-nucleon interaction
Nuclear ground state, 143
Nuclear Hamiltonian, 75, 128, 129, 131, 132, 169, 173, 242
Nuclear many-body problem, 129
Nuclear matter, 118, 168, 214, 283, 303, 368, 369, 372, 377
Nuclear matter distribution, 225, 226. *See also* Nuclear density
Nuclear reaction theory, 304
Nuclear structure and meson interactions, 347-378
Nuclear structure factor, 338
Nuclear targets, saturation of, in spin and isospin, 137, 192, 209
Nucleon current, *see* Current operators
Nucleon motion, 71, 254
Nucleon-nucleon interaction, 1, 19, 34, 57-60, 365-367
Nucleon-nucleus interactions, 5, 341

Odd isospin terms, 11, 230
Off-diagonal nuclear matrix elements, 136, 137, 148, 151
Off-energy-shell scattering, 12-15, 45-52, 194-198
Off-mass-shell effects, 15-17, 367
ω-meson, 2, 32, 271, 348, 351, 365, 368
One-boson-exchange (OBE), 58, 60, 352, 362, 368
One-meson-exchange, *see* One-boson-exchange (OBE)
One-nucleon reactions, *see* Reactions, one-nucleon (π,N) or (N, π)
One-nucleon transfer reactions, 330
One-pion-exchange (OPE), 58, 352, 361, 365, 368
On-mass-shell kinematics, 102
On-mass-shell scattering, 16, 17, 41, 56
Open channels, 275, 277, 281, 284, 286, 291
Optical absorption, 115, 305, 330, 342

Optical distortion, *see* Distorted wave impulse approximation; Wave distortion
Optical limit, 186
Optical potential, 3, 113-147, 154-177, 186-188, 191-227, 234-240, 304
  first -order, 135-137, 154, 159, 163, 203-206, 209, 211, 227, 238, 243
  second-order, 139, 141, 167, 205-207, 238
Optical theorem, 24, 115, 229, 235, 243, 252
Optics, 113-125, 213, 214, 218
Orthogonality, 33, 150, 349, 350, 353
Out field, 282
Outgoing waves, 27, 43, 121, 125, 275
Overlap brackets, 84

Parity inversion invariance, 9
Partially conserved axial current (PCAC), 55, 344, 361, 362, 376
Partial waves, 7-27, 47-53
  in three-body systems, 81-85
Particle-hole excitations, 304-325
Pauli correlations, 164, 207, 236
Pauli principle, 129, 168, 172, 213, 218, 248, 249, 301
Penetration factor, 20
PCAC, *see* Partially conserved axial current
Phase parameter, 122, 123, 178, 183, 186
Phase shifts, 10, 19, 178, 233
Phase transition, 369, 371
ϕ-meson, 32
Photodisintegration of deuteron, 374
Photon, 366
Photon spectrometer, 241
Photoproduction, *see* Radiative absorption
Physical cut, 230
π-meson, *see* Pion
Pion, 1, 5, 8, 271, 284, 306, 347, 348, 351
Pion absorption, 346
πd→NN reaction, 87, 236, 321
Pion condensation, 214, 221, 367, 369-373
Pion decay, 361
Pion-deuteron scattering, 54, 62-105, 112, 126, 268, 321
Pion-deuteron scattering lengths, 89-91
Pion exchange, *see* One-pion-exchange (OPE); Two-pion exchange (TPE)
Pionic atoms, 89, 214, 218, 221, 233-238
Pion-nucleon coupling constant, *see* Coupling constants

Pion-nucleon inelastic scattering, 22
Pion-nucleon interactions, 17-24, 34-57
Pion-nucleon scattering lengths, 21, 55
Pion-nucleon scattering volumes, 21
Pion-nucleon vertex, see Vertex function
Pion-nucleus interaction, 86, 124, 135,
   189, 228, 287, 378
Pion-nucleus vertex, 230
Pion-pion states, 57
Pion pole dominance, 361
Pion scattering, 190-260, 340, 363
Plane wave Born approximation (PWBA),
   327, 331, 337
Plane wave impulse approximation (PWIA),
   333, 336
Plane waves, 6, 8, 314
Polarizability, 214
Polarization function, 300, 372
Polarization measurements, 374
Polarized protons for ($\vec{p}$,π), 333
Pole approximation, 325, 343
Poles in scattering amplitudes, 41, 42, 56,
   230, 231, 233, 343
Pomeranchuk's theorem, 231
Potential theory, 27-34
Production, 2, 36, 110, 305-346
Production of pions, pp$\rightarrow$$\pi$$^+$d, 106, 321
Profile functions, 179, 180, 184, 187
Projectile, 5
Projectile-nucleon amplitude, 176
Projection methods, 284-295, 349, 350
Projection operators, 31, 40, 142, 144,
   153, 215, 219, 275, 276, 287, 296
Propagation of resonances, 295-304
Propagation vector, 115, 118
Propagator,28,29,100-102,268-272,279,372
Proper self-energy, 283
Proton density, 225, 235, 243
Proton radii, 225, 226
Pseudoscalar coupling, 35, 38, 361
Pseudoscalar current, 318
Pseudoscalar mesons, 34, 35
Pseudovector coupling, 35
p-wave πN effects, 42, 52, 56, 88-95, 191-
   235, 344

Quantum field theory, see Field theory
Quantum numbers, exchange of, in
   scattering, 240-251

Quarks, 3, 24, 176, 375
Quasi-deuteron model, 342
Quasi-free mechanism, 249, 250, 256-259
Quenching effects, 213

Radiative absorption, 63, 64, 306, 307, 343,
   344
Random phase approximation (RPA), 304
Rare meson decay modes, 4
Reaction cross section, see Absorption
   cross section
Reactions:
   direct, 326, 334
   (γ,p), 329
   one-nucleon (π,N) or (N, π), 327-333
   one-nucleon transfer, 330
   polarized protons in ($\vec{p}$, π), 333
   (p , 2p), 346
   two-nucleon: (π,2N), 33-342
Reaction theory, 304
Recoil, 211
Recoil energy, 210, 249
Recoupling of angular momentum, 49
Reduced mass, 9, 73, 81, 91, 235
Reduction formula, 283
References, 60, 111, 188, 260, 304, 346,
   379-393
Reflections, 166, 167, 172, 216, 225, 296
Relative momentum, 73, 81
Relativistic considerations, 3, 12, 95, 202,
   226, 262, 268-271
Relativistic energies, 6
Relativistic kinematics, 15
Relativistic three-body theories, 98-102,
   106
Removal energy, 210
Renormalization of axial coupling, 376-378
Renormalized coupling constant, 37,39,
   47, 57, 351, 376-378
Renormalized nucleon states, 351
Resonance propagation, 304
Resonances, see Delta-resonance,
   Λ(1405)-resonance
Resonances below threshold, 94
Resonances, in $\overline{K}$N system, 22
Resonances in πN system, 22
Resonant energy dependence, 94
Retarded potential, 33
ρ-exchange, 60, 110, 365

ρ-meson, 2, 32, 55, 58-60, 219, 247, 271, 286, 302, 347, 348, 351
ρNN vertex, 60, 219, 220
Right-hand cut, 104, 105
Rotational invariance, 9

Saturation of nuclear targets in spin and isospin, 157, 192, 209
Scalar variables, 14
Scattering lengths, 10, 21, 25, 26, 55, 89-91, 93, 95, 234
Scattering volumes, 10, 21
s-channel transfer, 104
Schroedinger pictures, 34
Schroedinger theory, 27-34, 261-304
Second-order optical potential, see Optical potential, second-order
Selection rules, 346
Selection rules in inelastic pion-nucleus scattering, 253
Self-consistent potential, 248
Self-energy, 264, 283
Semiclassical approximations, 113-124
Semirelativistic approach, 270
Separable interaction, 31, 32, 47, 50-52, 72, 84, 86, 87, 92, 93, 111, 197, 224, 295, 323
Separation energy, 210
Separation of projectile and nucleon variables, 125, 128, 173
Shell model, 248
Shift in delta-resonance position, 198, 199
Short-range correlations, see Correlations
Short-range phenomena, 367. See also Correlations
Siegert's theorem, 359
σ-meson, 2 56, 58, 59, 365, 368
Σ^* (1385)-resonance, 273
Single-channel Watson theory, 273
Single charge exchange, see Charge exchange
Single-nucleon mechanism for (π,N) or (N, π), 333
Single-nucleon reactions, see Reactions, one-nucleon (π,N) or (N,π)
Single-particle states, 210
Single scattering, 71, 92, 98
Single-scattering approximation, 67
s-matrix, 7, 10
Soft pion limit, 376

Spatial symmetry, 250
Spectral function, 101
Spectroscopic factor, 311, 329, 330
Spherical vector components, 34, 35
Spin, 35-40, 55, 83
Spin and isospin, saturation of nuclear targets in, 157, 192, 209
Spin density, 157, 369
Spin-isospin excitations, 372
Spin-orbit potential, 30
Spinors, 17, 358
Spreading potential, 211
Square integrability, 74, 75
Stability line of nuclei, 246
Standing spin-isospin waves, 369
Static approximation, see Static model in Chew-Low theory
Static interaction, 58
Static model in Chew-Low theory, 42, 46-56
Statistical methods, 312, 325, 326
Strangeness, 5,8,12,24,60,62,240,284,307
Strangeness analog resonance, see Strangeness analog state
Strangeness analog state, 242, 248, 250
Strangeness exchange reactions, 4, 106, 241, 247-251, 307, 310, 345
Subtraction in dispersion relation, 229, 231, 233
Supersymmetric state, 250
Surface thickness, 226

Talmi-Moshinsky transformation, 337
Tamm-Dancoff approximation (TDA), 303, 349
t-channel transfer, 104
Tensor coupling, 219, 365
Tensor force, 2, 89
Tensor interaction in nucleon-nucleon force, see Tensor force
Thermal emission of particles, 313
Thermal neutron capture cross section, 365
Three-body channels, 26
Three-body forces, 362, 368
Three-body systems, 62-111, 128, 171, 172, 239, 322
Three-particle correlations, 166, 207, 216
Three-particle terms, 165
Tibell plot, 252

Time-of-flight method, 334
Time-orderd product, 282
Time reversal, 327
T-matrix, 7, 13, 14, 27-29, 269, 277,
    279, 285, 293, 313-315
Total cross sections, 227
Total cross section for NN scattering, 18,
    19[
Total cross section for πN scattering, 18, 19
Total momentum, 8, 14, 15, 73, 74, 82
Transformation brackets, 84
Transformation of scattering amplitudes,
    201, 235
Transition amplitudes, *see* *t*-matrix
Transition density, 3
Transition matrix, *see* *T*-matrix
Transition operator, *see* *T*-matrix
Transition potentials, 353, 366
Translational invariance, 14, 15, 73
Transport equations, 326
Traveling spin-isospin waves, 369
Treiman-Yang angle, 335, 336
Treiman-Yang distribution, 339
Triple scattering, 92
True absorption, 115, 124, 129, 159, 188,
    205, 226, 227, 235, 236, 258, 305, 376
True correlation functions, 167, 207
Two-body problem, 72-74, 112, 129
Two-hole states, 324, 334, 346
Two-meson exchange, 352
Two-nucleon mechanism for (π,N) or
    (N,π), 352
Two-nucleon reactions, *see* Reactions,
    two-nucleon (π,2n)
Two-nucleon potential, *see*
    Nucleon-nucleon interaction
Two particle correlations, *see*
    Correlations
Two pion exchange, 2, 365

Uncertainty principle, 225
Unitarity, 10, 92, 101, 102
    multiparticle, 100

Unitarity condtition, 102
Unitary transform method, 356
Unphysical cut, 230
Unphysical region, 230, 231
Unrenormalized masses, 36

V-A theory of weak interaction, 358
Vector coupling, 365
Vector current, 358, 360, 361
Vector meson, 32, 272, 365
Vector meson dominance, 362
Velocity effects, 255
Vertex for absorption, 346
Vertex function, 37, 38, 52, 198, 200, 217-
    219, 317, 318, 353, 373, 377
Vertex operators, 55
Virtual decay, 239
Virtual meson, 347, 362
Virtual states, 32, 263

Watson multiple scattering theory,
    see Multiple scattering theory
Watson series, *see* Multiple scattering theory
Wave:
    external, 115
    extinction of, 115
Wave distortion, 148, 243, 244, 252, 256,
    258, 314, 330, 333, 336, 341, 377
    see also Distorted wave impulse
    approximation
Wave extinction, 115
Weak interactions, 63, 347, 348, 353, 356,
    357, 362, 376-378
Weisskopf units, 257
Width of resonance, 20

Yamaguchi potential, 87
Y*=(y*=resonance), 273, 295
Yukawa interaction, 34-38, 43, 45, 52, 56,
    57, 261-265, 271, 276, 293, 316, 351
Yukawa potential, 1, 55, 272, 352, 353

Zero-range assumption, 340